Dystonia and Dystonic Syndromes

Petr Kanovsky • Kailash P. Bhatia
Raymond L. Rosales
Editors

Dystonia and Dystonic Syndromes

Springer

Editors

Petr Kanovsky
Department of Neurology
Palacky University Medical School
University Hospital Olomouc
Olomouc
Czech Republic

Raymond L. Rosales
Department of Neurology and Psychiatry
The Royal and Pontifical University of
Santo Tomas
Manila
Philippines

Kailash P. Bhatia
Sobell Department of Motor
Neuroscience and Movement Disorders
University College London
Institute of Neurology
London
United Kingdom

ISBN 978-3-7091-1966-2 ISBN 978-3-7091-1516-9 (eBook)
DOI 10.1007/978-3-7091-1516-9

Springer Wien Heidelberg New York Dordrecht London

Softcover reprint of the hardcover 1st edition 2015

Printed on acid-free paper

Springer-Verlag GmbH Wien is part of Springer Science+Business Media (www.springer.com)

Contents

Contributors

Giovanni Abbruzzese Section of Neurology, DINOGMI, University of Genoa, Genoa, Italy

Laura Avanzino Section of Human Physiology, DIMES, University of Genoa, Genoa, Italy

Marek Baláž 1st Department of Neurology, Faculty of Medicine, St. Anne's University Hospital, Masaryk University Medical School, Brno, Czech Republic

Multimodal and Functional Imaging Research Group, CEITEC – Central European Institute of Technology, Masaryk University, Brno, Czech Republic

Martin Bareš 1st Department of Neurology, Faculty of Medicine, St. Anne's University Hospital, Masaryk University Medical School, Brno, Czech Republic

Kailash P. Bhatia Sobell Department of Motor Neuroscience and Movement Disorders, Institute of Neurology, University College London, London, UK

Carlo Colosimo Department of Neurology and Psychiatry, Sapienza University of Rome, Rome, Italy

Dirk Dressler Movement Disorders Section, Department of Neurology, Hanover University Medical School, Hannover, Germany

Mark J. Edwards Sobell Department of Motor Neuroscience and Movement Disorders, Institute of Neurology, University College London, London, UK

Hubert Fernandez Movement Disorders Program, Center for Neurological Restoration, Cleveland Clinic, Cleveland, OH, USA

Pavel Filip 1st Department of Neurology, Faculty of Medicine, St. Anne's University Hospital, Masaryk University Medical School, Brno, Czech Republic

Behavioural and Social Neuroscience Research Group, CEITEC – Central European Institute of Technology, Masaryk University, Brno, Czech Republic

Jorge Hernandez Franco Department of Neurological Rehabilitation, National Institute of Neurology and Neurosurgery, Mexico City, Mexico

Claudia M. Hendrix Department of Neurology, University of Minnesota, Minneapolis, MN, USA

Petr Kanovsky Department of Neurology, Palacky University Medical School, Olomouc, Czech Republic

Tarannum S. Khan Department of Neurology, Movement Disorders Program, Cleveland Clinic Florida, Weston, FL, USA

Christine Klein Department of Neurology, University of Lübeck Medical School, Lübeck, Germany

Antonella Macerollo Sobell Department of Motor Neuroscience and Movement Disorders, Institute of Neurology, University College London, London, UK

Michal Maršálek National Psychiatric Hospital Bohnice, Prague, Czech Republic

Department of Psychiatry, 1st Faculty of Medicine, Charles University, Prague, Czech Republic

Luca Marsili Department of Neurology and Psychiatry, Sapienza University of Rome, Rome, Italy

Kateřina Menšíková Department of Neurology, Palacky University Medical School, Olomouc, Czech Republic

Elisa Pelosin Section of Neurology, DINOGMI, University of Genoa, Genoa, Italy

Ana Claudia Saraiva Ribeiro Department of Neurology, Centro Hospitalar e Universitǎrio de Coimbra, Coimbra, Portugal

Raymond L. Rosales Department of Neurology and Psychiatry, The Royal and Pontifical University of Santo Tomas, Manila, Philippines

Erica Samantha A. Teleg Department of Neurology and Psychiatry, The Royal and Pontifical University of Santo Tomas, Manila, Philippines

Jerrold L. Vitek Department of Neurology, University of Minnesota, Minneapolis, MN, USA

Jürgen Voges Department of Neurosurgery, Otto von Guericke University Medical School, Magdeburg, Germany

Ana Westenberger Department of Neurology, University of Lübeck Medical School, Lübeck, Germany

Part I

Pathophysiology and Nosology of Dystonia

Dystonia: The Syndrome, Its Term, Concept and Their Evolution

1

Petr Kanovsky, Raymond L. Rosales, and Kailash P. Bhatia

The book fully dedicated to *dystonia* should start with this word. Nevertheless, going into history, this term has been invented relatively late. What we should mention are the paths of the syndrome in the medical and literary history. Notorically known are the artistic descriptions of torticollis by Dante (La Divina Commedia) and Rabelais (Gargantua and Pantagruel). The first medical depiction of dystonia was probably that of Gowers' from 1888, where the movement disorder was named "tetanoid chorea" (Fig. 1.1) [1]. Destarac used the term "torticolis spasmodique" to describe the cervical dystonia [2]. Schwalbe used in his doctoral thesis the term "tonic cramps" (Fig. 1.2) [3]. The word "dystonia" has been used for the first time by Oppenheim in 1911 in the form "dystonia musculorum deformans" (Fig. 1.3) [4]. Flatau and Sterling in the same year tried to name the disorder in a more innovative manner and used the term "progressive torsion spasms" (Fig. 1.4) [5]. The term "dystonia" prevailed in the neurological literature and has been used in the form "torsion dystonia" after Meige [6]. Its nature remained practically unknown for more than 50 years. At the end of the 1970s, dystonia was seen as something peculiar, balancing on the edge between organic disorder and psychiatric condition. The everlasting enigma of dystonia has been breached for the first time by David Marsden, who attributed the clinical syndrome to a disorder of basal ganglia

P. Kanovsky (✉)
Department of Neurology, Palacky University Medical School, Olomouc, Czech Republic
e-mail: petr.kanovsky@fnol.cz

R.L. Rosales, MD, PhD
Department of Neurology and Psychiatry, The Royal and Pontifical University of Santo Tomas, Manila, Philippines
e-mail: rlrosales@mnl.ust.edu.ph

K.P. Bhatia
Sobell Department of Motor Neuroscience and Movement Disorders, Institute of Neurology, University College London, Queen Square, London, UK
e-mail: k.bhatia@ion.ucl.ac.uk

P. Kanovsky et al. (eds.), *Dystonia and Dystonic Syndromes*,
DOI 10.1007/978-3-7091-1516-9_1

Fig. 1.1 William Richard Gowers (1845–1915)

Fig. 1.2 Gustav Schwalbe (1844–1916)

Fig. 1.3 Hermann Oppenheim (1858–1919)

Fig. 1.4 Edward Flatau (1868–1932)

Fig. 1.5 C. David Marsden (1938–1998)

functioning. As he postulated in the seminal work regarding the natural history and clinical presentation of dystonia, dystonia is a syndrome of sustained involuntary muscle contraction, frequently causing twisting or repetitive movements or abnormal postures (Fig. 1.5) [7, 8]. Dystonia has been described as an involuntary movement, which was (in principle) independent of the patient's aim to prime the movement (or rest) of a certain body part. Although a new definition of dystonia has been recently adopted [9], Marsden's concept is still a cornerstone of it.

One can undoubtedly characterise the dystonic movement as a result of volitional (albeit unconscious) motor action [10]. There is enough evidence that dystonia is the "normal" motor action abnormally accompanied by the action of antagonists, or by the co-contractions of inappropriate muscles, and by the defective reciprocal inhibition of other muscles. As a result, there is a dystonic movement or dystonic posture appearing as final vector of that deviant muscle activity.

Dystonia does not look like a chaotic medley of contractions and muscle relaxation, as do a majority of other hyperkinetic movement disorders. Dystonia rather looks like a highly (albeit aberrantly) organised motor performance. The motor action seen in patients with dystonia is apparently not a result of several abnormal contractions or muscle jerks. It is complex dyskinesia, which only sometimes changes character over time. This statement in a few words only rewrites more than 20-year-old Rondot's concept of dystonia. Named "a shadow of movement", dystonia was proposed as a primitive "kind of movement", similar to that which can still be seen in other primates. The motor action itself was seen as a "dystonic way back" in the phylogenetic history of motor action. In Rondot's view, dystonia is something that is constantly present in the mature human motor system, but which can only be seen in particular circumstances, for instance, in the case of basal ganglia lesion (or other basal ganglia involvement, as in the case of idiopathic dystonia). Nevertheless, in Rondot's motor hierarchy was the basal ganglia structure that modulates all motor outputs, and only a basal ganglia disorder can prime a dystonic motor performance (Fig. 1.6) [11].

Not surprisingly, the pathophysiological concept of dystonia is not complete only after the next 25 years. Certainly, it is now widely accepted that somatosensory input plays a substantial role in the priming of dystonia. The key role of the gamma system, namely, gamma motor neurons, at the spinal level has been confirmed and known since the mid of 1990s [12–15]. Gamma system has to adequately interact with the alpha system to achieve any alpha-gamma co-activation. Such a disorder must not be only localised at the spinal level. Normally, the gamma system acts under the strong control of the afferent somatosensory synapses. However, in the presence of any somatosensory disorder at any level of the central nervous system, the activation of the gamma system is abnormal and badly coordinated with the action of the alpha system. The well-known clinical evidence of the somatosensory system participation on the mechanism of dystonia is the existence of the "sensory tricks", or "geste antagoniste" [16], then successful treatment of dystonia by a local application of anaesthetics to the dystonic muscles [17–19] and, last but not least, the presence of dystonic dyskinesias in regional pain syndrome (RPS) or reflex sympathetic dystrophy [20].

However, it must be stressed here that the somatosensory abnormality can be seen as one of the characteristics of dystonia. The somatosensory theory does not fully apply to the existence of task-specific dystonia. It would be very difficult to

Fig. 1.6 Pierre Rondot (* 1923)

defend a concept in which task-specific dystonia is caused by the stereotyped abnormal somatosensory afferentation, which is present only during highly specialised, long-term trained motor acts and is not present during other, similarly specialised and long-term trained motor acts performed by the same small muscles or muscle groups (for instance, writing vs. drawing or piano playing vs. guitar playing).

In our view is the whole motor act initiated by the start of the motor programme which is retrieved from the "motor memory" [21, 22] and is continued by the update of the motor programme in both motor loops, direct and indirect. The substrate of the motor memory is not known, but there is an increasing evidence that the "motor programmes" may be stored in the dorsolateral anterior areas of the frontal lobe [23, 24].

Human individuals form a variety of stereotyped motor programmes during their life, from the intrauterine period to late senium; their processing is driven by the above-mentioned somatosensory feedback loops. Kaji proposed the term "motor routines" for primitive motor programmes and the term "motor subroutines" for the very short simple stereotyped programmes which are used for the generation of stereotyped and frequently used specialised movements [25, 26]. The "motor routines" could be seen as the motor programmes specific for given effectors and the "motor subroutines" as the motor programmes specialised for the dedicated complex motor action of these effectors (typing, writing, playing a musical instrument, mimicking movements).

There is an increasing evidence for that hypothesis coming from the field of clinical and experimental neurophysiology. Let us start with the well-known phenomenon of reciprocal inhibition. Its abnormality is a typical example of chaotic interaction between the alpha and gamma systems exemplified by the abnormal H-reflex tests done in the forearm muscles of patients with dystonia [27–30]. Defective reciprocal inhibition of muscle activity in the neck has been repeatedly found in polymyographic examinations of patients with torticollis dystonia [27–30]. Also, the patterns of active muscles and their mutual inhibition changed following an efficacious botulinum toxin treatment in dystonia [31–35].

Reciprocal inhibition is tightly connected with muscle spindles, and these apparently behave abnormally in dystonia [36]. The long-latency reflex which also depends on the function of muscle spindles is equally abnormal in dystonia [37]. It seems that the abnormal muscle spindle functions (which virtually are controlled by the gamma neurons) are probably the seminal source of defective functioning of the whole Ia proprioceptive afferent system. Further, the phenomena that ultimately depend on the normal functioning of Ia afferents (and then on the normal function of muscle spindles) are cortical excitability and intracortical inhibition. Abnormal functioning of muscle spindles and consequent abnormal afferent flow through the Ia afferents result in the disorders of both cortical functions. The disorder of cortical excitability is reflected in the abnormalities of cortical components of somatosensory evoked potentials (SEPs). The scalp amplitude of (either medial or tibial) SEP has been particularly studied, and it has been found increased in dystonia [32, 33, 38–41], confirming the fact of abnormally increased cortical excitability in dystonia. Similarly, the disorder of cortico-cortical inhibition has been shown in numerous studies with paired transcranial magnetic stimulation (pTMS) of the brain cortex, and this type of inhibition has been found to be abnormal in dystonia [42–44]. Both these disorders coexist in dystonia, and it seems that they also reflect the lateralised pattern, which is usual in the focal form [45].

However, one considers the hypothesis of the defective sorting of subroutines and routines into the supposed motor programme; some evidence of abnormality in the early phase of motor action should be present. In our opinion, this might be found in the experiments dealing with slow cortical potentials, which are seen by most authors as a correlate of the "motor programming" process. Bereitschaft or readiness potential (BP or RP) has been found to be abnormal in dystonia, and its abnormality has also been found lateralised depending on the side of dystonic symptomatology [46–49]. Contingent negative variation (CNV), another potential connected with the process of motor preparation, has also been found to be abnormal and lateralised in dystonia [25, 26, 50, 51]. Both potentials (RP and CNV) are presumed to be generated in the cortex; however, they were also repeatedly recorded in subcortical structures [52–56].

The normalisation of cortical disorder following botulinum toxin injections shows that the manipulation of proprioceptive afferent input has a substantial impact on the disorder at the cortical level [57, 58]. Based on this evidence, it can be assumed that the abnormalities of RP, CNV and ERD can indicate (with a high level of probability) a disorder in the motor programming process in dystonia, which is followed by the poor motor performance, reflected in the abnormalities of reciprocal inhibition, long-latency reflex, cortical excitability and intracortical inhibition.

All the above-mentioned physiological evidence must lead to the conclusion that the abnormal process in dystonia is the sensorimotor integration. This phenomenon is repeatedly mentioned in numerous papers [59–64]. The sensorimotor integration in the physiological perspective involves all parts of the motor and sensory system, including the motor circuits, in which the basal ganglia and the premotor and motor cortex are the principal components.

It has only recently been hypothesised that sensorimotor integration is in fact a function of brain plasticity. Then the disorder of plasticity should be present in dystonia, and this has been witnessed in several reports employing transcranial stimulation methods [65–68]. The most appropriate tool to investigate the brain plasticity, the functional magnetic resonance imaging (fMRI), has been used even more recently. We have seen the changes which are typical for the altered brain plasticity in torticollis patients (when compared to healthy individuals) and their "normalisation" following successful treatment with botulinum toxin A [69]. It is our belief that further studies will show us that the "sensorimotor integration" or "brain plasticity" represents the process of "motor preparation" itself, even in the expert motor performances [70–72].

All these recent and current indices are pieces of a puzzle. We believe that once completed, this jigsaw will show idiopathic dystonia in the frame of defective programming and organisation of motor action due to defective somatosensory flow, consequent defective sensorimotor integration or brain plasticity, in which the movement itself is perfectly performed. This aberrantly planned and perfectly performed movement is in our opinion of "pseudo-volitional" origin. There will probably be little doubt about this character of dystonic movement in task-specific dystonias. The difference between the most commonly seen dystonic disorders, which are classically described as involuntary (blepharospasm, laryngeal dystonia, oromandibular dystonias, cervical dystonia) and task-specific dystonias, should be seen in the sense of "unconscious voluntary movement" and "conscious voluntary movement". The above-mentioned dystonias are in fact representatives of the most commonly performed, automatic and unconscious motor acts: blinking, mimicking, voice performing, twisting or tilting of the head. From this point of view, dystonia is not simple, but a complex, highly organised movement. According to our belief, this movement is aberrantly programmed and, according to such a programme, precisely performed. Nevertheless, in the context of the aforementioned discussion, one should be aware that the phenomenology of dystonia should be gleaned carefully, especially in conditions where overlaps may occur (e.g. dystonia-parkinsonism) and pathophysiology may differ.

References

1. Gowers WR. A manual of diseases of the nervous system. 2nd ed. London: Churchill; 1888.
2. Destarac M. Torticollis spasmodique et spasmes fonctionnelles. Rev Neurol. 1901;9:591–7.
3. Schwalbe G. Lehrbuch der Neurologie. Jena: Springer; 1881.
4. Oppenheim H. Über eine eigenartige Krampfkrankheit des kindlichen und jugendlichen Alters (dysbasia lordotica progressiva, dystonia musculorum deformans). Neurologische Zbl. 1911;30:1090–107.
5. Flatau E, Sterling W. Torsionspasms bei kindern. Z Gesamte Neurol Psychiatr. 1911;7:586–612.
6. Meige H. Les convulsions de la face, une forme clinique de convulsion faciale, bilaterale et mediane. Rev Neurol. 1910;20:437–43.
7. Marsden CD. Dystonia: the spectrum of the disease. Res Publ Assoc Res Nerv Ment Dis. 1976;55:351–67.
8. Marsden CD, Rothwell JC. The physiology of idiopathic dystonia. Can J Neurol Sci. 1987;14 Suppl 3:521–7.
9. Albanese A, Bhatia K, Bressmann S, et al. Phenomenology and classification of dystonia: a consensus update. Mov Disord. 2013;29:863–73.
10. Kaňovský P. Dystonia: disorder of motor preparation or motor performance? Mov Disord. 2002;17:1143–8.
11. Rondot P. The shadow of movement. J Neurol. 1991;238:411–9.
12. Berardelli A, Rothwell JC, Hallet M, Thompson PD, Manfredi M, Marsden CD. The pathophysiology of primary dystonia. Brain. 1998;121:1195–212.
13. Hallet M. Is dystonia a sensory disorder? Ann Neurol. 1995;38:139–40.
14. Hallet M. The neurophysiology of dystonia. Arch Neurol. 1998;55:601–3.
15. Hallet M. Disorder of movement preparation in dystonia. Brain. 2000;123:1765–6.
16. Kaji R, Murase N. Sensory function of basal ganglia. Mov Disord. 2001;16:593–4.
17. Jankovic J. Post-traumatic movement disorders: central and peripheral mechanisms. Neurology. 1994;44:2006–14.
18. Rondot P, Korn H, Scherrer J. Suppression of an entire limb tremor by anaesthetising a selective muscular group. Arch Neurol. 1968;19:421–9.
19. Walshe F. Observation on the nature of the muscular rigidity of paralysis agitans, and its relationship with tremor. Brain. 1924;47:159–77.
20. Jankovic J, Van der Linden C. Dystonia and tremor induced by peripheral trauma: predisposing factors. J Neurol Neurosurg Psychiatry. 1988;51:1512–9.

21. Hallet M. Physiology of basal ganglia disorders: an overview. Can J Neurol Sci. 1993;20:177–83.
22. Young A, Penney J. Biochemical and functional organisation of the basal ganglia. In: Jankovic J, Tolosa E, editors. Parkinson's disease and movement disorders. 2nd ed. Philadelphia: Williams and Wilkins; 1993. p. 1–12.
23. Boyd LA, Lindsell MA. Excitatory repetitive transcranial magnetic stimulation to left dorsal premotor cortex enhances motor consolidation of new skills. BMC Neurosci. 2009;10:72.
24. Carlsson G, Hufnagel M, Jansen O, Claviez A, Nabavi A. Rapid recovery of motor and cognitive functions after resection of a right frontal lobe meningioma in a child. Childs Nerv Syst. 2010;26:105–11.
25. Kaji R, Ikeda A, Ikeda T, Kubori T, Mezaki T, Kohara A, Kanda M, Nagamine T, Honda M, Rothwell J, Shibasaki H, Kimura J. Physiological study of cervical dystonia. Task-specific abnormality in contingent negative variation. Brain. 1995;118:511–22.
26. Kaji R, Shibasaki H, Kimura J. Writer's cramp: a disorder of motor subroutine? Ann Neurol. 1995;38:837–8.
27. Day BL, Marsden CD, Obeso JA, Rothwell JC. Reciprocal inhibition between the muscles of the human forearm. J Physiol. 1984;349:519–34.
28. Deuschl G, Heinen F, Kleedorfer B, Wagner M, Lücking CH, Poewe W. Clinical and polymyographic investigation of spasmodic torticollis. J Neurol. 1992;239:9–15.
29. Deuschl G, Seifert C, Heinen F, Illert M, Lücking CH. Reciprocal inhibition of forearm flexor muscles in spasmodic torticollis. J Neurol Sci. 1992;113:85–90.
30. Priori A, Berardelli A, Mercuri B, Manfredi M. Physiological effects produced by botulinum toxin treatment of upper limb dystonia. Changes in reciprocal inhibition between forearm muscles. Brain. 1995;118:801–7.
31. Gelb D, Yoshimura D, Olney R, Lowenstein D, Aminoff M. Change in pattern of muscle activity following botulinum toxin injections for torticollis. Ann Neurol. 1991;29:370–6.
32. Kaňovský P, Dufek J, Halačková H, Rektor I. Change in the pattern of cervical dystonia might be the cause of benefit loss during botulinum toxin treatment. Eur J Neurol. 1997;4:79–84.
33. Kaňovský P, Streitová H, Dufek J, Rektor I. Lateralization of the P22/N30 component of the median nerve in patients with cervical dystonia. Mov Disord. 1997;12:553–60.
34. Marín C, Marti MJ, Tolosa E, Alvarez R, Montserrat L, Santamaría J. Modification of muscle activity after BOTOX injections in spasmodic torticollis. Ann Neurol. 1992;32:411–2.
35. Marín C, Martí MJ, Tolosa E, Alvarez R, Montserrat L, Santamaría J. Muscle activity changes in spasmodic torticollis after botulinum toxin treatment. Eur J Neurol. 1995;1:243–7.
36. Grünewald R, Yoneda Y, Shipman J, Sagar H. Idiopathic focal dystonia: a disorder of muscle spindle afferent processing? Brain. 1997;120:2179–85.
37. Naumann M, Reiners K. Long-latency reflexes of hand muscles in idiopathic focal dystonia and their modification by botulinum toxin. Brain. 1997;120:409–16.
38. Jones SJ, Sheean G, Ceballos-Baumann AO, Marsden CD. Evidence for enhanced sensorimotor cortex excitability in dystonia (abstract). Electroencephalogr Clin Neurophysiol. 1995;94 Suppl 1:S233.
39. Kaňovský P, Streitová H, Dufek J, Znojil V, Daniel P, Rektor I. Lateralization of the P22/N30 precentral cortical component of the median nerve somatosensory evoked potentials is different in patients with a tonic or tremulous form of cervical dystonia. Mov Disord. 1999;14:642–51.
40. Reilly JA, Hallet M, Cohen LG, Tarkka I, Dang N. The N30 component of somatosensory evoked potentials in patients with dystonia. Electroencephalogr Clin Neurophysiol. 1992;84:243–7.
41. Tinazzi M, Frasson E, Polo A, Tezzon F, Bovi P, Deotto L, et al. Evidence for an abnormal cortical sensory processing in dystonia: selective enhancement of lower limb P37/N50 somatosensory evoked potential. Mov Disord. 1999;14:473–80.
42. Berardelli A. Transcranial magnetic stimulation in movement disorders. Electroencephalogr Clin Neurophysiol. 1999;51 Suppl 1:276–80.
43. Ridding M, Sheean G, Rothwell JC, Inzelberg R, Kujirai T. Changes in the balance between motor cortical excitation and inhibition in focal, task–specific dystonia. J Neurol Neurosurg Psychiatry. 1995;59:493–8.

44. Rona S, Berardelli A, Vacca L, Inghilleri M, Manfredi M. Alterations of motor cortical inhibition in patients with dystonia. Mov Disord. 1998;13:118–24.
45. Kaňovský P, Bareš M, Streitová H, Klajblová H, Daniel P, Rektor. Abnormalities of cortical excitability and cortical inhibition in cervical dystonia evidence from somatosensory evoked potentials and paired transcranial magnetic stimulation recordings. J Neurol. 2003;250:41–50.
46. Deuschl G, Toro C, Matsumoto J, Hallet M. Movement-related cortical potentials in writer's cramp. Ann Neurol. 1995;38:862–8.
47. Féve A, Bathien A, Rondot P. Abnormal movement-related potentials in patients with lesions of basal ganglia and anterior thalamus. J Neurol Neurosurg Psychiatry. 1994;57:100–4.
48. Podivinsky F. Dissociation between motor potential and contingent negative variation after lesions of some basal ganglia in man: electrophysiological study of torticollis. Act Nerv Super. 1978;20:312–8.
49. Van der Kamp W, Rothwell J, Thompson P, Day B, Marsden C. The movement-related cortical potential is abnormal in patients with idiopathic torsion dystonia. Mov Disord. 1995;10:630–3.
50. Hamano T, Kaji R, Katayama M, Kubori T, Ikeda A, Shibasaki H, Kimura J. Abnormal contingent negative variation in writer's cramp. Clin Neurophysiol. 1999;110:508–15.
51. Ikeda A, Shibasaki H, Kaji R, Terada K, Nagamine T, Honda M, Hamano T, Kimura J. Abnormal sensorimotor integration in writer's cramp: study of contingent negative variation. Mov Disord. 1996;11:683–90.
52. Bareš M. Parallel processing of cognitive evoked potentials in the frontal cortex and basal ganglia. An intracerebral study. Homeostasis. 2001;41:55–7.
53. Bareš M, Rektor I. Basal ganglia involvement in sensory and cognitive processing. A depth electrode CNV study in human subjects. Clin Neurophysiol. 2001;112:2022–30.
54. Rektor I, Bareš M, Kaňovský P, Kukleta M. Intracerebral recording of readiness potential induced by a complex motor task. Mov Disord. 2001;16:698–704.
55. Rektor I, Bareš M, Kubová D. Movement-related potentials in the basal ganglia: a SEEG readiness potential study. Clin Neurophysiol. 2001;112:2146–53.
56. Rektor I, Kaňovský P, Bareš M, Louvel J, Lamarche M. Evoked potentials, ERP, CNV, readiness potential, and movement accompanying potential recorded from the posterior thalamus in human subjects. A SEEG study. Clin Neurophysiol. 2001;31:253–61.
57. Gilio F, Curra A, Lorenzano C, Modugno A, Mangredi M, Berardelli A. Effects of botulinum toxin type A on intracortical inhibition in patients with dystonia. Ann Neurol. 2000;48: 20–6.
58. Kaňovský P, Streitová H, Dufek J, Rektor I. Change in lateralization of the P22/N30 cortical component of median nerve somatosensory evoked potentials in patients after successful treatment with botulinum toxin A. Mov Disord. 1998;13:101–12.
59. Abbruzzese G, Marchese R, Buccolieri A, Gasparetto B, Trompetto C. Abnormalities of sensorimotor integration in dystonia. A transcranial magnetic stimulation study. Brain. 2001;124:537–45.
60. Cohen LG. A window into the role of inhibitory and excitatory mechanisms of perception? J Physiol. 2000;529(2):461–8.
61. Frasson E, Priori A, Bertolasi L, Mauguière F, Fiaschi A, Tinazzi M. Somatosensory disinhibition in dystonia. Mov Disord. 2001;16:674–82.
62. Hanajima R, Ugawa Y. Intracortical inhibition of the motor cortex in movement disorders. Brain Dev. 2000;22 Suppl 1:132–5.
63. Murase M, Kaji R, Shimazu H, Katayama–Hirota M, Ikeda A, Kohara N, Kimura J, Shibasaki H, Rothwell JC. Abnormal premovement gating of somatosensory input in writer's cramp. Brain. 2000;123:1813–29.
64. Rosenkranz K, Altenmüller E, Sigelkow S, Dengler R. Alteration of sensorimotor integration in musician's cramp: impaired focusing of proprioception. Clin Neurophysiol. 2000;111:2040–5.
65. Meunier S, Garnero I, Ducorps A, et al. Human brain mapping in dystonia reveals both endophenotypic traits and adaptive organisation. Ann Neurol. 2001;50:521–7.

66. Nelson AJ, Blake DT, Chen R. Digit-specific aberrations in the primary somatosensory cortex in writer's cramp. Ann Neurol. 2009;66:146–54.
67. Quartarone A, Rizzo V, Bagnato S. Homeostatic-like plasticity of the primary motor hand area is impaired in focal hand dystonia. Brain. 2005;128:1943–50.
68. Quartarone A, Siebner HR, Rothwell JC. Task-specific hand dystonia: can too much plasticity be bad for you? Trends Neurosci. 2006;29:192–9.
69. Opavský R, Hluštík P, Otruba P, Kaňovský P. Sensorimotor network in cervical dystonia and the effect of botulinum toxin treatment: a functional MRI study. J Neurol Sci. 2011;306:71–5.
70. Milton J, Solodkin A, Hluštík P, Small S. The mind of expert motor performance is cool and focused. Neuroimage. 2007;35:804–13.
71. Marsden CD. Motor disorders in basal ganglia disease. Hum Neurobiol. 1984;2:245–50.
72. Stamelou M, Edwards MJ, Hallett M, Bhatia KP. The non-motor syndrome of primary dystonia: clinical and pathophysiological implications. Brain. 2012;145:1668–81.

Physiology of Dystonia

2

Giovanni Abbruzzese, Elisa Pelosin, and Laura Avanzino

2.1 Introduction

Dystonia is defined as a neurological syndrome characterized primarily by involuntary, sustained, and patterned contractions of opposite muscles, frequently causing twisting and repetitive movements or abnormal postures [1]. Dystonia is observed in many different clinical conditions and may be classified as "primary or idiopathic" (with unknown or genetic causes, no gross morphological changes or other associated symptoms) or "secondary or symptomatic" (with identifiable lesions or causes and additional symptoms) [2]. Dystonia can present with a variable topographical distribution, from focal to generalized, but some clinical aspects (co-contraction of agonist-antagonist muscles, excess of movement with loss of selectivity and overflow in muscle activation) are usually present across the different forms. However, it is still uncertain whether all the types of dystonia recognize a common physiology [3]. Indeed, most of physiological data have been obtained in "adult-onset focal dystonias" which are likely to be all related to each other [4], and the physiology of these disorders is likely to be shared.

In addition, a number of physiological abnormalities have been documented also in non-affected body regions suggesting that dystonia is characterized by a widespread impairment of motor control. For instance, in patients with cervical dystonia, an impaired activation of specific brain regions was reported during movement imagination (bilateral parietal, left premotor, and cingulate cortex regions) and movement execution (ipsilateral putamen, insula, and cingulate cortex) of a normal hand [5]. Further supporting the idea that a general disorganization of cerebral motor control occurs in focal (cervical) dystonia, we demonstrated that kinematic

G. Abbruzzese (✉) • E. Pelosin
Section of Neurology, DINOGMI, University of Genoa, Genoa, Italy
e-mail: giabbr@unige.it

L. Avanzino
Section of Human Physiology, DIMES, University of Genoa, Genoa, Italy

P. Kanovsky et al. (eds.), *Dystonia and Dystonic Syndromes*,
DOI 10.1007/978-3-7091-1516-9_2

parameters of reaching movements performed with non-dystonic upper limbs are abnormal in patients with cervical dystonia [6].

Neurophysiological studies have provided important information on the pathophysiological mechanisms underlying dystonia. The main findings are reviewed in this chapter.

2.2 Loss of Inhibition

Common features in dystonia are the presence of sustained simultaneous co-contraction of agonist and antagonist muscles (which can be triggered or worsened by task-specific actions and relieved by some sensory tricks) and the excess of movement with involvement of adjacent or distant muscles ("overflow"). Both these phenomena are likely to depend on an unbalance between excitatory and inhibitory circuits because of defective inhibitory mechanisms operating at various levels of the central nervous system.

Loss of inhibition in dystonia was first demonstrated in spinal and brainstem reflexes. Reciprocal inhibition (the spinal process by which a muscle is inhibited when its antagonist is activated) is abnormal in patients with focal hand dystonia [7]; the second phase of reciprocal inhibition is reduced or absent because of impaired presynaptic inhibition of muscle afferent inputs as a result of defective descending motor control. Abnormalities of blink reflex recovery have been documented in patients with blepharospasm [8]. Similar findings can also be found in patients with generalized dystonia [9, 10].

Loss of inhibition has been reported also at the cortical level by using various techniques of transcranial magnetic stimulation (TMS). Different TMS protocols allow evaluating inhibitory circuits within the motor cortex by testing the activity of specific populations of inhibitory interneurons. Several studies have reported alterations of many types of cortical inhibition in patients with dystonia [11, 12].

Short-interval intracortical inhibition (SICI) following paired-pulse TMS (with a conditioning-test paradigm) largely reflects the effect of inhibitory influences within the cortex mediated by GABA-A receptors. The motor evoked potential (MEP) produced by a suprathreshold TMS stimulus is inhibited by a conditioning subthreshold TMS stimulus preceding the test one by less than 5 ms. SICI was found reduced in patients with focal hand dystonia [13] and afterward in other types of dystonia [14–17]. Other inhibitory mechanisms (long-interval intracortical inhibition, cortical silent period, short- and long-afferent inhibition, interhemispheric inhibition), related to the activity of different intracortical receptors (GABA-B, cholinergic), are also abnormal in dystonic patients [11, 12]. Some of these inhibitory mechanisms may contribute to "surround inhibition": when the motor cortex produces an accurate voluntary movement, there is a widespread inhibition of muscles not involved in the task [18]. Surround inhibition is reduced in patients with focal hand dystonia making finger movements [19] or even imagining a voluntary movement [20]. These results might explain the difficulty of dystonic patients to focus their voluntary motor activity and why dystonic movements can be worsened by the mere

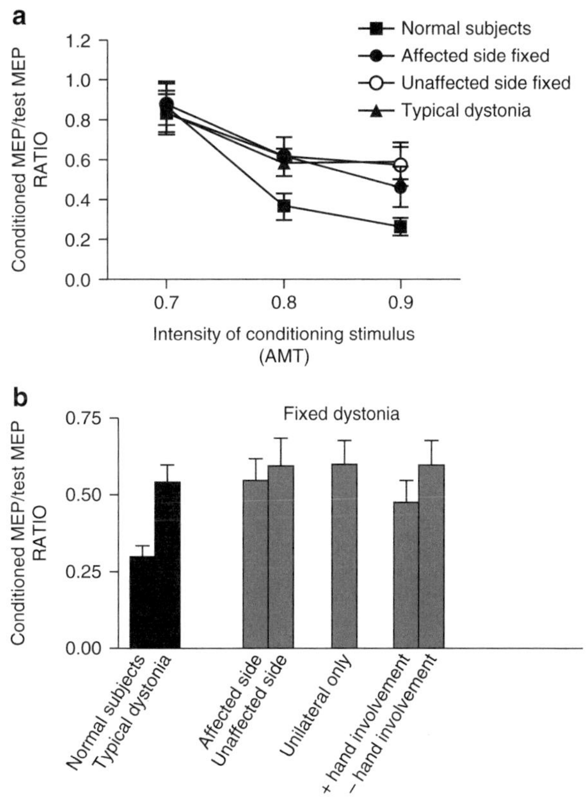

Fig. 2.1 Short-interval intracortical inhibition (*SICI*) in patients with "typical" dystonia, in healthy subjects, and in patients with psychogenic "fixed" dystonia. SICI (the size of the conditioned response is expressed as a ratio to the unconditioned test response) is significantly reduced in patients with "typical" dystonia, compared to healthy subjects, as well as on both sides of patients with "fixed" dystonia. (**a**) Effect of conditioning stimulus intensity (**b**) Combined data from 80 % and 90 % conditioning stimulus (Reproduced from: Avanzino et al. [17])

intention to move (see, for instance, patients with writer's cramp). Interestingly, both SICI [21] and surround inhibition [22] are specifically dependent on the activation of GABA-A receptors, and reduced levels of GABA in both the basal ganglia and the sensory motor cortex have been demonstrated in dystonic patients [23].

However, it should be pointed out that most of the alterations of inhibitory mechanisms are rather nonspecific: they can be present in non-affected body regions and have also been observed in various other neurological conditions, including psychogenic dystonia (Fig. 2.1) [15, 16]. Consequently, the clinical correlate of impaired inhibitory mechanisms is not obvious. Defective inhibition might just represent a predisposing background, and other causative factors are necessary to induce dystonic manifestations.

2.3 Sensory and Sensorimotor Integration Abnormalities

Although dystonia is generally regarded as a pure "motor" disorder, a number of "non-motor" features have been also reported [24]. In particular, sensory abnormalities are not uncommon. Some patients may complain of mild sensory symptoms before dystonic symptoms develop [25–27]. Conventional sensory examination is usually normal, but kinesthesia and perception of vibratory stimulation are defective [25–27]. Finally, psychophysical studies demonstrated that temporal and spatial discrimination are impaired in dystonia [11]. Interestingly, the abnormalities were documented also in non-affected body regions [28], in the unaffected carriers of the DYT1 mutation [29], and in psychogenic dystonia [30]. Using voxel-based morphometry analysis in unaffected relatives of patients with adult-onset primary torsion dystonia, a significant relationship between abnormal temporal discrimination and a bilateral increase in putaminal gray matter was documented [31]. Altogether, these findings suggest that abnormalities of temporal and spatial discrimination might represent a neurophysiological (endophenotypic) trait predisposing to develop a dystonic posture triggered by additional factors.

What is the explanation for these sensory abnormalities? The clinical phenomenon of the "sensory trick" (a tactile or proprioceptive stimulation which serves to temporarily interrupt dystonia), although its physiology is not fully understood, indicates the possible influence of peripheral sensory inputs onto motor circuits. Indeed, several neurophysiological evidences suggest that patients with dystonia have an abnormal processing of sensory information within the lemniscal system and defective sensorimotor integration. Studies with somatosensory evoked potentials (SEPs) [32] showed a loss of lateral inhibition [12], and we demonstrated that long-afferent inhibition is selectively impaired in patients with focal hand dystonia: motor cortical excitability (as tested with TMS) was not reduced 200–1,000 ms after a conditioning stimulation of the contralateral median nerve, and inhibition was converted into facilitation (Fig. 2.2) [33].

The correct execution of voluntary movements is tightly connected to the assistance of peripheral sensory feedback. Therefore, we may postulate that impaired processing of sensory information contributes to changes in the functional organization of cortical motor circuits resulting in abnormal motor programs (co-contraction, overflow).

Such hypothesis is further supported by some experimental data. Studies with EEG [34], MEG [35], and fMRI [36] demonstrated an abnormal somatotopy of finger representation within the primary motor cortex of dystonic subjects even in the non-dystonic hand [35]. Interestingly, it has been shown that treatment with botulinum toxin (BT) can transiently modify the enlarged cortical maps [37]. This is not surprising as BT is able to change intrafusal feedback in dystonia [38], and the resulting modification of sensory inputs can indirectly promote a plastic cortical rearrangement. Some of the abnormalities in kinematic parameters that we described in patients with cervical dystonia [6] partially disappeared 3 weeks after BT injections in the cervical muscles, indicating an influence on motor reorganization of clinically not affected body segments. Similarly, in a recent study Kojovic et al. [39] reported that BT injected into neck muscles in patients with cervical dystonia decreases sensorimotor associative plasticity in the hand area by reducing afferent input from neck muscles.

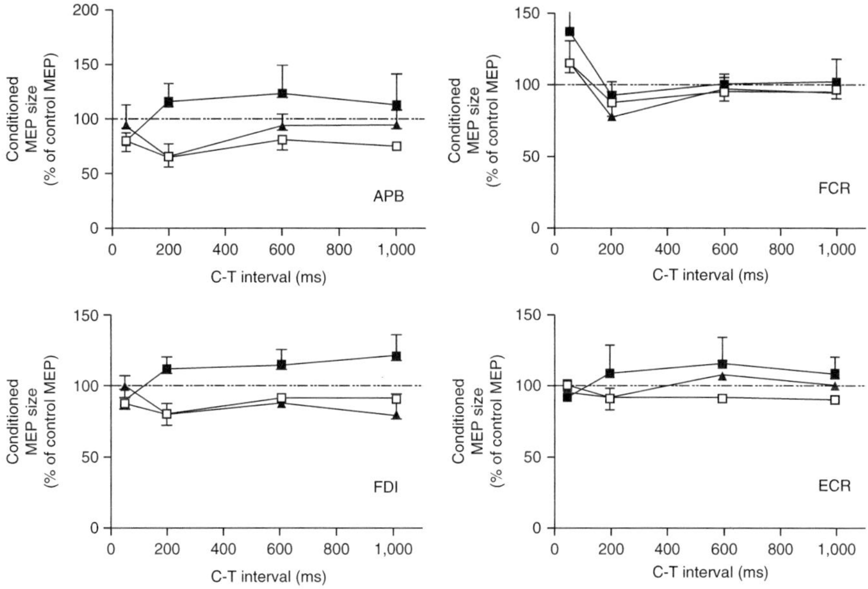

Fig. 2.2 Effect of conditioning stimulation of the median nerve on the test motor evoked potential amplitude in normal controls, in patients with cervical dystonia, and in patients with hand dystonia. Motor cortical excitability is significantly reduced to 200–1,000 ms after conditioning stimulation of the contralateral median nerve in normal individuals but not in patients with focal hand dystonia (Reproduced from: Abbruzzese et al. [33])

2.4 Changes of Cortical Plasticity

Another important pathophysiological abnormality in dystonia is that of plasticity. A number of neurophysiological methods (using TMS) have been developed to explore plasticity in humans. In particular, studies using the paired associative stimulation (PAS) protocol showed abnormal plasticity in patients with primary dystonia, even in unaffected muscles [40, 41]. The PAS protocol combines low-frequency peripheral nerve electrical stimulation and magnetic stimulation applied to the contralateral motor cortex. The peripheral excitatory input preceding the cortical stimulation increases the synaptic strength reflecting a mechanism of long-term potentiation (LTP). PAS-induced effects have a topographical specificity being limited to the muscles innervated by the stimulated peripheral nerve.

In patients with focal hand dystonia, two main abnormalities were documented: first, the amount of MEP facilitation is larger than in healthy subjects; second, the spatial specificity is lost (Fig. 2.3). Such excessive motor cortex plasticity is generalized across the entire sensorimotor system; it may favor maladaptive plasticity and is now considered an endophenotypic trait of the disease [41]. It should be noticed that the circuit tested by PAS is the same engaged in motor learning;

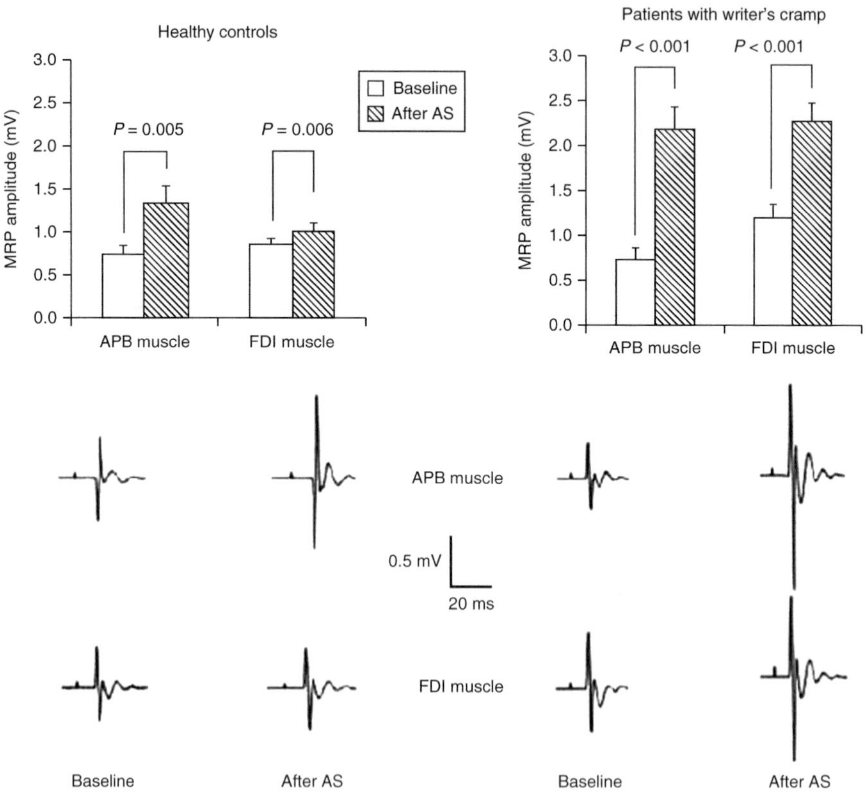

Fig. 2.3 Effect of paired associative stimulation (*PAS*) on the size of motor evoked potentials (*MEPs*) in healthy controls and in patients with writer's cramp. PAS induced a significantly larger facilitation in patients with writer's cramp compared to healthy controls. The spatial specificity is lost in patients with writer's cramp since the increase in MEP size can be observed both in the APB and in the FDI muscles (Reproduced from: Quartarone et al. [40])

not surprisingly abnormalities in the learning of sequential movements have been shown in DYT1 mutation carriers [42].

Another TMS paradigm used to explore plasticity is the "theta burst stimulation" (TBS) [43]. TMS pulses are applied in high-frequency bursts (three pulses at 50 Hz, repeated five times per second) either intermittently causing facilitation or continuously leading to suppression. Edwards et al. [44] investigated the effects of TBS in DYT1 carriers (with or without manifest dystonia) and in patients with sporadic cervical dystonia. The effects of TBS were enhanced in DYT1 gene carriers with dystonia and in subjects with sporadic torticollis but not in DYT1 gene carriers without dystonia.

This study demonstrated that an excessive cortical plasticity was present only in subjects with clinical dystonia suggesting that the propensity to undergo plastic changes may affect the development of symptoms in genetically susceptible individuals.

Some of the aspects of abnormal plasticity may as well reflect an impairment of inhibitory mechanisms (particularly, lateral or surround inhibition). However, the mutual relationship between impaired intracortical inhibitory mechanism and abnormal plasticity is controversial.

Avanzino et al. [17] showed that SICI was significantly reduced in both hemispheres of patients with "typical" and (psychogenic) "fixed dystonia" (FD) compared to healthy subjects suggesting that the abnormal cortical excitability observed in FD might represent an underlying trait predisposing to different clinical forms of dystonia. Quartarone et al. [45] confirmed the presence of abnormal motor cortical inhibition in both organic and psychogenic groups. However, they found that plasticity (as tested with PAS) was abnormally high only in the organic group. They concluded that abnormal plasticity is a hallmark of organic dystonia and is not a consequence of reduced inhibition since the latter is seen in psychogenic patients who have normal plasticity.

Ruge et al. [46] investigated longitudinally the effects of deep brain stimulation (DBS) to the internal globus pallidus on short-interval intracortical inhibition (SICI) and long-term potentiation-like synaptic plasticity (with PAS). DBS is an effective treatment for primary dystonia, and the optimal clinical effect often occurs only weeks to months after starting stimulation. Before surgery, inhibition was reduced and plasticity increased in dystonic patients compared with healthy controls. Following surgery, SICI increased toward normal levels over the following months with the same monotonic time course as the patients' clinical benefit. In contrast, synaptic plasticity changed rapidly, following a nonmonotonic time course. The authors postulated that DBS induces a disruption of abnormal basal ganglia signals, resulting in the absent response to paired associative stimulation at 1 month. However, the clinical benefit is delayed because engrams of abnormal movement persist and take time to normalize.

2.5 The Role of Basal Ganglia

Primary dystonia is traditionally regarded as a basal ganglia disorder although gross structural abnormalities are not present (hence the name "primary"). However, imaging and clinicopathological studies provided several evidences for basal ganglia abnormalities in "secondary" dystonia. To further support the functional anatomy of dystonia, we recently described the possible correlations between clinico-radiological features and pathophysiological mechanisms in patients with dystonia secondary to focal brain lesions [47].

By analogy with the dystonic manifestations observed in patients with focal lesions of the basal ganglia (putamen and caudate) [48], it is largely assumed that dystonia may result from dysfunction of the circuits connecting the basal ganglia with the cortical motor areas. However, it is not completely clear how the basal ganglia abnormalities could produce dystonia. According to the traditional model of basal ganglia functional organization, the "direct" pathway (striatum-internal globus pallidus-thalamus-cortex) would provide a facilitatory influence on the desired movements, while the "indirect" pathway (striatum-external globus pallidus-subthalamic nucleus-internal globus pallidus-thalamus-cortex) exerts an inhibitory effect on unwanted movements (Fig. 2.4) [18]. The imbalance in the direct and indirect pathways should lead to the excessive movement peculiar to dystonia.

However, recent imaging and physiological studies in patients with focal hand dystonia documented an abnormal connectivity within the sensorimotor system between the basal ganglia and the dorsal premotor cortex [49, 50]. Therefore, modern views indicate that primary dystonia might be regarded as a neurodevelopmental circuit disorder, involving several functional circuits

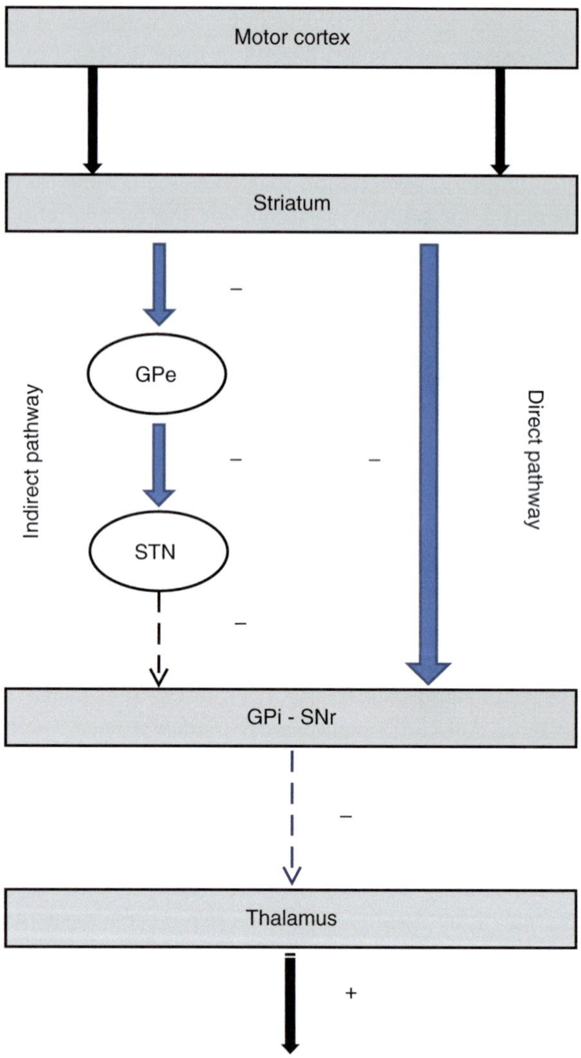

Fig. 2.4 Abnormal function of basal ganglia circuits in dystonia. The imbalance in the "direct" and "indirect" pathways increases the cortical motor tonic drive leading to excessive movement of dystonic patients

(the cortico-striato-pallido-thalamo-cortical and the cerebello-thalamo-cortical pathways) [51, 52].

2.6 The Role of Cerebellum

Several pieces of evidence suggest a possible involvement of the cerebellum in the pathophysiology of dystonia. Neuropathological studies of primary dystonia are extremely limited in number, and their results have been inconsistent; in particular, there was no specific exploration of cerebellar pathology or cerebellar-basal ganglia projections in brains of dystonic patients. However, the evidence that damage to the cerebellum can result in dystonia is provided by some forms of secondary dystonia [53]. Although in primary dystonia there is an absence of cerebellar signs on clinical examination, a number of case reports do link dystonia to structural lesions of the cerebellum [54]. Further, patients with degenerative cerebellar disorders (SCA3, DYTCA) commonly present dystonia as a part of their clinical phenotype [55].

In humans, some neurophysiological studies have provided evidence supporting the role of the cerebellum in the pathophysiology of dystonia. Brighina et al. showed a reduced cerebellar modulation on motor cortex excitability in patients with focal upper limb dystonia [56]. Reduced or absent cerebellar modulation has also been reported in patients with ataxia of various origins [57] or with lesions of the cerebellum or the dentate-thalamocortical pathways and in patients with focal cerebellar lesions and hemicerebellectomy [58], suggesting that dysfunctioning Purkinje cells in the cerebellar cortex might affect the inhibitory drive to the dentate-thalamocortical pathways. The eyeblink conditioning paradigm is a test strongly dependent on the olivocerebellar circuit and represents a good neurophysiological indicator of cerebellar function [59]. In 2008, Teo et al. showed that healthy controls produced more conditioned eyeblink responses than patients with focal dystonia indicating an abnormality of associative learning in this patient population [60].

Finally, it is noteworthy that dystonic patients present abnormalities in motor control that go beyond the sole movement execution and involve cognitive aspects of motor control, such as the processing of sensory input and imagination of movement [20, 33, 61–63]. The cerebellum plays a crucial role in integrating signals deriving from different sensory systems with the aim of contributing to a correct motor planning and might contribute to motor control abnormalities in dystonia [64].

However, so far, the exact role of the cerebellum in the pathophysiology of dystonia is far from being clarified [54, 64]. Three different hypotheses can be postulated: (1) both the cerebellum and basal ganglia might be abnormal in functioning (via separate loops toward the motor cortex); (2) abnormal function of one of the two structures (basal ganglia or the cerebellum) might in turn induce an abnormal functioning of the other since the cerebellum and basal ganglia are anatomically and functionally connected; (3) damage to one of the two structures may lead to compensatory reactions of the other since the two motor systems present somewhat overlapping functions.

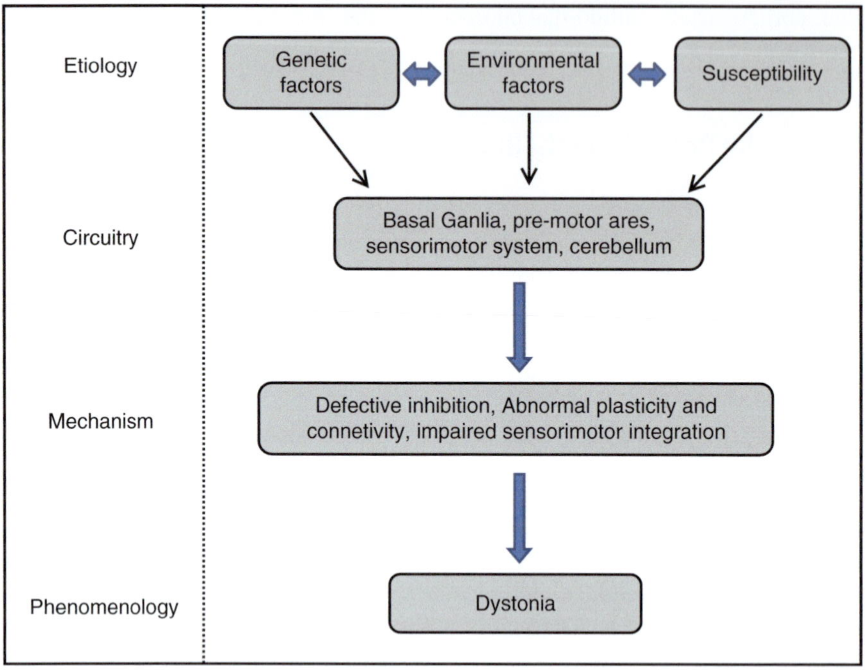

Fig. 2.5 The multistep process underlying dystonia (Modified from: Albanese and Lalli [66])

Conclusions

Although physiological studies documented a number of abnormalities (deficit of inhibitory mechanisms, increased cortical plasticity, abnormal sensorimotor integration, abnormal functional connectivity), the definite mechanisms underlying dystonia are still unclear, and a unifying theory is lacking.

We may hypothesize that genetic influences play a relevant role making subjects more susceptible to develop abnormalities of motor control mechanisms. In particular, the defective inhibition is likely to contribute to altered sensorimotor representations within the central nervous system. Environmental factors (such as repetitive training or peripheral nervous system injuries) may act on individuals with a genetic background triggering a "maladaptive plasticity" [65] or an abnormal connectivity. Therefore, the pathophysiology of dystonia can be seen as a multistep process (Fig. 2.5).

References

1. Fahn S, Marsden CD, Calne DB. Classification and investigation of dystonia. In: Marsden CD, Fahn S, editors. Movement disorders, vol. 2. London: Buttherworths; 1987. p. 332–58.
2. Albanese A, Asmus F, Bhatia KP, et al. EFNS guidelines on diagnosis and treatment of primary dystonias. Eur J Neurol. 2011;18:5–18.

3. Breakefield XO, Blood AJ, Li Y, et al. The pathophysiological basis of dystonias. Nat Rev Neurosci. 2008;9:222–34.
4. Defazio G, Berradelli A, Hallett M. Do primary adult-onset focal dystonias share aetiological factors? Brain. 2007;130:1183–93.
5. de Vries PM, Johnson KA, de Jong BM, et al. Changed patterns of cerebral activation related to clinically normal hand movement in cervical dystonia. Clin Neurol Neurosurg. 2008;110:120–812.
6. Pelosin E, Bove M, Marinelli L, Abbruzzese G, Ghilardi MF. Cervical dystonia affects aimed movements of nondystonic segments. Mov Disord. 2009;24:1955–61.
7. Nakashima K, Rothwell JC, Day BL, et al. Reciprocal inhibition between forearm muscles in patients with writer's cramp and other occupational cramps, symptomatic hemidystonia and hemiparesis due to stroke. Brain. 1989;112:681–97.
8. Berardelli A, Rothwell JC, Day BL, Marsden CD. Pathophysiology of blepharospasm and oromandibular dystonia. Brain. 1985;108:593–608.
9. Nakashima K, Rothwell JC, Thompson PD, et al. The blink reflex in patients with idiopathic torsion dystonia. Arch Neurol. 1990;47:413–6.
10. Tisch S, Limousin P, Rothwell JC, et al. Changes in forearm reciprocal inhibition following pallidal stimulation for dystonia. Neurology. 2006;66:1091–3.
11. Tinazzi M, Squintani G, Berardelli A. Does neurophysiological testing provide the information we need to improve the clinical management of primary dystonia? Clin Neurophysiol. 2009;120:1424–232.
12. Hallett M. Neurophysiology of dystonia: the role of inhibition. Neurobiol Dis. 2011;42:177–84.
13. Ridding MC, Sheean G, Rothwell JC, Inzelberg R, Kujirai T. Changes in the balance between motor cortical excitation and inhibition in focal, task specific dystonia. J Neurol Neurosurg Psychiatry. 1995;59:493–8.
14. Gilio F, Curra A, Lorenzano C, et al. Effects of botulinum toxin type A on intracortical inhibition in patients with dystonia. Ann Neurol. 2000;48:20–6.
15. Edwards MJ, Huang YZ, Wood NW, Rothwell JC, Bhatia KP. Different patterns of electrophysiological deficits in manifesting and non-manifesting carriers of the DYT-1 gene mutation. Brain. 2003;126:2074–80.
16. Espay AJ, Morgante F, Purzner J, et al. Cortical and spinal abnormalities in psychogenic dystonia. Ann Neurol. 2006;59:825–34.
17. Avanzino L, Martino D, van de Warrenburg BPC, et al. Cortical excitability is abnormal in patients with the "Fixed Dystonia" syndrome. Mov Disord. 2008;23:646–52.
18. Mink JW. The basal ganglia: focused selection and inhibition of competing motor programs. Prog Neurobiol. 1996;50:381–425.
19. Sohn YH, Hallett M. Disturbed surround inhibition in focal hand dystonia. Ann Neurol. 2004;56:595–9.
20. Quartarone A, Bagnato S, Rizzo V, et al. Corticospinal excitability during motor imagery of a simple tonic finger movement in patients with writer's cramp. Mov Disord. 2005;20:1488–95.
21. Di Lazzaro V, Oliviero A, Saturno E, et al. Effects of lorazepam on short latency afferent inhibition and short latency intracortical inhibition in humans. J Physiol. 2005;564:661–8.
22. Matsumura M, Sawaguchi T, Kubota K. GABAergic inhibition of neuronal activity in the primate motor and premotor cortex during voluntary movement. J Neurophysiol. 1992;68:692–702.
23. Levy LM, Hallett M. Impaired brain GABA in focal dystonia. Ann Neurol. 2002;51:93–101.
24. Stamelou M, Edwards MJ, Hallett M, Bhatia KP. The non-motor syndrome of primary dystonia: clinical and pathophysiological implications. Brain. 2012;135:1668–81.
25. Abbruzzese G, Berardelli A. Sensorimotor integration in movement disorders. Mov Disord. 2003;18:231–40.
26. Tinazzi M, Rosso T, Fiaschi A. Role of the somatosensory system in primary dystonia. Mov Disord. 2003;18:605–22.
27. Tinazzi M, Fiorio M, Fiaschi A, Rothwell JC, Bhatia KP. Sensory functions in dystonia: insights from behavioral studies. Mov Disord. 2009;24:1427–36.

28. Molloy FM, Carr TD, Zeuner KE, et al. Abnormalities of spatial discrimination in focal and generalized dystonia. Brain. 2003;126:2175–82.
29. Fiorio M, Gambarin M, Valente EM, et al. Defective temporal processing of sensory stimuli in DYT1 mutation carriers: a new endophenotype of dystonia? Brain. 2007;130:134–42.
30. Morgante F, Tinazzi M, Squintani G, et al. Abnormal tactile temporal discrimination in psychogenic dystonia. Neurology. 2011;77:1191–7.
31. Bradley D, Whelan R, Walsh R, et al. Temporal discrimination threshold: VBM evidence for an endophenotype in adult onset primary torsion dystonia. Brain. 2009;132:2327–35.
32. Kanovský P. Dystonia: a disorder of motor programming or motor execution? Mov Disord. 2002;17:1143–7.
33. Abbruzzese G, Marchese R, Buccolieri A, et al. Abnormalities of sensorimotor integration in focal dystonia: a transcranial magnetic stimulation study. Brain. 2001;124:537–45.
34. Bara-Jimenez W, Catalan MJ, Hallett M, et al. Abnormal somatosensory homunculus in dystonia of the hand. Ann Neurol. 1998;44:828–31.
35. Meunier S, Garnero L, Ducorps A, et al. Human brain mapping in dystonia reveals both endophenotypic traits and adaptive reorganization. Ann Neurol. 2001;50:521–7.
36. Butterworth Butterworth S, Francis S, Kelly E. Abnormal cortical sensory activation in dystonia: an fMRI study. Mov Disord. 2003;18:673–82.
37. Thickbroom GW, Byrnes ML, Stell R, Mastaglia FL. Reversible reorganisation of the motor cortical representation of the hand in cervical dystonia. Mov Disord. 2003;18:395–402.
38. Trompetto C, Currà A, Buccolieri A, Suppa A, Abbruzzese G, Berardelli A. Botulinum toxin changes intrafusal feedback in dystonia: a study with the tonic vibration reflex. Mov Disord. 2006;21:777–82.
39. Kojovic M, Caronni A, Bologna M, Rothwell JC, Bhatia KP, Edwards MJ. Botulinum toxin injections reduce associative plasticity in patients with primary dystonia. Mov Disord. 2011;26:1282–9.
40. Quartarone A, Bagnato S, Rizzo V, et al. Abnormal associative plasticity of the human motor cortex in writer's cramp. Brain. 2003;126:2586–96.
41. Quartarone A, Morgante F, Sant' Angelo A, et al. Abnormal plasticity of sensorimotor circuits extends beyond the affected body part in focal dystonia. J Neurol Neurosurg Psychiatry. 2008;79:985–90.
42. Ghilardi MF, Carbon M, Silvestri G, et al. Impaired sequence learning in carriers of the DYT1 dystonia mutation. Ann Neurol. 2003;54:102–9.
43. Huang YZ, Edwards MJ, Rounis E, Bhatia KP, Rothwell JC. Theta burst stimulation of the human motor cortex. Neuron. 2005;45:201–6.
44. Edwards MJ, Huang YZ, Mir P, Rothwell JC, Bhatia KP. Abnormalities in motor cortical plasticity differentiate manifesting and nonmanifesting DYT1 carriers. Mov Disord. 2006;21:2181–6.
45. Quartarone A, Rizzo V, Terranova C, et al. Abnormal sensorimotor plasticity in organic but not in psychogenic dystonia. Brain. 2009;132:2871–7.
46. Ruge D, Tisch S, Hariz MI, et al. Deep brain stimulation effects in dystonia: time course of electrophysiological changes in early treatment. Mov Disord. 2011;26:1913–21.
47. Trompetto C, Avanzino L, Marinelli L, et al. Corticospinal excitability in patients with secondary dystonia due to focal lesions of the basal ganglia and thalamus. Clin Neurophysiol. 2012;123:808–14.
48. Bhatia KP, Marsden CD. The behavioural and motor consequences of focal lesions of the basal ganglia in man. Brain. 1994;117:859–76.
49. Jin SH, Lin P, Auh S, Hallett M. Abnormal functional connectivity in focal hand dystonia: mutual information analysis in EEG. Mov Disord. 2011;26:1274–81.
50. Delnooz C, Helmich RC, Toni I, van de Warrenburg BPC. Reduced parietal connectivity with a premotor writing area in writer's cramp. Mov Disord. 2012;27:1425–31.
51. Niethammer M, Carbon M, Argyelan M, Eidelberg D. Hereditary dystonia as a neurodevelopmental circuit disorder: evidence from neuroimaging. Neurobiol Dis. 2011;42:202–9.

52. Poston KL, Eidelberg D. Functional brain networks and abnormal connectivity in the movement disorders. Neuroimage. 2012;62:2261–70.
53. Ledoux MS, Brady KA. Secondary cervical dystonia associated with structural lesions of the central nervous system. Mov Disord. 2003;18:60–9.
54. Sadnicka A, Hoffland BS, Bhatia KP, van de Warrenburg BP, Edwards MJ. The cerebellum in dystonia. Help or hindrance? Clin Neurophysiol. 2012;123:65–70.
55. Münchau A, Dressler D, Bhatia KP, Vogel P, Zühlke C. Machado-Joseph disease presenting as severe generalised dystonia in a German patient. J Neurol. 1999;246:840–2.
56. Brighina F, Romano M, Giglia G, Saia V, Puma A, Giglia F, et al. Effects of cerebellar TMS on motor cortex of patients with focal dystonia: a preliminary report. Exp Brain Res. 2009;192:651–6.
57. Ugawa Y, Terao Y, Hanajima R, Sakai K, Furubayashi T, Machii K, Kanazawa I. Magnetic stimulation over the cerebellum in patients with ataxia. Electroencephalogr Clin Neurophysiol. 1997;104:453–8.
58. Di Lazzaro V, Restuccia D, Nardone R, Leggio MG, Oliviero A, Profice P, Tonali P, Molinari M. Motor cortex changes in a patient with hemicerebellectomy. Electroencephalogr Clin Neurophysiol. 1995;97:259–63.
59. Gerwig M, Kolb FP, Timmann D. The involvement of the human cerebellum in eyeblink conditioning. Cerebellum. 2007;6:38–57. Review.
60. Teo JT, van de Warrenburg BP, Schneider SA, Rothwell JC, Bhatia KP. Neurophysiological evidence for cerebellar dysfunction in primary focal dystonia. J Neurol Neurosurg Psychiatry. 2009;80:80–3.
61. Fiorio M, Tinazzi M, Aglioti SM. Selective impairment of hand mental rotation in patients with focal hand dystonia. Brain. 2006;129:47–54.
62. Fiorio M, Weise D, Önal-Hartmann C, Zeller D, Tinazzi M, Classen J. Impairment of the rubber hand illusion in focal hand dystonia. Brain. 2011;134:1428–37.
63. Castrop F, Dresel C, Hennenlotter A, Zimmer C, Haslinger B. Basal ganglia-premotor dysfunction during movement imagination in writer's cramp. Mov Disord. 2012;27:1432–9.
64. Avanzino L, Abbruzzese G. How does the cerebellum contribute to the pathophysiology of dystonia? Basal Ganglia. 2012;2:231–5.
65. Quartarone A, Siebner HR, Rothwell JC. Task-specific hand dystonia: can too much plasticity be bad for you? Trends Neurosci. 2006;29:192–9.
66. Albanese A, Lalli S. Update on dystonia. Curr Opin Neurol. 2012;25:483–90.

Genetics of Dystonia

3

Ana Westenberger and Christine Klein

The first dystonia gene, *GCH1*, was reported more than 20 years ago [1]. Since that time, 15 new dystonia genes have been discovered (Fig. 3.1, Table 3.1). Recent technological advances have considerably accelerated the process of gene identification through the advent of next-generation sequencing, as demonstrated by the discovery of a half of these dystonia genes within a six-year (2008–2013) period (Fig. 3.1).

Clinically and according to the recent consensus revision [28], dystonia is defined as a movement disorder characterized by sustained or intermittent muscle contractions causing abnormal, often repetitive, movements, postures, or both. Dystonic movements are typically patterned, twisting, and may be tremulous. Dystonia is often initiated or worsened by voluntary action and associated with overflow muscle activation.

More than a 100 years ago, in 1911, the German neurologist Hermann Oppenheim established the term dystonia in his report of four young patients that displayed a generalized abnormality of muscle tone with coexisting hypo- and

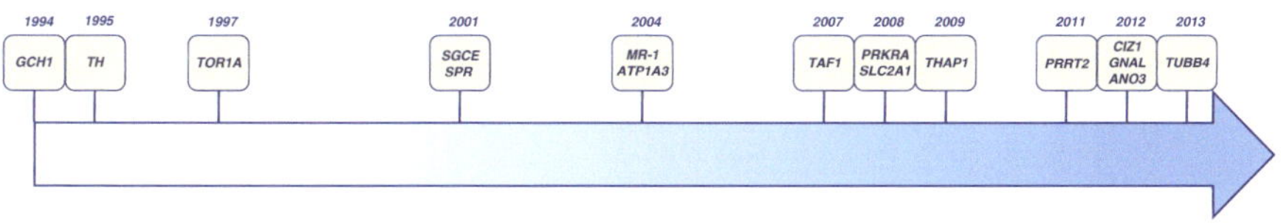

Fig. 3.1 Novel dystonia genes reported in each of the past 20 years. Names of the genes are shown within *yellow boxes*

A. Westenberger, PhD (✉) • C. Klein, MD
Institute of Neurogenetics, University of Lübeck, Lübeck, Germany
e-mail: ana.westenberger@neuro.uni-luebeck.de;
christine.klein@neuro.uni-luebeck.de

P. Kanovsky et al. (eds.), *Dystonia and Dystonic Syndromes*,
DOI 10.1007/978-3-7091-1516-9_3

Table 3.1 Monogenic forms of dystonia (DYTs)

Designation	Dystonia type	Mode of inheritance	Gene locus	Gene	OMIM number	Reference
DYT1	Early-onset generalized	Autosomal dominant	9q	*TOR1A*	128100	[2]
DYT2	Early-onset generalized	Autosomal recessive	Unknown	Unknown	224500	[3]
DYT3	X-linked dystonia-parkinsonism; "lubag"	X-linked recessive	Xq	*TAF1*	314250	[4]
DYT4	Whispering dysphonia	Autosomal dominant	19p	*TUBB4*	128101; 602662	[5]
DYT5a/DYT14	Dopa-responsive dystonia; Segawa syndrome	Autosomal dominant	14q	*GCH1*	128230	[1, 6]
DYT5b	Dopa-responsive dystonia; Segawa syndrome	Autosomal recessive	11p	*TH*	605407	[7]
DYT6	Adolescent-onset mixed phenotype	Autosomal dominant	8p	*THAP1*	602629	[8]
DYT7	Adult-onset focal	Autosomal dominant	18p	Unknown	602124	[9]
DYT8	Paroxysmal nonkinesigenic dyskinesia 1	Autosomal dominant	2q	*PNKD*	118800	[10, 11]
DYT10	Paroxysmal kinesigenic dyskinesia 1	Autosomal dominant	16p-q	*PRRT2*	128200	[12]
DYT11	Myoclonus-dystonia	Autosomal dominant	7q	*SGCE*	159900	[13]
DYT12	Rapid-onset dystonia-parkinsonism	Autosomal dominant	19q	*ATP1A3*	128235	[14]
DYT13	Adolescent-onset multifocal/ segmental	Autosomal dominant	1p	Unknown	607671	[15]
DYT15	Myoclonus-dystonia	Autosomal dominant	18p	Unknown	607488	[16]
DYT16	Early-onset generalized with parkinsonism	Autosomal recessive	2p	*PRKRA*	603424	[17]
DYT17	Adolescent onset	Autosomal recessive	20pq	Unknown	612406	[18]

DYT18/DYT9	Paroxysmal exertion-induced dyskinesia	Autosomal dominant	1p	*SLC2A1 (GLUT1)*	612126	[19, 20]
DYT19	Paroxysmal kinesigenic dyskinesia 2	Autosomal dominant	16q	Unknown	611031	[21]
DYT20	Paroxysmal nonkinesigenic dyskinesia 2	Autosomal dominant	2q	Unknown	607488	[22]
DYT21	Adult-onset generalized/multifocal	Autosomal dominant	2q	Unknown	614588	[23]
DYT23	Adult-onset cervical dystonia	Autosomal dominant	9q	*CIZ1*	614860	[24]
DYT24	Adult-onset craniocervical dystonia	Autosomal dominant	11p	*ANO3*	615034	[25]
DYT25	Adult-onset cervical dystonia	Autosomal dominant	18p	*GNAL*	615073	[26]
No DYT designation yet	Dopa-responsive dystonia	Autosomal recessive	2p	*SPR*	612716	[27]

Adapted from www.ncbi.nlm.nih.gov/omim/

DYT22 has not yet been published

hypertonia [29]. In this fascinating paper Oppenheim noted that all of his patients were of Jewish descent, mostly of Russian origin, and that "hereditary burden likely plays a major role" [29]. Importantly, Oppenheim insisted that their disorder was of "organic nature" and not psychogenic (hysteria) as was commonly claimed by his peers.

In this chapter we will discuss the current knowledge and understanding of the molecular genetics of dystonia. In brief, we will review the genetic and clinicogenetic classification of dystonia and summarize the known molecular mechanisms leading to the major monogenic forms of dystonia. We will also describe state-of-the-art methods for the identification of dystonia genes and implications for genetic testing.

3.1 Classification of Dystonia

Based on the distribution of symptoms, dystonias may be focal (affecting only a single part of the body), segmental (affecting several adjacent body parts), generalized (affecting most of the body), or restricted to one side of the body (hemidystonia).

Monogenic forms of dystonia, caused by mutations in a single gene, are overall rare. In the current dystonia genetics nomenclature, 23 different chromosomal regions (loci) or genes are termed DYT and numbered in chronological order of their identification (Table 3.1). However, this listing of DYTs has several inconsistencies, and thus it should serve as an orientation and not as a classification in the true sense of the word. Rather, the monogenic forms can be clinicogenetically grouped as "isolated" and "combined" dystonias. In isolated dystonias (DYT1, DYT2, DYT4, DYT6, DYT7, DYT13, DYT17, DYT21, DYT23, DYT24, and DYT25) dystonia is the only sign of the disease, with the possible exception of tremor. In combined dystonias, that can have a persistent (DYT3, DYT5a, DYT5b, DYT11, DYT12, DYT15, DYT16, and *SPR*) or paroxysmal (episodic) temporal pattern (DYT8–10 and DYT18–20) (Fig. 3.2), an additional movement disorder or a motor symptom other than tremor accompanies the dystonia.

3.1.1 Isolated Dystonias

Causative genes have been known for a number of years for the early onset, mostly generalized forms of dystonia DYT1 and DYT6 [2, 8, 17]. In addition, *GNAL* mutations were recently found to be responsible for rare cases of the adult-onset form of focal dystonia (DYT25). In 2012 and 2013, mutations in the *TUBB4* gene (DYT4), *CIZ1* (DYT23), and *ANO3* (DYT24) have also been linked to adult-onset isolated dystonias in single reports and await independent validation [5, 24–26]. Among isolated dystonias, DYT1 is best studied, and it seems to be the

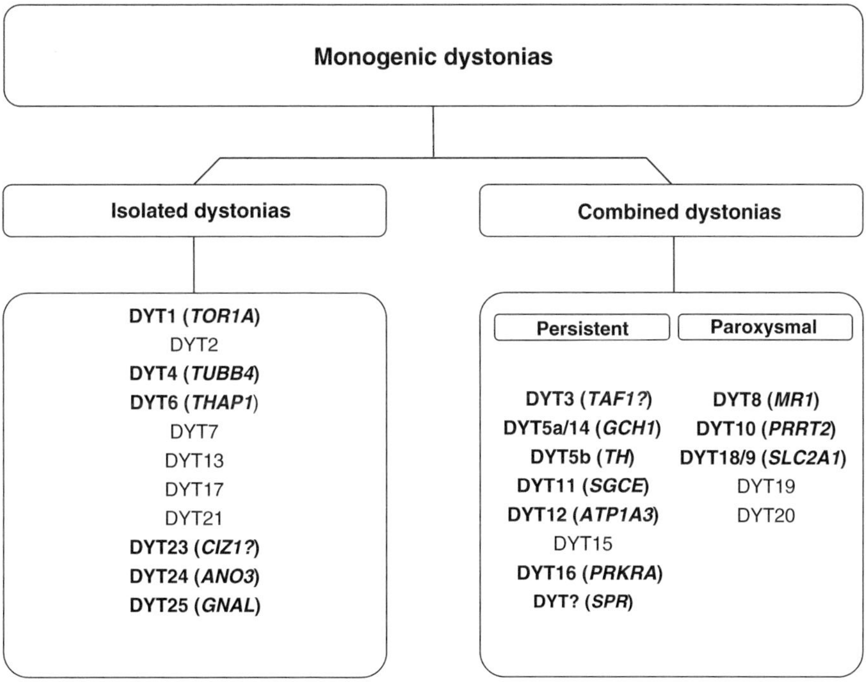

Fig. 3.2 Scheme of dystonia classification. The DYT loci within which dystonia-associated genes are known are shown in *bold*. Question marks denote loci with assumed but uncertain genetic causes (i.e. *TAF1, CIZ1*) or genes currently without DYT designation (i.e. *SPR*)

most prevalent. DYT1 is particularly frequent in Ashkenazi Jews, in whom it accounts for about 80 % of early onset generalized dystonia [30]. DYT6 is the second most common form of isolated dystonia with more than 60 mutation-carrying families reported to date. It is probably too early to draw any firm conclusions about the frequency of mutations in the four recently reported genes [5, 24–26]. However, it seems that among those, *GNAL* mutations might be the most common cause of dystonia [31–35]. Thus far, DYT4 dystonia has been described in a single family and likely in one additional familial patient [5]. Possibly pathogenic variants in *CIZ1* and *ANO3* have been found in about 1 % of the investigated dystonia patients [24, 25], whereas *GNAL* mutations were identified in 19 % of analyzed families [26].

All above mentioned isolated dystonias with known genetic cause follow an autosomal dominant pattern of inheritance and have reduced penetrance. In addition, in all isolated forms of dystonia, type and severity of the symptoms vary among affected individuals, signifying variable disease expressivity.

In the following chapters we will list and review single isolated forms of dystonia in the order of their currently known frequency and significance.

3.1.1.1 *TOR1A* (DYT1)

DYT1 dystonia is characterized by an early onset with a mean age of onset in childhood or adolescence. The disease starts in a limb in 90 % of the cases and often generalizes with a caudo-rostral gradient, usually sparing the cranial muscles [36]. Nevertheless, in rare instances, DYT1 dystonia may have an only mild manifestation, such as writer's cramp [37].

Due to a founder mutation, DYT1 is particularly frequent among Ashkenazi Jews, as already pointed out by Oppenheim who likely provided the first detailed clinical description of DYT1 dystonia [29]. Penetrance of the disease is significantly reduced to about 30 %, and thus it affects only every third carrier of the disease-causing mutation. This mutation is identical in almost all cases, a GAG deletion in exon 5 of the *TOR1A* gene encoding the torsinA protein. This three-nucleotide deletion results in an in-frame deletion of glutamic acid. Three other putatively disease-causing mutations have been found in single cases only [38–40]. Interestingly, the frequency of a common polymorphism (torsinA p.D216H) is increased in non-manifesting carriers and decreased in carriers with dystonia, implying a possible protective effect [41].

TorsinA is a member of the AAA (ATPases Associated with various cellular Activities) family of molecular chaperones that mediate conformational changes in target proteins. Interestingly, while wild-type torsinA is predominantly found in the endoplasmic reticulum (ER), mutated protein is located in the nuclear envelope [42]. Transfection of mutant protein into cells with wild-type protein results in re-localization of wild-type protein to the nuclear envelope [42]. Thus, it seems that the mutant exhibits a dominant-negative effect by recruiting wild-type torsinA to the nuclear envelope. Neurons from both torsinA null and homozygous GAG mutant "knockin" mice contain severely abnormal nuclear membranes that are absent in non-neuronal cell types [43].

3.1.1.2 *THAP1* (DYT6)

In contrast to DYT1, the second early onset isolated dystonia, DYT6, often starts in the craniocervical region and has a tendency to spread to adjacent body regions showing a rostro-caudal gradient and involving the legs in only about 4 % of the cases [36]. Thus, speech involvement and dysphonia, which are rarely seen in DYT1, are common features in DYT6.

DYT6 was first described in five related Amish-Mennonite families [44] in whom the disease was present with reduced penetrance (to about 60 %). However, after the responsible gene was identified, it became clear that its role was not restricted only to families of Amish-Mennonite origin. The disease gene, *THAP1*, unlike *TOR1A*, shows significant mutational heterogeneity, and there are currently over 60 different missense and truncating *THAP1* mutations reported mainly in European but also in Chinese and Brazilian patients [45]. The *THAP1* gene codes for the THAP domain-containing apoptosis-associated protein 1 transcription factor that has a conserved DNA-binding domain (THAP domain) at its N-terminus and a nuclear localization signal (NLS) at its C-terminus. Regions encoding these two important domains are most frequently affected by mutations that disturb either the ability of the protein to bind DNA or its ability to be translocated into the nucleus.

Notably, a THAP-binding site was found in the promoter of the *TOR1A* gene, and in HeLa cells and human neuroblastoma (SH-SY5Y) cells, THAP1 specifically binds to that site and suppresses promoter activity in a concentration-dependent manner [46]. In addition, DYT6-associated *THAP1* mutations abolished THAP1-mediated repression of *TOR1A* to various degrees. Thus, *TOR1A* appears to be a target for the transcription factor activity of THAP1, suggesting a molecular link between DYT1 and DYT6.

3.1.1.3 *GNAL* (DYT25)

Recently, the first and long-awaited gene convincingly associated with adult-onset (segmental) dystonia – *GNAL* – has been identified using exome sequencing in two unrelated families [26]. Mutational screening of *GNAL* in 39 additional families revealed six additional likely pathogenic mutations in six different families [26]. This was followed by multiple independent studies identifying *GNAL* mutation carriers among German, Serbian, Chinese, Japanese, and African-American patients [31, 33–35]. The average age of dystonia onset in the mutation carriers was 31.3 years (range: 7–54 years). More than 80 % of the mutation carriers had onset in the neck, and in most of them the dystonia progressed to other body sites. In addition, 44 % of the patients had speech involvement. Of note, this type of dystonia shows almost complete penetrance, given that out of the eight families, only a single mutation carrier in one family was not affected. No phenotype-genotype correlation could be drawn with respect to different mutations, and, moreover, phenotypes varied within single families.

GNAL encodes the stimulatory α-subunit, $G\alpha_{olf}$, first identified as a guanine nucleotide-binding protein (G protein) that mediates odorant signaling in the olfactory epithelium [49]. G proteins are heterodimers that consist of α-, β-, and γ-subunits and usually serve as a molecular link between transmembrane receptor and downstream effector molecules. Heterodimer containing $G\alpha_{olf}$ associates with dopamine type 1 receptors, and *GNAL* mutations significantly impair the ability of $G\alpha_{olf}$ to couple with β- and γ-subunits.

3.1.1.4 *CIZ1* (DYT23)

Using linkage analysis and exome sequencing, an exonic splicing enhancer mutation in the *CIZ1* gene was identified in a large Caucasian kindred with focal (cervical) dystonia with five definitely and five possibly affected members [24]. None of the patients was affected in a body region beyond the neck [50]. In all definitely affected individuals, the first symptoms appeared in adulthood (mean age of onset: 35 years, range: 18–49 years). Upon follow-up examination, the dystonia remained focal, i.e., it was restricted to the neck and shoulder. Screening of an additional 300 cervical dystonia patients revealed only another two likely pathogenic missense changes. *CIZ1* encodes the Cip1-interacting zinc finger protein 1 that is expressed in the brain and involved in DNA synthesis and cell-cycle control [50].

As a general caveat regarding recent methods of gene identification, the utility of exome sequencing is still limited by a considerable amount of false-positive and false-negative results. False-positive results can subsequently be corrected through Sanger sequencing; however, false-negative results remain unrecognized. For example, in the original report of *CIZ1* mutations, 40 % of the exome was covered with low-read-depth sequencing only [24]. Therefore, as the authors rightfully stressed, the association between *CIZ1* and cervical dystonia should be interpreted with caution [24], and independent confirmation of this link is required [51].

3.1.1.5 *ANO3* (DYT24)

Recently, *ANO3* has been reported as another gene responsible for adult-onset cervical dystonia. Investigation of a three-generational family with five affected members by linkage analysis and exome sequencing revealed a missense mutation in exon 15 of the *ANO3* gene coding for anoctamin 3 [25]. All patients from this family had tremulous cervical dystonia with associated upper limb dystonic tremor. In addition, three affected individuals showed laryngeal involvement. Age of onset was between 19 and 39 years, although most of the patients developed their initial symptoms in their late 30s. Analysis of exon 15 in additional 384 familial and sporadic patients revealed another family with an *ANO3* missense mutation and an almost identical phenotype; however, the index patient had had disease onset in her early teens. Screening of the entire gene in 188 cervical dystonia cases resulted in the discovery of additional three missense and one 5'UTR variant [25].

Anoctamin 3 belongs to the family of proteins that encode Ca^{2+}-activated Cl^{-} channels and is highly expressed in the striatum [25]. Functional studies of a fibroblast culture carrying the original exon 15 mutation revealed abnormalities in Ca^{2+} signaling, possibly caused by a defect in ER-related Ca^{2+} handling in these cells [25].

3.1.1.6 *TUBB4* (DYT4)

Hereditary "whispering dysphonia," later designated DYT4, was first described in an Australian family with 20 affected family members [52]. The phenotype of this disease is clinically characterized by craniocervical dystonia with prominent spasmodic dysphonia and shows variable expressivity within the family. The dystonia frequently generalizes and is at least partially responsive to alcohol and propranolol [53]. In addition, affected individuals have characteristic clinical features beyond the motor phenotype, i.e., a thin face and body habitus that completely co-segregated with the DYT4 motor phenotype [53]. DYT4 dystonia follows an autosomal dominant mode of inheritance with seemingly high penetrance based on lack of unaffected obligate carriers and a high number of affected individuals in large sibships. Notably, two recent studies of the same original DYT4 family independently identified a novel mutation (p.R2G) in the *tubulin beta-4* (*TUBB4*) gene as the likely genetic cause of "whispering dysphonia" [5, 54]. *TUBB4* expression was subsequently analyzed in three different cell types revealing that the amount of the TUBB4 protein in the heterozygous mutation carrier was significantly reduced compared to controls. Furthermore, a second missense mutation was found in an independent patient with familial craniocervical dystonia [5].

Interestingly, the spectrum of *TUBB4* mutations has now been shown to be much broader and includes hypomyelination with atrophy of the basal ganglia and cerebellum (H-ABC) and similar phenotypes [55, 56], thereby challenging the classification of DYT4 as an isolated form of dystonia.

3.1.2 Combined Dystonias

In the combined dystonias, an additional movement disorder or a motor symptom other than tremor accompanies the dystonia. Notably, the dystonia and the associated feature are usually permanently present; however, they may also occur in an episodic fashion induced by a trigger. Thus, based on their temporal expression pattern, combined dystonias may be subgrouped into persistent combined dystonias (DYT3, DYT5a, DYT5b, DYT11, DYT12, DYT15, DYT16, and *SPR*) or paroxysmal (episodic) combined dystonias (DYT8–10 and DYT18–20). Persistent combined dystonia commonly have parkinsonism (DYT3, DYT5a, DYT5b, DYT12, DYT16, and *SPR*) or myoclonus (DYT11 and DYT15) as the additional feature. Below we will list and discuss the combined forms of dystonia with known genetic causes based on their frequency.

3.1.2.1 Persistent Combined Dystonias with Parkinsonism

GCH1 (DYT5a), *TH* (DYT5b), and *SPR* (no DYT Designation)

The typical phenotype of DYT5 or dopa-responsive dystonia (DRD) includes childhood-onset leg dystonia. In later-onset patients parkinsonism is usually more prominent [57]. DRD is characterized by worsening of the symptoms as the day progresses, improvement after sleep, and excellent response to L-dopa therapy, hence the name. In the case of DRD, effects of the causative mutations are very well understood. The autosomal dominant form of DRD occurs due to the mutations in the gene coding for GTP cyclohydrolase 1 (*GCH1*) [1], while changes in the *tyrosine hydroxylase* (*TH*) [7] gene cause an autosomal recessive form. Both of these enzymes are required for dopamine synthesis. Mutations in a gene encoding another enzyme in the dopamine synthesis pathway, *sepiapterin reductase* (*SPR*), are responsible for a second autosomal recessively inherited DRD phenotype [27]. However, this gene has not been assigned a DYT designation. Importantly, mutations in *TH* and *SPR* cause a much more severe clinical phenotype than dopa-responsive dystonia due to *GCH1* mutations and resemble the phenotype observed in the rare carriers of homozygous *GCH1* mutations [58].

To date, more than 100 different mutations, spread across the entire *GCH1* coding region, have been reported and include missense, nonsense, and splice site mutations, small and large (whole-exon or whole-gene) deletions, and mutations in the untranslated regions. In contrast, mutations in *TH* and *SPR* are much less frequent ($n<20$). Of note, the DYT14 locus had been designated erroneously because of lack of gene dosage analysis of the *GCH1* gene in the proband and initial clinical misclassification of one family member impacting on the genetic linkage analysis. This family with clinically typical DRD was later discovered to harbor a heterozygous deletion of exons 3–6 of the *GCH1* gene [6].

ATP1A3 (DYT12)

In DYT12 or rapid-onset dystonia-parkinsonism (RDP), onset of dystonia is abrupt, within hours or weeks, and the dystonia typically follows a rostro-caudal gradient (face > arm > leg). The first symptoms usually appear in adolescence or in the early 20s and are frequently triggered by physical or emotional stress. Within 1 month from onset, stabilization may occur; however, abrupt worsening often follows later in life. Disease-causing mutations were identified in the alpha 3 subunit of a sodium pump that uses ATP hydrolysis to exchange Na^+ and K^+ across the cell membrane [14]. Functional biochemical studies with several pathogenic mutations showed reduced Na^+ affinity [59] suggesting that a defect in handling of Na^+ may be a general pathophysiological feature of the disease. Pharmacological replication of the loss-of-function sodium pump mutations in the basal ganglia of mice caused parkinsonism-like symptoms, but no involuntary dystonic movements in the treated animals [60]. However, treatment of the cerebellum resulted initially in ataxia that over time transformed into generalized dystonia [60]. Thus, these data suggest involvement of both the basal ganglia and the cerebellum in the generation of RDP symptoms. Combining the pharmacological treatment with stress increased the severity of dyskinesia and resulted in mild to severe dystonia [60].

The spectrum of phenotypes associated with mutations in *ATP1A3* has been recently expanded. Namely, it has been shown that the *ATP1A3* mutations are the cause of 74 % of alternating hemiplegia of childhood (AHC) cases [61]. AHC is a severe neurodevelopmental syndrome characterized by recurrent hemiplegic episodes and distinct neurological manifestations. Notably, most AHC cases are caused by one of seven recurrent *ATP1A3* mutations, one of which was observed in 35 % of the patients [61].

PRKRA (DYT16)

Clinically, DYT16 frequently starts in the legs but spreads extensively and includes prominent bulbar involvement, dysphonia, dysarthria, and dysphagia. The 10 out of 12 DYT16 patients described in the literature showed a generalized phenotype with varying severity [17, 47].

DYT16 is an autosomal recessive form of combined dystonias and to date has been described only in two Brazilian families, one Polish family, and four sporadic patients [17, 47, 48]. While the three families and one sporadic patient carried homozygous missense mutations in the *PRKRA* gene [17, 48], a single heterozygous protein-truncating mutation was found in a sporadic case from Germany [47]. In addition, two more heterozygous missense changes were found in two Polish patients [48]. These two mutations in accord with the first described heterozygous frameshift mutation [47] further raise the possibility that heterozygous *PRKRA* variants might constitute a risk for dystonia. Probably due to the apparent rarity of mutations in this gene, few *PRKRA* screening studies have been performed. Therefore, the mutational and associated phenotypic spectrum is likely largely unknown. Likewise, little is known about the DYT16 disease mechanism. The *PRKRA* gene is involved in cellular stress response; however, it is unclear how this may lead to dystonia.

TAF1 (DYT3)

In several ways, DYT3 is unique in comparison to other forms of dystonia: (1) this X-linked disorder is recessively inherited, thus affecting predominantly males; (2) it is the only neurodegenerative form of the DYTs; and (3) it has currently been confirmed, due to a genetic founder effect, only in individuals of Filipino ancestry [62]. Thus, although probably rare worldwide, DYT3 is prevalent in the Philippines, especially on Panay island (5.74 per 100,000), where the genetic founder originated [62]. Initial symptoms of DYT3, focal dystonia or less frequently parkinsonism, usually emerge in the fourth decade of life. After several years, parkinsonism appears or becomes more prominent, and dystonia generalizes, incapacitating the patient and leading to premature death from pneumonia or starvation. If, however, the patient survives for approximately 15 years after the onset of symptoms, he will enter the parkinsonian phase in which dystonia subsides and parkinsonism is the only feature. Thus, DYT3 is also known as X-linked dystonia-parkinsonism (XDP), and it may even be considered a form of parkinsonism.

A large retrotransposon insertion within intron 32 of the *TAF1* gene coding for the largest subunit of the transcription factor IID is currently considered to be the disease-causing change [63]. As a pathogenic mechanism, it was proposed that this insertion interrupts a potential neuron-specific cis-regulatory element within intron 32, thereby downregulating the expression of a neuron-specific TAF1 isoform. This may, in turn, downregulate transcription of neuron-specific proteins [63].

However, the genetic cause and molecular pathways leading to XDP are currently not completely understood, given that within the 427-kb-linked region, there are six additional sequence variants (five disease-specific single-nucleotide changes (DSCs 1, 2, 3, 10, and 12) and one 48-bp deletion)) found exclusively in XDP patients. These changes co-segregate due to linkage disequilibrium. Therefore, which one, if any, of the disease-specific changes is pathogenic currently remains a matter of debate. To test this, we analyzed the seven changes in 166 patients and 473 Filipino controls from the island of Panay. The aim of this approach was to detect recombination events in patients or single occurrences of any of the genetic variants in controls, which would indicate through segregation analysis which variant/s is/are truly specific to the phenotype. Surprisingly, all seven reported changes co-segregate in a single haplotype [64]. Therefore, although the genetic cause of XDP remains unsolved, these results could indicate that the original XDP proband, the first person to ever be affected, has come from another population and was not of Filipino ethnic origin. Using the next-generation sequencing approach, we identified six single-nucleotide variants that flank the DSCs. The subsequent genotyping of these SNVs in all of the patient and control samples allowed for definition of the boundaries of the disease-specific haplotype and narrowing the disease locus to a 294-kb region [64] that includes four known genes (*TAF1*, *OGT*, *ACRC*, and *CXCR3*).

3.1.2.2 Persistent Combined Dystonias with Myoclonus

SGCE (DYT11)

DYT11 or myoclonus-dystonia (MD) is a childhood-onset disorder in which dystonia is accompanied by brief lightening-like myoclonic jerks. The first symptoms occur earlier in girls than in boys [65]. In many patients, myoclonic jerks affect the neck, trunk, and upper limbs, and they show an excellent response to alcohol [66]. Various types of mutations have been described in the *SGCE* gene coding for epsilon-sarcoglycan. The mode of inheritance of MD is autosomal dominant, and penetrance is reduced due to imprinting of the gene, an epigenetic phenomenon that results in the selective silencing of one of the two parental alleles. This silencing occurs through *SGCE* promoter methylation [67]. Consequently, disease expression depends on the transmitting parent, and almost all individuals who inherit an *SGCE* mutation from their father will develop the disease.

Most *SGCE* mutations are thought to result in loss of function of the protein. Interestingly large deletions of the entire *SGCE* gene are frequently accompanied by a deletion of the neighboring gene *COL1A2* (collagen type I alpha 2) coding for the fibrillar collagen found in cartilage. Thus, MD patients carrying large deletions within the DYT11 locus may have associated phenotypes such as delayed skeletal development and severe osteoporosis [68, 69].

Epsilon-sarcoglycan is a member of the sarcoglycan family of transmembrane proteins that form a complex linking the actin cytoskeleton to the extracellular matrix. In cellular overexpression studies disease-associated epsilon-sarcoglycan missense mutants were unable to reach the plasma membrane [70]. Instead, they were polyubiquitinated and rapidly degraded intracellularly. Interestingly, this proteosomal degradation is augmented by binding of torsinA to the misfolded epsilon-sarcoglycan mutants [70], providing a possible link between DYT1 and DYT11. Additional confirmation of this association came from a recent study showing the presence of the abnormal nuclear envelopes, normally seen in torsinA knockout mice, in the striatal medium spiny neurons in *SGCE* knockout mice [71]. These mice show myoclonus, motor deficits, and alterations of emotional responses and monoamine contents in the striatum, suggesting that functional alterations in the striatum may contribute to DYT11 [71]. However, in animals in which *SGCE* was selectively inactivated in the striatum, there were no nuclear envelope abnormalities, and the mice did not exhibit myoclonus [71]. Thus, loss of epsilon-sarcoglycan in the striatum alone does not cause myoclonus, and epsilon-sarcoglycan functions in multiple brain regions and likely contributes to their different roles.

3.1.2.3 Paroxysmal Combined Dystonias

Paroxysmal dystonias are a heterogeneous group of disorders characterized by sudden and transient attacks of involuntary movements that can be precipitated by (1) alcohol and caffeine in the nonkinesigenic form, DYT8; (2) physical exercise or emotional stress in case of the exercise-induced form, DYT18/DYT9; or (3) sudden movements in the kinesigenic form, DYT10. All three forms of paroxysmal dystonia are inherited in an autosomal dominant manner and have a slightly reduced penetrance.

PNKD (DYT8)

In addition to alcohol and caffeine, DYT8 (also known as paroxysmal nonkinesigenic dyskinesia) attacks can be precipitated by stress, hunger, fatigue, and tobacco. They usually consist of a combination of dystonia, chorea, athetosis, or ballismus; last from minutes to hours; and in the most severe cases, may occur several times daily.

Two missense mutations (p.A7V and p.A9V) in the *PNKD* (paroxysmal nonkinesigenic dyskinesia) gene, previously known as myofibrillogenesis regulator 1 (*MR-1*), are the cause of the disease in about 15 families reported to date [10, 11, 72–74]. Another mutation, p.A33P, was found to co-segregate with the disease phenotype in a single family [75]. PNKD has three isoforms, a long (PNKD-L), a medium (PNKD-M), and a short form (PNKD-S). While the M and the S forms are ubiquitously expressed, PNKD-L was found only in the brain. PNKD-L is a plasma membrane-associated protein [76]. The N-terminus of wild-type PNKD-L is normally cleaved off from the mature protein [77]. In cellular protein overexpression studies, both mutant forms (p.A7V and p.A9V) conferred resistance to this cleavage and were therefore degraded faster in comparison to the wild-type PNKD-L [77]. Interestingly, transgenic mice expressing the two human PNKD-L mutants recapitulated the DYT8 phenotype upon caffeine and ethanol intake and after prolonged handling of stress [76]. The phenotype was not observed in mice expressing human wild-type PNKD-L or in the knockout mice, thus arguing that PNKD-L mutations are gain-of-function alleles. Finally, mice carrying a mutated PNKD-L transgene displayed lower striatal dopamine levels under basal conditions in comparison to their wild-type littermates [76]. However, their dopamine release significantly increased in response to stress of caffeine. Thus, PNKD mice show alterations of dopamine signaling in the striatum.

SLC2A1 (DYT18/DYT9)

Paroxysmal exercise-induced dystonia (PED) or DYT18 is precipitated by exercise, exertion, or stress and is characterized by transient (10–40 min) abnormal involuntary movements, such as dystonia, choreoathetosis, and ballismus affecting the exercised limbs [78]. The attacks usually start in childhood, and patients may have associated phenotypes such as epilepsy, migraine, mild mental retardation, and anemia [19, 79, 80].

DYT18 (and as later shown also DYT9) is caused by various missense and truncating mutations in the solute carrier family 2 member 1 (*SLC2A1, GLUT1*) gene [20, 79]. *SLC2A1* encodes a glucose transporter that is expressed on endothelial cells of the blood-brain barrier and facilitates glucose entry into the brain. All functionally tested mutations showed a decrease in glucose uptake. After prolonged physical activity, energy demands of the organism are increased, and therefore PED is probably caused by reduced glucose transport into the brain under these circumstances. Thus, a low-carbohydrate (ketogenic) diet, that directs the body towards burning fat rather than carbohydrates, often results in marked clinical improvement of the motor and seizure symptoms [81].

PRRT2 (DYT10)

Paroxysmal kinesigenic dyskinesia (PKD) usually starts in childhood or adolescence and is triggered by sudden movements [82]. Attacks usually last several minutes and may appear up to 100 times per day. They mostly consist of dystonic and choreoathetotic movements. PKD has been clinically and genetically linked to a variety of conditions including benign familial infantile seizures (BFIS); the syndrome of rolandic epilepsy, paroxysmal exercise-induced dyskinesia, and writer's cramp; and even the trait of wet ear wax (cerumen) [83]. Indeed, the co-occurrence of BFIS and PKD in families has resulted in the identification of another illness, infantile convulsions, and paroxysmal choreoathetosis (ICPC).

Recently, missense and truncating mutations in the proline-rich transmembrane protein 2 (*PRRT2*) gene were identified as the cause of PKD [12]. Subsequently, it was shown that coexisting ICPC is caused by the same mutations [83]. Thus, these two apparently clinically distinct syndromes are allelic conditions caused by the same mutation in *PRRT2* [83]. PRRT2 is expressed in the brain and spinal cord, and it localizes to axons in primary neuronal cultures [84]; however, its role in the central nervous system is unknown. It interacts with the synaptic protein SNAP25 [84, 85] raising the possibility that, similar to PNKD, PKD may be caused by synaptic dysfunction. Functional studies of four PRRT2 truncating mutants overexpressed in HEK293T cells showed that these proteins were extremely unstable and present either in very low amounts or completely absent [84]. However, they did not have a dominant-negative effect on wild-type protein levels.

3.2 Inconsistencies of DYT Designation

As previously mentioned, the current dystonia genetics nomenclature, embodied in the list of DYT symbols, features a number of inconsistencies that hinder its usefulness [86].

1. *Erroneous linkage.* The DYT14 locus was erroneously designated due to initial clinical misclassification of one family member and lack of comprehensive mutational analysis of previously known genes. The family was later found to have a deletion within the *GCH1* gene.
2. *Duplication of loci.* DYT18 has been assigned to *GLUT1* deficiency due to *SLC2A1* mutations. Phenotypic similarity and localization of the DYT18 and DYT9 in the same chromosomal region prompted reexamination of the DYT9-linked family. Affected members of this family were later also found to carry *SLC2A1* changes. Thus, DYT18 and DYT9 were found to be identical.

3. *Duplication of DYT symbols*. Usage of a chronological numbering system is particularly problematic at a time of rapid gene identification. This is illustrated by the fact that in the period immediately following their discovery, both *CIZ1* and *ANO3* carried the DYT23 designation.
4. *Missing gene locus*. Sometimes DYT designations serve to indicate a family with an apparently inherited disorder, however, in the absence of any linkage to a chromosomal region, resulting in a locus symbol with no related locus, such as in the case of DYT2.
5. *Missing locus symbol*. Some established disease-associated genetic loci have not been granted a DYT designation. One example is the *SPR* gene, mutations in which can cause DRD.
6. *Missing gene locus and locus symbol*. Currently, the DYT22 designation has only been reserved, but neither the putative clinical phenotype nor the gene locus/disease-causing mutations have been published.
7. *Complex phenotype*. Some conditions include more than one prominent movement disorder. For example, DYT3 was originally included as a member of the DYT list given that it clinically starts with dystonia. However, based on the neuropathological findings and disease progression, it may equally belong to the list of Parkinson disease-related disorders.

3.3 Identification of New Dystonia Genes

Historically, new dystonia-linked genes or risk factors were identified by gene mapping or candidate gene approaches. The gene underlying any heritable form of human disease can be mapped and identified by linkage analysis if DNA samples from a sufficient number of affected and unaffected family members are available. By this method, genomic regions shared by all the affected members are identified and sequenced in search for disease-causing mutations. Once the gene is identified, the molecular mechanism of the disease can be hypothesized and further investigated. Alternatively, based on their known function, levels of expression, or mode of interaction (candidate gene approach), some genes can be considered plausible candidates and as such tested for in cohorts of patients and controls.

The advent of recent sequencing technology, most notably next-generation sequencing (NGS) of the whole genome or whole exome, i.e., the coding part of the genome, greatly facilitates gene identification. It is now possible to obtain the entire sequence of a patient's genome in a fast and cost-efficient manner and to compare it to reference genomes. Although the analysis of such large amounts of data is not an easy task, NGS has already been proven useful in identifying dystonia-causing mutations in *TUBB4*, *PRRT2*, *ANO3*, and *GNAL*. Thus, this approach will likely replace classical linkage analysis in the future.

3.4 Genetic Testing: Whom to Test and How?

Genetic testing is a very useful tool for establishing diagnosis, minimizing further patient workup, clarifying treatment approaches, and assisting with future family planning. Unlike many other genetic conditions, several forms of dystonia (e.g., DRD and PED), when correctly diagnosed, can be successfully treated (dopamine and ketogenic diet).

Currently, diagnostic genetic tests for all of the known dystonia genes are commercially available. Certified laboratories offering such diagnostic tests can be found at www.genetests.org. However, lack of formal testing guidelines for genetic forms of dystonia, with the exception of DYT1 dystonia [87], poses a challenge to clinicians when trying to decide whether to test a patient and which test to order.

General clinical features pointing towards a monogenic form of dystonia are a positive family history, an early age of onset, a specific clinical picture, and a specific ethnic or cultural origin. As an example, a male patient of Filipino origin with an adult-onset dystonia with signs of parkinsonism should first be tested for the DYT3-associated genotype. Similarly, a patient with a childhood-onset leg dystonia without response to levodopa is likely to carry a GAG deletion in the *TOR1A* gene. Also, a phenotype of myoclonus-dystonia in combination with impaired skeletal development and osteoporosis offers a clear clue that *SGCE* exonic or gene deletions are to be investigated.

It also has to be kept in mind that dystonia patients without an apparent family history may nevertheless suffer from an inherited form of dystonia. Common reasons for a pseudo-negative family history are non-paternity, adoption, early death of a parent (before the onset of symptoms), de novo mutations, reduced penetrance (not all individuals with a mutation are affected), and variable expressivity (not all individuals with a mutation show the expected phenotype). One example explaining reduced penetrance is imprinting of the *SGCE* gene in DYT11. In addition, as discussed in the *TOR1A* (DYT1) chapter, the common polymorphism p.D216H seems to be associated with reduced penetrance of DYT1 dystonia and thus, through an as yet unknown mechanism, protects against the clinical manifestation of the GAG deletion.

As it is difficult to keep up with the rapid advances in dystonia genetics and the wealth of clinical and genetic clues and exceptions thereof, it may be helpful to contact a movement disorders center with specific expertise in diagnosing genetic dystonia for specific advice.

In conclusion, the past 20 years of dystonia genetics research markedly improved our understanding of the genetic and molecular mechanisms of dystonia. In addition, the knowledge and experience gathered to date highlight the importance of careful clinical-genetic studies enabling us to appreciate the full phenotypic spectrum associated with each of the dystonia genes, as well as any potential phenotype-genotype correlations.

References

1. Ichinose H, Ohye T, Takahashi E, Seki N, Hori T, Segawa M, Nomura Y, Endo K, Tanaka H, Tsuji S, et al. Hereditary progressive dystonia with marked diurnal fluctuation caused by mutations in the GTP cyclohydrolase I gene. Nat Genet. 1994;8(3):236–42.
2. Ozelius LJ, Hewett JW, Page CE, Bressman SB, Kramer PL, Shalish C, de Leon D, Brin MF, Raymond D, Corey DP, Fahn S, Risch NJ, Buckler AJ, Gusella JF, Breakefield XO. The early-onset torsion dystonia gene (DYT1) encodes an ATP-binding protein. Nat Genet. 1997;17(1):40–8.
3. Gimenez-Roldan S, Delgado G, Marin M, Villanueva JA, Mateo D. Hereditary torsion dystonia in gypsies. Adv Neurol. 1988;50:73–81.
4. Lee LV, Pascasio FM, Fuentes FD, Viterbo GH. Torsion dystonia in Panay, Philippines. Adv Neurol. 1976;14:137–51.
5. Lohmann KWR, Winkler S, et al. Whispering dysphonia (DYT4 dystonia) is caused by a mutation in the TUBB4 gene. Ann Neurol. 2013;73(4):537–45.
6. Wider C, Melquist S, Hauf M, Solida A, Cobb SA, Kachergus JM, Gass J, Coon KD, Baker M, Cannon A, Stephan DA, Schorderet DF, Ghika J, Burkhard PR, Kapatos G, Hutton M, Farrer MJ, Wszolek ZK, Vingerhoets FJ. Study of a Swiss dopa-responsive dystonia family with a deletion in GCH1: redefining DYT14 as DYT5. Neurology. 2008;70(16 Pt 2):1377–83.
7. Ludecke B, Dworniczak B, Bartholome K. A point mutation in the tyrosine hydroxylase gene associated with Segawa's syndrome. Hum Genet. 1995;95(1):123–5.
8. Fuchs T, Gavarini S, Saunders-Pullman R, Raymond D, Ehrlich ME, Bressman SB, Ozelius LJ. Mutations in the THAP1 gene are responsible for DYT6 primary torsion dystonia. Nat Genet. 2009;41(3):286–8.
9. Leube B, Rudnicki D, Ratzlaff T, Kessler KR, Benecke R, Auburger G. Idiopathic torsion dystonia: assignment of a gene to chromosome 18p in a German family with adult onset, autosomal dominant inheritance and purely focal distribution. Hum Mol Genet. 1996;5(10):1673–7.
10. Lee HY, Xu Y, Huang Y, Ahn AH, Auburger GW, Pandolfo M, Kwiecinski H, Grimes DA, Lang AE, Nielsen JE, Averyanov Y, Servidei S, Friedman A, Van Bogaert P, Abramowicz MJ, Bruno MK, Sorensen BF, Tang L, Fu YH, Ptacek LJ. The gene for paroxysmal non-kinesigenic dyskinesia encodes an enzyme in a stress response pathway. Hum Mol Genet. 2004;13(24):3161–70.
11. Rainier S, Thomas D, Tokarz D, Ming L, Bui M, Plein E, Zhao X, Lemons R, Albin R, Delaney C, Alvarado D, Fink JK. Myofibrillogenesis regulator 1 gene mutations cause paroxysmal dystonic choreoathetosis. Arch Neurol. 2004;61(7):1025–9.
12. Chen WJ, Lin Y, Xiong ZQ, Wei W, Ni W, Tan GH, Guo SL, He J, Chen YF, Zhang QJ, Li HF, Murong SX, Xu J, Wang N, Wu ZY. Exome sequencing identifies truncating mutations in PRRT2 that cause paroxysmal kinesigenic dyskinesia. Nat Genet. 2011;43(12):1252–5.
13. Zimprich A, Grabowski M, Asmus F, Naumann M, Berg D, Bertram M, Scheidtmann K, Kern P, Winkelmann J, Muller-Myhsok B, Riedel L, Bauer M, Muller T, Castro M, Meitinger T, Strom TM, Gasser T. Mutations in the gene encoding epsilon-sarcoglycan cause myoclonus-dystonia syndrome. Nat Genet. 2001;29(1):66–9.
14. de Carvalho AP, Sweadner KJ, Penniston JT, Zaremba J, Liu L, Caton M, Linazasoro G, Borg M, Tijssen MA, Bressman SB, Dobyns WB, Brashear A, Ozelius LJ. Mutations in the Na+/K+−ATPase alpha3 gene ATP1A3 are associated with rapid-onset dystonia parkinsonism. Neuron. 2004;43(2):169–75.
15. Valente EM, Bentivoglio AR, Cassetta E, Dixon PH, Davis MB, Ferraris A, Ialongo T, Frontali M, Wood NW, Albanese A. DYT13, a novel primary torsion dystonia locus, maps to chromosome 1p36.13–36.32 in an Italian family with cranial-cervical or upper limb onset. Ann Neurol. 2001;49(3):362–6.

16. Grimes DA, Han F, Lang AE, St George-Hyssop P, Racacho L, Bulman DE. A novel locus for inherited myoclonus-dystonia on 18p11. Neurology. 2002;59(8):1183–6.
17. Camargos S, Scholz S, Simon-Sanchez J, Paisan-Ruiz C, Lewis P, Hernandez D, Ding J, Gibbs JR, Cookson MR, Bras J, Guerreiro R, Oliveira CR, Lees A, Hardy J, Cardoso F, Singleton AB. DYT16, a novel young-onset dystonia-parkinsonism disorder: identification of a segregating mutation in the stress-response protein PRKRA. Lancet Neurol. 2008;7(3):207–15.
18. Chouery E, Kfoury J, Delague V, Jalkh N, Bejjani P, Serre JL, Megarbane A. A novel locus for autosomal recessive primary torsion dystonia (DYT17) maps to 20p11.22-q13.12. Neurogenetics. 2008;9(4):287–93.
19. Suls A, Dedeken P, Goffin K, Van Esch H, Dupont P, Cassiman D, Kempfle J, Wuttke TV, Weber Y, Lerche H, Afawi Z, Vandenberghe W, Korczyn AD, Berkovic SF, Ekstein D, Kivity S, Ryvlin P, Claes LR, Deprez L, Maljevic S, Vargas A, Van Dyck T, Goossens D, Del-Favero J, Van Laere K, De Jonghe P, Van Paesschen W. Paroxysmal exercise-induced dyskinesia and epilepsy is due to mutations in SLC2A1, encoding the glucose transporter GLUT1. Brain. 2008;131(Pt 7):1831–44.
20. Weber YG, Kamm C, Suls A, Kempfle J, Kotschet K, Schule R, Wuttke TV, Maljevic S, Liebrich J, Gasser T, Ludolph AC, Van Paesschen W, Schols L, De Jonghe P, Auburger G, Lerche H. Paroxysmal choreoathetosis/spasticity (DYT9) is caused by a GLUT1 defect. Neurology. 2011;77(10):959–64.
21. Tomita H, Nagamitsu S, Wakui K, Fukushima Y, Yamada K, Sadamatsu M, Masui A, Konishi T, Matsuishi T, Aihara M, Shimizu K, Hashimoto K, Mineta M, Matsushima M, Tsujita T, Saito M, Tanaka H, Tsuji S, Takagi T, Nakamura Y, Nanko S, Kato N, Nakane Y, Niikawa N. Paroxysmal kinesigenic choreoathetosis locus maps to chromosome 16p11.2-q12.1. Am J Hum Genet. 1999;65(6):1688–97.
22. Spacey SD, Adams PJ, Lam PC, Materek LA, Stoessl AJ, Snutch TP, Hsiung GY. Genetic heterogeneity in paroxysmal nonkinesigenic dyskinesia. Neurology. 2006;66(10): 1588–90.
23. Norgren N, Mattson E, Forsgren L, Holmberg M. A high-penetrance form of late-onset torsion dystonia maps to a novel locus (DYT21) on chromosome 2q14.3-q21.3. Neurogenetics. 2011;12(2):137–43.
24. Xiao J, Uitti RJ, Zhao Y, Vemula SR, Perlmutter JS, Wszolek ZK, Maraganore DM, Auburger G, Leube B, Lehnhoff K, LeDoux MS. Mutations in CIZ1 cause adult onset primary cervical dystonia. Ann Neurol. 2012;71(4):458–69.
25. Charlesworth G, Plagnol V, Holmstrom KM, Bras J, Sheerin UM, Preza E, Rubio-Agusti I, Ryten M, Schneider SA, Stamelou M, Trabzuni D, Abramov AY, Bhatia KP, Wood NW. Mutations in ANO3 cause dominant craniocervical dystonia: Ion channel implicated in pathogenesis. Am J Hum Genet. 2012;91(6):1041–50.
26. Fuchs T, Saunders-Pullman R, Masuho I, Luciano MS, Raymond D, Factor S, Lang AE, Liang TW, Trosch RM, White S, Ainehsazan E, Herve D, Sharma N, Ehrlich ME, Martemyanov KA, Bressman SB, Ozelius LJ. Mutations in GNAL cause primary torsion dystonia. Nat Genet. 2012;45(1):88–92.
27. Bonafe L, Thony B, Penzien JM, Czarnecki B, Blau N. Mutations in the sepiapterin reductase gene cause a novel tetrahydrobiopterin-dependent monoamine-neurotransmitter deficiency without hyperphenylalaninemia. Am J Hum Genet. 2001;69(2):269–77.
28. Albanese A, Bhatia K, Bressman SB, Delong MR, Fahn S, Fung VS, Hallett M, Jankovic J, Jinnah HA, Klein C, Lang AE, Mink JW, Teller JK. Phenomenology and classification of dystonia: a consensus update. Mov Disord. 2013;28(7):863–73.
29. Oppenheim H. Über eine eigenartige Krampfkrankheit des kindlichen und jugendlichen Alters (Dysbasia lordotica progressiva, dystonia musculorum deformans). Neurologisches Centralblatt. 1911;30:1090–107.
30. Bressman SB, de Leon D, Kramer PL, Ozelius LJ, Brin MF, Greene PE, Fahn S, Breakefield XO, Risch NJ. Dystonia in Ashkenazi Jews: clinical characterization of a founder mutation. Ann Neurol. 1994;36(5):771–7.

31. Dobricic V, Kresojevic N, Westenberger A, Svetel M, Tomic A, Ralic V, Petrovic I, Lukic MJ, Lohmann K, Novakovic I, Klein C, Kostic VS. De novo mutation in the GNAL gene causing seemingly sporadic dystonia in a Serbian patient. Mov Disord. 2014;29(9):1190–3.
32. Ziegan J, Wittstock M, Westenberger A, Dobricic V, Wolters A, Benecke R, Klein C, Kamm C. Novel GNAL mutations in two German patients with sporadic dystonia. Mov Disord. 2014;29(14):1833–4.
33. Vemula SR, Puschmann A, Xiao J, Zhao Y, Rudzinska M, Frei KP, Truong DD, Wszolek ZK, LeDoux MS. Role of Galpha(olf) in familial and sporadic adult-onset primary dystonia. Hum Mol Genet. 2013;22(12):2510–9.
34. Kumar KR, Lohmann K, Masuho I, Miyamoto R, Ferbert A, Lohnau T, Kasten M, Hagenah J, Bruggemann N, Graf J, Munchau A, Kostic VS, Sue CM, Domingo AR, Rosales RL, Lee LV, Freimann K, Westenberger A, Mukai Y, Kawarai T, Kaji R, Klein C, Martemyanov KA, Schmidt A. Mutations in GNAL: a novel cause of craniocervical dystonia. JAMA Neurol. 2014;71(4):490–4.
35. Miao J, Wan XH, Sun Y, Feng JC, Cheng FB. Mutation screening of GNAL gene in patients with primary dystonia from Northeast China. Parkinsonism Relat Disord. 2013;19(10):910–2.
36. Ozelius LJ, Lubarr N, Bressman SB. Milestones in dystonia. Mov Disord. 2011;26(6):1106–26.
37. Gasser T, Windgassen K, Bereznai B, Kabus C, Ludolph AC. Phenotypic expression of the DYT1 mutation: a family with writer's cramp of juvenile onset. Ann Neurol. 1998;44(1):126–8.
38. Kabakci K, Hedrich K, Leung JC, Mitterer M, Vieregge P, Lencer R, Hagenah J, Garrels J, Witt K, Klostermann F, Svetel M, Friedman J, Kostic V, Bressman SB, Breakefield XO, Ozelius LJ, Pramstaller PP, Klein C. Mutations in DYT1: extension of the phenotypic and mutational spectrum. Neurology. 2004;62(3):395–400.
39. Calakos N, Patel VD, Gottron M, Wang G, Tran-Viet KN, Brewington D, Beyer JL, Steffens DC, Krishnan RR, Zuchner S. Functional evidence implicating a novel TOR1A mutation in idiopathic, late-onset focal dystonia. J Med Genet. 2010;47(9):646–50.
40. Zirn B, Grundmann K, Huppke P, Puthenparampil J, Wolburg H, Riess O, Muller U. Novel TOR1A mutation p.Arg288Gln in early-onset dystonia (DYT1). J Neurol Neurosurg Psychiatry. 2008;79(12):1327–30.
41. Kock N, Naismith TV, Boston HE, Ozelius LJ, Corey DP, Breakefield XO, Hanson PI. Effects of genetic variations in the dystonia protein torsinA: identification of polymorphism at residue 216 as protein modifier. Hum Mol Genet. 2006;15(8):1355–64.
42. Goodchild RE, Dauer WT. Mislocalization to the nuclear envelope: an effect of the dystonia-causing torsinA mutation. Proc Natl Acad Sci U S A. 2004;101(3):847–52.
43. Goodchild RE, Kim CE, Dauer WT. Loss of the dystonia-associated protein torsinA selectively disrupts the neuronal nuclear envelope. Neuron. 2005;48(6):923–32.
44. Almasy L, Bressman SB, Raymond D, Kramer PL, Greene PE, Heiman GA, Ford B, Yount J, de Leon D, Chouinard S, Saunders-Pullman R, Brin MF, Kapoor RP, Jones AC, Shen H, Fahn S, Risch NJ, Nygaard TG. Idiopathic torsion dystonia linked to chromosome 8 in two Mennonite families. Ann Neurol. 1997;42(4):670–3.
45. Blanchard A, Ea V, Roubertie A, Martin M, Coquart C, Claustres M, Beroud C, Collod-Beroud G. DYT6 dystonia: review of the literature and creation of the UMD Locus-Specific Database (LSDB) for mutations in the THAP1 gene. Hum Mutat. 2011;32(11):1213–24.
46. Kaiser FJ, Osmanoric A, Rakovic A, Erogullari A, Uflacker N, Braunholz D, Lohnau T, Orolicki S, Albrecht M, Gillessen-Kaesbach G, Klein C, Lohmann K. The dystonia gene DYT1 is repressed by the transcription factor THAP1 (DYT6). Ann Neurol. 2010;68(4):554–9.
47. Seibler P, Djarmati A, Langpap B, Hagenah J, Schmidt A, Bruggemann N, Siebner H, Jabusch HC, Altenmuller E, Munchau A, Lohmann K, Klein C. A heterozygous frameshift mutation in PRKRA (DYT16) associated with generalised dystonia in a German patient. Lancet Neurol. 2008;7(5):380–1.

48. Zech M, Castrop F, Schormair B, Jochim A, Wieland T, Gross N, Lichtner P, Peters A, Gieger C, Meitinger T, Strom TM, Oexle K, Haslinger B, Winkelmann J. DYT16 revisited: exome sequencing identifies PRKRA mutations in a European dystonia family. Mov Disord. 2014;29(12):1504–10.
49. Jones DT, Reed RR. Golf: an olfactory neuron specific-G protein involved in odorant signal transduction. Science. 1989;244(4906):790–5.
50. Uitti RJ, Maraganore DM. Adult onset familial cervical dystonia: report of a family including monozygotic twins. Mov Disord. 1993;8(4):489–94.
51. Klein C, Konig IR, Lohmann K. Exome sequencing for gene discovery: time to set standard criteria. Ann Neurol. 2012;72(4):627–8.
52. Parker N. Hereditary whispering dysphonia. J Neurol Neurosurg Psychiatry. 1985;48(3):218–24.
53. Wilcox RA, Winkler S, Lohmann K, Klein C. Whispering dysphonia in an Australian family (DYT4): a clinical and genetic reappraisal. Mov Disord. 2011;26(13):2404–8.
54. Hersheson J, Mencacci N, Davis M, MacDoanld HM, Trabzuni D, Ryten M, Pittman A, Paudel R, Kara E, Fawcett K, Plagnol V, Bhatia KP, Medlar A, Stanescu HC, Hardy J, Kleta R, Wood NW, Houlden H. Mutations in the autoregulatory domain of β-tubulin 4a cause hereditary dystonia. Ann Neurol. 2013;73(4):546–53.
55. Simons C, Wolf NI, McNeil N, Caldovic L, Devaney JM, Takanohashi A, Crawford J, Ru K, Grimmond SM, Miller D, Tonduti D, Schmidt JL, Chudnow RS, van Coster R, Lagae L, Kisler J, Sperner J, van der Knaap MS, Schiffmann R, Taft RJ, Vanderver A. A de novo mutation in the β-tubulin gene TUBB4A results in the leukoencephalopathy hypomyelination with atrophy of the basal ganglia and cerebellum. Am J Hum Genet. 2013;92(5):767–73.
56. Erro R, Hersheson J, Ganos C, Mencacci NE, Stamelou M, Batla A, Thust SC, Bras JM, Guerreiro RJ, Hardy J, Quinn NP, Houlden H, Bhatia KP. H-ABC syndrome and DYT4: Variable expressivity or pleiotropy of TUBB4 mutations? Mov Disord. 2014 [Epub ahead of print].
57. Tadic V, Kasten M, Bruggemann N, Stiller S, Hagenah J, Klein C. Dopa-responsive dystonia revisited: diagnostic delay, residual signs, and nonmotor signs. Arch Neurol. 2012;69(12):1558–62.
58. Bruggemann N, Spiegler J, Hellenbroich Y, Opladen T, Schneider SA, Stephani U, Boor R, Gillessen-Kaesbach G, Sperner J, Klein C. Beneficial prenatal levodopa therapy in autosomal recessive guanosine triphosphate cyclohydrolase 1 deficiency. Arch Neurol. 2012;69(8):1071–5.
59. Blanco-Arias P, Einholm AP, Mamsa H, Concheiro C, Gutierrez-de-Teran H, Romero J, Toustrup-Jensen MS, Carracedo A, Jen JC, Vilsen B, Sobrido MJ. A C-terminal mutation of ATP1A3 underscores the crucial role of sodium affinity in the pathophysiology of rapid-onset dystonia-parkinsonism. Hum Mol Genet. 2009;18(13):2370–7. ddp170 [pii].
60. Calderon DP, Fremont R, Kraenzlin F, Khodakhah K. The neural substrates of rapid-onset Dystonia-Parkinsonism. Nat Neurosci. 2011;14(3):357–65.
61. Heinzen EL, Swoboda KJ, Hitomi Y, Gurrieri F, Nicole S, de Vries B, Tiziano FD, Fontaine B, Walley NM, Heavin S, Panagiotakaki E, Fiori S, Abiusi E, Di Pietro L, Sweney MT, Newcomb TM, Viollet L, Huff C, Jorde LB, Reyna SP, Murphy KJ, Shianna KV, Gumbs CE, Little L, Silver K, Ptacek LJ, Haan J, Ferrari MD, Bye AM, Herkes GK, Whitelaw CM, Webb D, Lynch BJ, Uldall P, King MD, Scheffer IE, Neri G, Arzimanoglou A, van den Maagdenberg AM, Sisodiya SM, Mikati MA, Goldstein DB. De novo mutations in ATP1A3 cause alternating hemiplegia of childhood. Nat Genet. 2012;44(9):1030–4.
62. Lee LV, Rivera C, Teleg RA, Dantes MB, Pasco PM, Jamora RD, Arancillo J, Villareal-Jordan RF, Rosales RL, Demaisip C, Maranon E, Peralta O, Borres R, Tolentino C, Monding MJ, Sarcia S. The unique phenomenology of sex-linked dystonia parkinsonism (XDP, DYT3, "Lubag"). Int J Neurosci. 2011;121 Suppl 1:3–11.
63. Makino S, Kaji R, Ando S, Tomizawa M, Yasuno K, Goto S, Matsumoto S, Tabuena MD, Maranon E, Dantes M, Lee LV, Ogasawara K, Tooyama I, Akatsu H, Nishimura M, Tamiya G.

Reduced neuron-specific expression of the TAF1 gene is associated with X-linked dystonia-parkinsonism. Am J Hum Genet. 2007;80(3):393–406.

64. Domingo A, Westenberger A, Lee LV, Brænne I, Liu T, Vater I, Rosales R, Jamora RD, Pasco PM, Cutiongco-Dela Paz EM, Freimann K, Schmidt TG, Dressler D, Kaiser FJ, Bertram L, Erdmann J, Lohmann K, Klein C. New insights into the genetics of X-linked dystonia-parkinsonism (XDP, DYT3). Eur J Hum Genet. 2015. [Epub ahead of print].
65. Raymond D, Saunders-Pullman R, de Carvalho AP, Schule B, Kock N, Friedman J, Harris J, Ford B, Frucht S, Heiman GA, Jennings D, Doheny D, Brin MF, de Leon BD, Multhaupt-Buell T, Lang AE, Kurlan R, Klein C, Ozelius L, Bressman S. Phenotypic spectrum and sex effects in eleven myoclonus-dystonia families with epsilon-sarcoglycan mutations. Mov Disord. 2008;23(4):588–92.
66. Kyllerman M, Forsgren L, Sanner G, Holmgren G, Wahlstrom J, Drugge U. Alcohol-responsive myoclonic dystonia in a large family: dominant inheritance and phenotypic variation. Mov Disord. 1990;5(4):270–9.
67. Muller B, Hedrich K, Kock N, Dragasevic N, Svetel M, Garrels J, Landt O, Nitschke M, Pramstaller PP, Reik W, Schwinger E, Sperner J, Ozelius L, Kostic V, Klein C. Evidence that paternal expression of the epsilon-sarcoglycan gene accounts for reduced penetrance in myoclonus-dystonia. Am J Hum Genet. 2002;71(6):1303–11.
68. Grunewald A, Djarmati A, Lohmann-Hedrich K, Farrell K, Zeller JA, Allert N, Papengut F, Petersen B, Fung V, Sue CM, O'Sullivan D, Mahant N, Kupsch A, Chuang RS, Wiegers K, Pawlack H, Hagenah J, Ozelius LJ, Stephani U, Schuit R, Lang AE, Volkmann J, Munchau A, Klein C. Myoclonus-dystonia: significance of large SGCE deletions. Hum Mutat. 2008;29(2):331–2.
69. Asmus F, Hjermind LE, Dupont E, Wagenstaller J, Haberlandt E, Munz M, Strom TM, Gasser T. Genomic deletion size at the epsilon-sarcoglycan locus determines the clinical phenotype. Brain. 2007;130(Pt 10):2736–45.
70. Esapa CT, Waite A, Locke M, Benson MA, Kraus M, McIlhinney RA, Sillitoe RV, Beesley PW, Blake DJ. SGCE missense mutations that cause myoclonus-dystonia syndrome impair epsilon-sarcoglycan trafficking to the plasma membrane: modulation by ubiquitination and torsinA. Hum Mol Genet. 2007;16(3):327–42.
71. Yokoi F, Dang MT, Zhou T, Li Y. Abnormal nuclear envelopes in the striatum and motor deficits in DYT11 myoclonus-dystonia mouse models. Hum Mol Genet. 2012; 21(4):916–25.
72. Pons R, Cuenca-Leon E, Miravet E, Pons M, Xaidara A, Youroukos S, Macaya A. Paroxysmal non-kinesigenic dyskinesia due to a PNKD recurrent mutation: report of two Southern European families. Eur J Paediatr Neurol. 2012;16(1):86–9.
73. Djarmati A, Svetel M, Momcilovic D, Kostic V, Klein C. Significance of recurrent mutations in the myofibrillogenesis regulator 1 gene. Arch Neurol. 2005;62(10):1641.
74. Chen DH, Matsushita M, Rainier S, Meaney B, Tisch L, Feleke A, Wolff J, Lipe H, Fink J, Bird TD, Raskind WH. Presence of alanine-to-valine substitutions in myofibrillogenesis regulator 1 in paroxysmal nonkinesigenic dyskinesia: confirmation in 2 kindreds. Arch Neurol. 2005;62(4):597–600.
75. Ghezzi D, Viscomi C, Ferlini A, Gualandi F, Mereghetti P, DeGrandis D, Zeviani M. Paroxysmal non-kinesigenic dyskinesia is caused by mutations of the MR-1 mitochondrial targeting sequence. Hum Mol Genet. 2009;18(6):1058–64.
76. Lee HY, Nakayama J, Xu Y, Fan X, Karouani M, Shen Y, Pothos EN, Hess EJ, Fu YH, Edwards RH, Ptacek LJ. Dopamine dysregulation in a mouse model of paroxysmal nonkinesigenic dyskinesia. J Clin Invest. 2012;122(2):507–18.
77. Shen Y, Lee HY, Rawson J, Ojha S, Babbitt P, Fu YH, Ptacek LJ. Mutations in PNKD causing paroxysmal dyskinesia alters protein cleavage and stability. Hum Mol Genet. 2011;20(12):2322–32.
78. Munchau A, Valente EM, Shahidi GA, Eunson LH, Hanna MG, Quinn NP, Schapira AH, Wood NW, Bhatia KP. A new family with paroxysmal exercise induced dystonia and migraine: a clinical and genetic study. J Neurol Neurosurg Psychiatry. 2000;68(5):609–14.

79. Weber YG, Storch A, Wuttke TV, Brockmann K, Kempfle J, Maljevic S, Margari L, Kamm C, Schneider SA, Huber SM, Pekrun A, Roebling R, Seebohm G, Koka S, Lang C, Kraft E, Blazevic D, Salvo-Vargas A, Fauler M, Mottaghy FM, Munchau A, Edwards MJ, Presicci A, Margari F, Gasser T, Lang F, Bhatia KP, Lehmann-Horn F, Lerche H. GLUT1 mutations are a cause of paroxysmal exertion-induced dyskinesias and induce hemolytic anemia by a cation leak. J Clin Invest. 2008;118(6):2157–68.
80. Wang D, Pascual JM, Yang H, Engelstad K, Jhung S, Sun RP, De Vivo DC. Glut-1 deficiency syndrome: clinical, genetic, and therapeutic aspects. Ann Neurol. 2005;57(1):111–8.
81. Brockmann K. The expanding phenotype of GLUT1-deficiency syndrome. Brain Dev. 2009;31(7):545–52.
82. Bhatia KP. Paroxysmal dyskinesias. Mov Disord. 2011;26(6):1157–65.
83. Schmidt A, Kumar KR, Redyk K, Grunewald A, Leben M, Munchau A, Sue CM, Hagenah J, Hartmann H, Lohmann K, Christen HJ, Klein C. Two faces of the same coin: benign familial infantile seizures and paroxysmal kinesigenic dyskinesia caused by PRRT2 mutations. Arch Neurol. 2012;69(5):668–70.
84. Lee HY, Huang Y, Bruneau N, Roll P, Roberson ED, Hermann M, Quinn E, Maas J, Edwards R, Ashizawa T, Baykan B, Bhatia K, Bressman S, Bruno MK, Brunt ER, Caraballo R, Echenne B, Fejerman N, Frucht S, Gurnett CA, Hirsch E, Houlden H, Jankovic J, Lee WL, Lynch DR, Mohamed S, Muller U, Nespeca MP, Renner D, Rochette J, Rudolf G, Saiki S, Soong BW, Swoboda KJ, Tucker S, Wood N, Hanna M, Bowcock A, Szepetowski P, Fu YH, Ptacek LJ. Mutations in the novel protein PRRT2 cause paroxysmal kinesigenic dyskinesia with infantile convulsions. Cell Rep. 2012;1(1):2–12.
85. Stelzl U, Worm U, Lalowski M, Haenig C, Brembeck FH, Goehler H, Stroedicke M, Zenkner M, Schoenherr A, Koeppen S, Timm J, Mintzlaff S, Abraham C, Bock N, Kietzmann S, Goedde A, Toksoz E, Droege A, Krobitsch S, Korn B, Birchmeier W, Lehrach H, Wanker EE. A human protein-protein interaction network: a resource for annotating the proteome. Cell. 2005;122(6):957–68.
86. Marras C, Lohmann K, Lang A, Klein C. Fixing the broken system of genetic locus symbols: Parkinson disease and dystonia as examples. Neurology. 2012;78(13):1016–24.
87. Bressman SB, Sabatti C, Raymond D, de Leon D, Klein C, Kramer PL, Brin MF, Fahn S, Breakefield X, Ozelius LJ, Risch NJ. The DYT1 phenotype and guidelines for diagnostic testing. Neurology. 2000;54(9):1746–52.

4 Nosology and Classification of Dystonia

Ana Claudia Saraiva Ribeiro and Kailash P. Bhatia

4.1 Introduction

The first descriptions of dystonia date back to the nineteenth century, and since that time, its nosologic definition and classification have been continuously evolving after considerable scrutiny and debate [1, 2].

Gowers introduced the term "tetanoid chorea" to describe two siblings suffering from "tonic spasms which was continuous and varied by paroxysmal attacks of similar but more intense spasms." The true nature of this condition was later recognized as related to hepatic cirrhosis and acknowledged as probable hepatolenticular degeneration. Nonetheless, it is accounted as the first description of dystonia in medical literature [1, 3]. Destarac, in 1901, was probably the author of the first true description of what Oppenheim would call 10 years later "dystonia musculorum deformans" [1, 2]. The term dystonia was then introduced to define an organic disease of the nervous system consisting of an alternating state of hypertonia and hypotonia more pronounced on voluntary movements [1]. There were objections to this new term, mainly by Flatau and Sterling, which considered "progressive torsion spasm" has a more suitable description; hence, they considered there was no varying in muscle tone [4]. Notwithstanding, the term dystonia prevailed even though its phenomenology was by that time far from our current conceptions.

In 1929, at the Tenth Annual Neurological Reunion held in Paris with the main purpose to discuss the concept of "dystonia" as a morbid entity, Wimmer concluded that "Pathology has not been able to bring evidence that torsion dystonia is a disease entity. Dystonia as a syndrome is in no way pathognomonic," as it was recognized in

A.C. Saraiva Ribeiro
Department of Neurology, Centro Hospitalar e Universitãrio de Coimbra, Coimbra, Portugal

K.P. Bhatia (✉)
Sobell Department of Motor Neuroscience and Movement Disorders, Institute of Neurology, University College London, Queen Square, London, UK
e-mail: k.bhatia@ion.ucl.ac.uk

P. Kanovsky et al. (eds.), *Dystonia and Dystonic Syndromes*,
DOI 10.1007/978-3-7091-1516-9_4

other conditions such as Wilson's and Parkinson's disease, among others [5]. In the following years, there was an assumption that it was too early to decide whether dystonia only played a part as a syndrome, which could occur with various different diseases, or was perhaps a morbid entity by itself. These misconceptions were later clarified by Herz, after considering all available data and electromyography recordings, when he found that dystonia as an organic disease consisted in movements and postures and could only be diagnosed in case of "gradual development, without recognizable etiologic factors on onset." He further stated that the clinician should be able to separate those from other well-known conditions such as Wilson's disease [6]. This concept became firmly established after its hereditary trait was acknowledged by Zeman and Dyken in 1967 [7]. Marsden and Harrison elaborated further on Herz criteria for what they called "idiopathic torsion dystonia" to include normal perinatal history and development as well as the absence of evidence of other neurological symptoms or conditions other than dystonia itself and negative laboratory investigations [8]. The first consensus definition was later provided by the Ad Hoc Committee, in 1984, where dystonia was held "a syndrome of sustained muscle contractions, frequently causing twisting and repetitive movements, or abnormal postures" [9].

As concepts and misconceptions about dystonia evolved through time, the classification schemes and cathegories also underwent important changes in close relation to the increasing knowledge about its etiology. Their main purpose is to organize thinking in the light of new evidence and thus providing an essential aid to improve clinical diagnosis, genetic testing, and treatment of dystonia.

With this regard and to overcome some of the current difficulties, a panel of international experts introduced a new definition and classification scheme for dystonia in early 2013 [10].

In this new classification scheme, the utility of the cathegories *age of onset* and *body distribution* was acknowledged as clinically useful, and the cathegories were therefore included with minor changes. More significant refinements were done to the previous cathegory of *etiology*.

The proposal is based on the concept of two axes instead of the three formerly used. *Age of onset* and *body distribution* are now together under Axis 1 which addresses the clinical characteristics associated with dystonia and includes two other subcategories – *temporal pattern* and *associated features*. Axis 1 is aimed to facilitate the recognition of different syndromic associations, while Axis 2 is reserved for the *etiology* and relies on two main subcategories – *nervous system pathology* and division according to *inherited or acquired* forms.

4.2 Axis I: Clinical Characteristics of Dystonia

4.2.1 Classification by Age at Onset

The first attempt to classify dystonia was by age of onset as Herz observed that there were mainly three cathegories of distribution and accordingly early, juvenile, and late forms (Fig. 4.1) [6]. A few years later, Cooper also drew attention to the worse prognosis on the early-onset cathegory and differentiated these groups even further

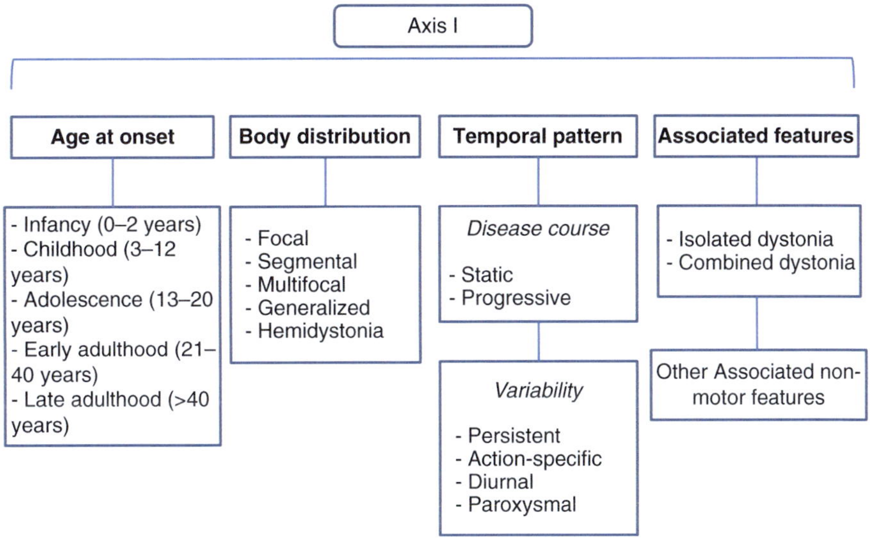

Fig. 4.1 Axis I scheme according to the new consensus update classification [10]

(Table 4.1) [11]. Marsden and Harrison hallmarked that two factors seemed to predict outcome in the patients they reviewed, one was age and second the site of onset. They found that both were closely related as the progression of dystonia to generalized forms was associated to a younger age and onset in the legs [8]. In a later discussion, they concluded that the site of origin and its progression rate were determined by age of onset and divided their data accordingly into two separate groups suggesting that dystonia has a bimodal overall distribution and stated that this division does not imply different etiologies (Table 4.1) [12]. Hence, the classification by age at onset has both diagnostic and prognostic values and, for this, was further acknowledged by the *Ad Hoc Committee* in 1984 and expanded in the new consensus update [9, 12]. In the new classification, the young onset is subdivided into infancy and childhood and the adult onset into early or late forms, accordingly (Table 4.1).

4.2.2 Classification by Body Distribution

To describe the body parts involved by dystonia, Marsden and colleagues coined the terms focal, segmental, and generalized in their review. The term *focal* was meant to be used when the presence of dystonia remained confined to the site of origin and involved only one single region such as the arm or neck. When dystonia spread to neighboring regions but remained limited to two or more contiguous regions, the word *segmental* would be applied. The term *generalized* included the patients in which dystonia has spread to involve all four limbs and axial structures [12]. The same authors also acknowledged that generalized dystonia was more likely related

Table 4.1 Evolution of classification according to age at onset

Classification of dystonia according to Herz [6] 1. Early: onset on or after birth 2. Juvenile: onset between the ages of 5 and 15 years 3. Late: onset after the age of 15
Classification of dystonia according to Cooper [11] 1. Childhood form: onset between 4 and 6 years of age; nearly always with foot inversion and rapid progression to generalized dystonia 2. Adolescent form: onset between 8 and 13 years of age; usually in the foot, sometimes in the arms, and rarely in axial structures; slower rate of progression and less disability 3. Adult form: pattern of axial dystonia, starting in the arm, slow progression, and sparing the extremities with even less disability
Classification of dystonia according to Marsden et al. [8, 12] 1. Early-onset dystonia: onset before the age of 20; the highest percentage of cases (70 %); onset in the leg > arm > axial; high progression to generalized dystonia and often inherited 2. Adult-onset dystonia: onset after the age of 20; low percentage of cases (30 %); onset in the arm > axial; focal or progression to segmental and usually not inherited
Classification of dystonia according to the Ad Hoc Committee [9] 1. Childhood: onset between birth and 12 years of age 2. Adolescent: onset between 13 and 20 years of age 3. Adult: onset above 20 years of age
Classification of dystonia according to the new consensus update 2013 [10] 1. Infancy (birth to 2 years) 2. Childhood (3–12 years) 3. Adolescence (13–20 years) 4. Early adulthood (21–40 years) 5. Late adulthood (>40 years)

with early-onset dystonia and a higher risk was seen with onset on the first decade. According to their observations, the group with onset after the second decade was more prone to develop segmental or focal dystonia, the latter also common in even later-onset cases (Table 4.1) [12]. In 1984, the Ad Hoc Committee expanded this terminology further in their consensus, and the same concepts were kept in the new classification scheme (Table 4.2) [9, 10].

The classification according to body distribution is valuable first as an indicator of severity and second regarding the choice of treatment. The restricted syndromes of focal or segmental dystonia can be offered treatment with *botulinum toxin* injections, while for the generalized forms, drugs and surgery remain the only options.

4.2.3 Classification According to Temporal Pattern

This new cathegory concerns two important issues related to the dystonia phenomenology. The first is the *disease course* which can be either static or progressive, the

Table 4.2 Classification of dystonia according to the body regions involved

1. Focal dystonia – a single body part affected Eyelids (blepharospasm) Mouth (oromandibular dystonia) Larynx (dystonic adductor dysphonia) Neck (torticollis) Arm (writer's cramp)
2. Segmental dystonia – two or more contiguous body parts affected Cranial (head and neck) Axial (trunk and neck) Brachial (one or both arms ± trunk ± neck) Crural (one or both legs ± trunk)
3. Generalized dystonia – a combination of affected segments including the trunk and two other sites
4. Multifocal dystonia – two or more regions affected, noncontiguous
5. Hemidystonia – ipsilateral arm and leg affected

Modified from Fahn et al. [13]

latter being frequently related with spreading to other body regions. The second is the *variability* and comprises four patterns – persistent, action specific, diurnal, and paroxysmal. Dystonia can be classified as persistent if it remains the same throughout the day, action specific if it occurs during a selected activity or task, with diurnal fluctuations if there is a circadian variation in its phenomenology, and paroxysmal if it happens only intermittently.

4.2.4 Recognition of Associated Features

Dystonia can present by itself or be associated with other recognizable features such as tremor, myoclonus, parkinsonism, or others. One of the first recognized conditions associated with idiopathic dystonia was tremor, and today it is considered to be a part of its phenotypical expression and thus not a true associated feature. Later, it was acknowledged that myoclonus and parkinsonism could coexist with dystonia, and therefore a new subcategory was created to accommodate these dystonia-plus syndromes. This new term was meant to comprise those entities which differed from the common inherited neurodegenerative conditions and were also distinct from the "pure" primary dystonias. Nonetheless, they shared some similar features with the latter group such as presence of dystonia as a main feature, lack of recognizable cause, and absence of brain pathology correlate. These considerations gave rise to a lot of debate and questioning of where to consider this new group under etiology (see below) [14].

In the new classification scheme, the same subcategories were maintained under a different terminology. The conditions with dystonia as its only recognizable feature will be considered under the term *isolated dystonia* (as before, with or without tremor), and *combined dystonia* will be used for those associated with parkinsonism, myoclonus, or others. Combined dystonia is now also applied to heredodegenerative syndromes and thus is a mere phenomenological concept and does not carry etiological implications [10]. For these reasons, it was considered under Axis 1 (Fig. 4.1).

The occurrence of other neurological or systemic nonmotor features is also considered under Axis 1 and should be noted [15].

The main purpose of Axis I is to allow the recognition of syndromic associations grouped according to their clinical phenotype. Dystonia is well known for its clinical and genetic variability, and Axis I is a guide to create cathegories according to their similarities and thus facilitate its diagnosis. There are four syndromic patterns well recognized – the childhood or adolescent-onset generalized isolated dystonia which can be due to DYT1 gene mutation with autosomal dominant transmission. The focal or segmental isolated dystonia with adult onset where one can account cervical dystonia as the most frequent (but also blepharospasm and writer's cramp among others). The other two patterns result from the combination of dystonia with either parkinsonism or myoclonus and can accommodate a large number of etiologies.

4.3 Axis II: Etiology

Since its early recognition, dystonia was believed to have a hereditary trait even before it was acknowledged as a true organic entity [16]. This concept of inheritance was also recognized by Flatau and Sterling as they observed that the "progressive torsion spasm in children," similar to Tay-Sachs disease, affected the Jewish descendants more [4]. Later, Santangelo described the same condition in a consanguineous non-Jewish family [17]. These results were further replicated by Zeman when he reported an undeniable hereditary pattern among Ashkenazi Jews [18].

As the interest about dystonia grew among clinicians and investigators and similar findings were replicated around the globe producing extensive literature descriptions, the inherence patterns became more evident. The first description of an autosomal dominant form was by Zeman and Dyken, and they also recognized the importance of examining carefully all family members as they found considerable variability on disease expression and documented *formes frustes* [19]. Additional studies revealed a second pattern of inheritance, recognized as autosomal recessive and strongly supported by the previous findings among Ashkenazi Jew descendants [20]. A third form, X-linked, was later described in the Philippines' Island of Panay where pedigree analysis revealed 28 families with only males affected [21].

Historically, the first comprehensive classification of dystonic syndromes was based on etiology and acknowledged a primary or idiopathic group, divided into hereditary and nonhereditary forms, a secondary group associated with heredodegenerative syndromes and environmental causes and a third group of psychological causes (Table 4.3) [22]. It became fairly accepted that in primary dystonia the only neurological abnormality was the presence of dystonia itself, not caused by other

Table 4.3 Classification of dystonia by etiology according to Fahn [22]

1. Primary torsion dystonia A. Hereditary Autosomal dominant Autosomal recessive X-linked B. Idiopathic Generalized Segmental Focal
2. Secondary dystonia A. Hereditary Wilson's disease Huntington's disease Hallervorden-Spatz disease Juvenile neuronal ceroid lipofuscinosis Dystonia with neural deafness Amyotrophic dystonia paraplegia Dystonia with subcapsular cataracts Myoclonic dystonia with nose malformation B. Environmental Perinatal cerebral injury Infection related Postinfectious Head trauma Focal cerebral vascular injury Brain tumor Toxins: manganese, carbon monoxide Drugs: phenothiazines
3. Psychological dystonia

Modified from Fahn [22]

recognized disease, and that in secondary forms the dystonia was often accompanied by other signs or symptoms suggestive of neurological disease. The definition into primary or secondary forms was thus based on an exclusion diagnosis. The primary group was also distinguishable from the secondary because of the lack of pathological correlate.

A few years later, Fahn, Marsden, and Calne introduced their new classification according to the three axes, namely, age of onset, body distribution, and etiology [13]. Subsequently, the classification further evolved to consider four subcategories in the etiology group – primary, dystonia-plus syndromes, secondary, and heredodegenerative [14]. This scheme was further refined by Bressman to include some of the new genetic achievements. She made a clear cut between a primary group mainly composed of genetic causes and a secondary group which included the dystonia-plus syndromes, degenerative diseases, and complex/unknown and acquired forms

[23]. The classification schemes further evolved in the revisions provided by the EFNS/MDS-ES Task Force in 2006 and later in 2011 (Table 4.4) [24, 25].

The evolution of the classification schemes mirrored the prominent advances of neurogenetics over these last decades. As dystonia genes and loci were discovered, a new nomenclature was introduced by the Human Genome Organization Gene Nomenclature Committee to label each mapped locus with a DYT number (Table 4.4) [26]. The main goal of this new genetic classification is to provide help with genetic testing guided though phenotype correlation and apparent mode of inheritance [27, 28].

Nonetheless, this genetic classification was not included in the new consensus update as it was considered limited due to its many inconsistencies. Firstly, the methods used are prone to technical errors as they are mainly based on statistical findings and assume that all the family members share the same genotype. Secondly, there are also issues regarding loci that have been found without gene mapping and without further replication studies that still persist on the classification. Some of the latter were based on a single family pedigree. There are also issues with locus replication giving rise to two DYT having to share the same gene (e.g., DYT 5a and 14) [29] (Table 4.4).

Since dystonia was firmly acknowledged, there have been considerable advances in regard to its etiology. The two main issues to consider when discussing the etiology are the presence of anatomical changes, identified by neuroimaging or pathology, or of an inheritance pattern. To address this, the new consensus classification expert panel divided etiology into two subcategories. The first comprises the *nervous system pathology* and can be further divided into degenerative (presence of an identified structural abnormality), static (comprising non-degenerative causes), or absence of lesions (degenerative or structural). The second acknowledges inherited, acquired, and idiopathic forms [10] (Fig. 4.2).

This new classification introduced in early 2013 is a reflection of how much the knowledge about dystonia grew over these last decades. The clinicians are therefore encouraged to use it in their daily practice as it promises to overcome the current terminology misconceptions, especially regarding the former primary and secondary cathegories, now presumably better classified and more easily understood.

Nonetheless, even though this new classification is clinically more descriptive, its application on the clinical setting may be difficult because it introduces several scattered cathegories that one must link to combine in a syndromic association. However, this better categorization of patients can be of interest from a research point of view, as it creates more homogenous populations. If one assumes that similar clinical or imagiological findings may share similar pathophysiology, then a more homogenous population may lead to a better understanding of disease mechanisms, diagnosis, and development of new treatments for dystonia patients.

There has been a considerable evolution regarding the concept of dystonia, from the early times when its organicity was put to question until this new era of genetic explosion (Table 4.5). There is great expectation that in a nearby future, as it has been happening with neurogenetics, science will finally be able to create more answers than questions regarding this interesting field of dystonia, and new classifications will surely be developed to illustrate it.

Table 4.4 Dystonia genes and loci

Gene	Locus	Inheritance	Phenotype	Gene product (gene)
DYT 1	9q32–34	AD	Young onset, generalized	Torsin A (TOR1A)
DYT 2	NM	AR	Young onset, generalized	
DYT 3	Xq13.1	XR	Filipino dystonia-parkinsonism	Gene transcription factor (TAF1)
DYT 4	NM	AD	Laryngeal ± limb dystonia (1 family)	
DYT 5a	14q22.1-2	AD	Young-onset, dopa-responsive dystonia-parkinsonism	GTP cyclohydrolase I (GTPCH1)
DYT 5b	11p15.5	AR		Tyrosine hydroxylase (TH)
DYT 6	8p11.21	AD	Young onset, craniocervical, or generalized	Thanatos-associated protein 1 (THAP1)
DYT 7	18p	AD	Adult-onset, focal dystonia (1 family)	Not identified
DYT 8	2q35	AD	PNKD1	Myofibrillogenesis regulator 1 (MR-1)
DYT 9	1p31	AD	EID1/episodic chorea or ataxia and spasticity	Glucose transporter 1 (SLC2A1)
DYT 10	16p11-q12	AD	PKD1	PRRT2
DYT 11	7q21.3	AD	Myoclonus dystonia	E-sarcoglycan (SGCE)
DYT 12	19q13.2	AD	Rapid-onset dystonia-parkinsonism	Na+/K+ATPase α3 subunit (ATP1A3)
DYT 13	1p36	AD	Young-onset segmental or generalized dystonia (1 family)	Not identified
DYT 14 = DYT 5a				
DYT 15	18p11	AD	Myoclonus dystonia (1 family)	Not identified
DYT 16	2q31.2	AR	Young-onset, generalized dystonia-parkinsonism	Stress-response protein (PRKRA)
DYT 17	20p11-q13	AR	Young-onset, mixed phenotype (1 family)	Not identified
DYT 18	1p34.2	AD	EID2	Glucose transporter 1 (SLC2A1)
DYT 19	16q13–22	AD	PKD2 (1 family)	Not identified
DYT 20	2q31	AD	PNKD2 (1 family)	Not identified
DYT 21	2q14–q21	AD	Adult-onset mixed phenotype (1 family)	Not identified
DYT 22	Reserved			
DYT 23	9q34.11	AD	Adult-onset cervical dystonia	Cip-1-interacting zinc protein
DYT 24	11p14.2	AD	Adult-onset craniocervical dystonia	Anoctamin
DYT 25	18p11	AD	Mixed phenotype	G protein subunit α_{olf}

AD autosomal dominant, *AR* autosomal recessive, *XR* X-linked recessive, *NM* not mapped, *PNKD* paroxysmal non-kinesigenic dystonia, *PKD* paroxysmal kinesigenic dystonia, *EID* exercise-induced dystonia

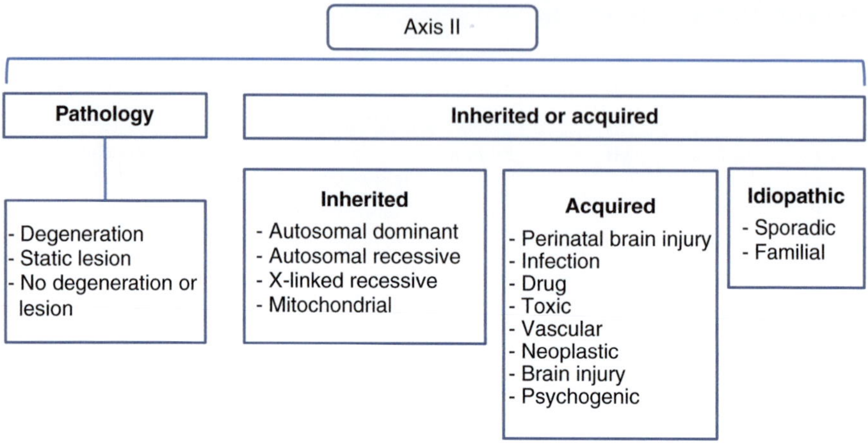

Fig. 4.2 Axis II scheme according to the new consensus update classification [10]

Table 4.5 Evolution of dystonia classification schemes

Fahn et al. [13]	Fahn et al. [14]	Bressman [23]	EFNS/MDS-ES Task Force 2006	EFNS/MDS-ES Task Force 2011
1. By etiology	1. By etiology	1. By etiology	1. By etiology	1. By etiology
(a) Idiopathic	(a) Primary	(a) Primary	(a) Primary	(a) Primary
Sporadic	Sporadic	(b) Secondary	(b) Dystonia-plus syndromes	Primary *pure* dystonias
Familial	Familial	Inherited	(c) Heredodegenerative	Primary *plus* dystonias
(b) Symptomatic	(b) Dystonia-plus syndromes	Dystonia-plus	(d) Secondary	Primary *paroxysmal* dystonias
	(c) Secondary	Degenerative	(e) Paroxysmal	(b) Heredodegenerative
	(d) Heredodegenerative diseases	Complex/unknown		(c) Secondary
		Acquired		
		(c) Dystonia as a feature in other conditions		
		(d) Pseudodystonia		
2. By age at onset	2. By age at onset	2. By age at onset	2. By age at onset	2. By age at onset
(a) Childhood (0–12 years)	(a) Childhood (0–12 years)	(a) Early (< 26 years)	(a) Early (< 20–30 years)	(a) Early (< 20–30 years)
(b) Adolescent (13–20)	(b) Adolescent (13–20)	(b) Late	(b) Late	(b) Late
(c) Adult (>20)	(c) Adult (>20)			
3. By distribution	3. By distribution	3. By distribution	3. By distribution	3. By distribution
(a) Focal	(a) Focal	(a) Focal	(a) Focal	(a) Focal
(b) Segmental	(b) Segmental	(b) Segmental	(b) Segmental	(b) Segmental
(c) Multifocal	(c) Multifocal	(c) Multifocal	(c) Multifocal	(c) Multifocal
(d) Generalized	(d) Generalized	(d) Generalized	(d) Generalized	(d) Generalized
(e) Hemidystonia	(e) Hemidystonia	(e) Hemidystonia	(e) Hemidystonia	(e) Hemidystonia

Modified from [13, 14, 23–25]

References

1. Gowers WR. A manual of diseases of the nervous system. 2nd ed. Philadelphia: P Blakiston, Son & Co.; 1893.
2. Destarac M. Torticollis spasmodique et spasmes fonctionels. Rev Neurol. 1901;9:591–7.
3. Gowers WR. On tetanoid chorea and its association with cirrhosis of the liver. Rev Neurol Psychiatry. 1906;4:249–58.
4. Flatau E, Sterling W. Progressiver torsionsspasmus bei Kindern. Z Die Gesamten Neurol Psychiatr. 1911;7(1):586–612.
5. Wimmer A. Le spasme de torsion. Reunion Neurol Int. 1929;36:904–15.
6. Herz E. Dystonia II. Clinical classification. Arch Neurol Psychiatr. 1944;51(4):319–55.
7. Zeman W, Dyken P. Dystonia musculorum deformans: clinical, genetic and pathoanatomical studies. Psychiatr Neurol Neurochir. 1967;70:77–121.
8. Marsden CD, Harrison MJG. Idiopathic torsion dystonia (dystonia musculorum deformans). A review of forty-two patients. Brain. 1974;97(1):793–810.
9. Fahn S, Marsden CD, Calne DB. Classification and investigation of dystonia. In: Marsden CD, Fahn S, editors. Movement disorders. Vol. 2. London: Butterworths; 1987. p. 332–58
10. Albanese A, Bhatia K, Bressman SB, De Long MR, Fahn S, Fung VSC et al. Phenomenology and classification of dystonia: a consensus update. Mov Disord. 2013;28:863–73
11. Cooper IS. Dystonia musculorum deformans: natural history and neurosurgical alleviation. J Pediatr. 1969;74(4):585–92.
12. Marsden CD, Harrison MJ, Bundey S. Natural history of idiopathic torsion dystonia. Adv Neurol. 1976;14:177.
13. Fahn S, Marsden CD, Calne DB. Classification and investigation of dystonia. Mov Disord. 1987;2:332–58.
14. Fahn S, Bressman SB, Marsden CD. Classification of dystonia. Adv Neurol. 1998;78:1.
15. Stamelou M, Edwards MJ, Hallett M, Bhatia KP. The non-motor syndrome of primary dystonia: clinical and pathophysiological implications. Brain. 2012;135(6):1668–81.
16. Schwalbe W. Eine eigentümliche tonische Krampfform mit hysterischen Symptomen (Doctoral dissertation). Friedrich-Wilhelms-Universität zu Berlin; 1908.
17. Santangelo G. Contributo clinico alla conoscenza delle forme familiari della dysbasia lordotica progressiva (spasmo di torsione). G Psychiatr Neuropathol. 1934;62:52–77.
18. Zeman W, Kaelbling R, Pasamanick B, Jenkins JT. Idiopathic dystonia musculorum deformans I. The hereditary pattern. Am J Hum Genet. 1959;11(2 Pt 1):188.
19. Zeman W, Dyken P. Dystonia musculorum deformans. In: Vinken PJ, Bruyn, editors. Handbook of clinical neurology, vol. 6. Amsterdam: North-Holland Publishing Company; 1968.
20. Eldridge R. The torsion dystonias: literature review and genetic and clinical studies. Neurology. 1970;20(11):1.
21. Lee LV, Pascasio FM, Fuentes FD, Viterbo GH. Torsion dystonia in Panay, Philippines. Adv Neurol. 1976;14:137.
22. Fahn S, Eldridge R. Definition of dystonia and classification of the dystonic states. Adv Neurol. 1976;14:1.
23. Bressman SB. Dystonia genotypes, phenotypes, and classification. Adv Neurol. 2004;94:101.
24. Albanese A, Barnes MP, Bhatia KP, Fernandez-Alvarez E, Filippini G, Gasser T, et al. A systematic review on the diagnosis and treatment of primary (idiopathic) dystonia and dystonia plus syndromes: report of an EFNS/MDS-ES task force. Eur J Neurol. 2006;13(5):433–44.
25. Albanese A, Asmus F, Bhatia KP, Elia AE, Elibol B, Filippini G, et al. EFNS guidelines on diagnosis and treatment of primary dystonias. Eur J Neurol. 2011;18(1):5–18.
26. Müller U, Steinberger D, Németh AH. Clinical and molecular genetics of primary dystonias. Neurogenetics. 1998;1(3):165–77.
27. Müller U. The monogenic primary dystonias. Brain. 2009;132(8):2005–25.
28. Schmidt A, Klein C. The role of genes in causing dystonia. Eur J Neurol. 2010;17:65–70.
29. Marras C, Lohmann K, Lang A, Klein C. Fixing the broken system of genetic locus symbols Parkinson disease and dystonia as examples. Neurology. 2012;78(13):1016–24.

Secondary Dystonia

5

Petr Kanovsky, Dirk Dressler, and Kateřina Menšíková

Classical taxonomical and nosological classification, done by Marsden and Fahn more than 30 year ago, introduced four categories of dystonia based on its aetiology: *primary dystonia*, *secondary* (*symptomatic*) *dystonia*, *dystonia-plus syndromes* and *heredodegenerative dystonia* [1, 2]. With the increasing evidence of the genetic, metabolic and (probable) environmental aetiology of many diseases manifesting with dystonia, these strictly splitted categories coupled back and even intermixed in the recent years [3–6]. Current trend is clearly into the definition of two groups: primary dystonias and non-primary (secondary) dystonias [7]. The second category is sometimes splitted into the dystonia-plus syndromes and symptomatic dystonias. Probably the best solution of this nosological and taxonomical confusion was offered by Schneider and Bhatia in 2010 when they described the wide group of secondary dystonias on the basis of syndromic associations and clinical clues [8]. They introduced the seminal clinical features suggestive of secondary dystonia (Table 5.1) and the list of most frequent syndromic associations (Table 5.2). For clinical purpose, this is probably the most useful diagnostic tool. Nevertheless, for the description of "secondary" dystonia, the current version of "splitted" aetiological classification will be used (Table 5.3).

P. Kanovsky (✉) • K. Menšíková
Department of Neurology, Palacky University Medical School, Olomouc, Czech Republic
e-mail: petr.kanovsky@fnol.cz

D. Dressler, MD, PhD
Department of Neurology, Hanover University Medical School, Hannover, Germany
e-mail: dressler.dirk@mh-hannover.de

P. Kanovsky et al. (eds.), *Dystonia and Dystonic Syndromes*,
DOI 10.1007/978-3-7091-1516-9_5

Table 5.1 Features suggestive of secondary dystonia [8]

Sudden onset and rapid progression
Hemidystonia
Cranial onset in childhood
Restriction to focal or segmental dystonia with childhood onset
Onset in legs in adults
Progression to generalized dystonia with adult onset
Prominent oro-bulbar involvement
Other neurological or systemic signs
Presence of painful spasms

Table 5.2 Syndromic associations and clinical clues in secondary dystonia [8]

Clinical features	Differential diagnosis to consider
Dystonia with prominent oromandibular involvement	Drug-induced dystonia PKAN PLA2G6-associated neurodegeneration Neuroacanthocytosis
Dystonia and peripheral neuropathy	Neuroferritinopathy Lesch-Nyhan disease Niemann-Pick type C Metachromatic leukodystrophy Friedreich's ataxia Ataxia telangiectasia
Dystonia and eye movement disorders	Spinocerebellar ataxias (e.g. SCA3) Niemann-Pick type C Huntington's disease Ataxia telangiectasia
Dystonia and retinitis pigmentosa	PKAN GM2 gangliosidosis
Dystonia and deafness	Metachromatic leukodystrophy Mitochondrial disease Mohr-Tranebjaerg syndrome Woodhouse-Sakati syndrome
Dystonia and parkinsonism	Dopa-responsive dystonia Young-onset parkinsonism Wilson's disease Kufor-Rakeb disease PLA2G6-associated neurodegeneration Atypical parkinsonism (PSP, CBD, MSA)
Dystonia with progressive dementia	GM1, GM2 gangliosidosis Glutaric acidemia Huntington's disease Huntington's disease look-alike disorders

Table 5.3 Secondary dystonia classified according to underlying pathological process [5]

I. Focal brain lesions
Vascular processes
Trauma
Tumour
Encephalitis

Table 5.3 (continued)

II. Diffuse brain damage
Impairment of neuronal energy metabolism
Ischaemia
Hypoxia
Mixed
Glucose
Impairment of systemic metabolism
Bilirubin
Calcium
Natrium
Increased intracranial pressure
Subdural haematoma
Normal pressure hydrocephalus
Drugs
Acute
Dopamine-blocking agents
Benzamides
Antihistamines
Dopamine agonists in Parkinson's disease
Tardive
Phenothiazines
Thioxanthenes
Butyrophenones
Diphenylbutyrylpiperidine
Dibenzazepine
Dibenzodiazepine
Thienobenzodiazepine
Substituted benzamides
Indolenes
Pyrimidinone
Tricyclic antidepressants
Chemical agents
Manganese
Copper
Methanol
Cyanide
Carbon monoxide
Ephedrone
Physical interactions
Electricity, radiation
Indirect central nervous effects
Peripheral trauma with reflex sympathetic dystrophy
Peripheral trauma without causalgia

5.1 Secondary (Symptomatic) Dystonia

Pure secondary, i.e. symptomatic, dystonias are those caused by the focal brain lesion (trauma, tumour, stroke, etc.), diffuse brain disorder, perinatal encephalopathy, infections and drugs or by the direct impact of chemical agents. Regarding the focal brain lesions, the dystonias – either focal, segmental or hemidystonia – are the most frequently caused movement disorder. This type of dystonia might appear in the case of lesion of sensitive brain structures, most frequently reported being the striatum (particularly putamen), pallidum, thalamus (VPL nucleus of thalamus), dorsolateral parietal cortex and cerebellar hemisphere [9–13]. It has been postulated that putaminal lesions cause a pure hemidystonia syndrome, and it is also believed that thalamic lesions cause a complex movement disorder with the predominant limb dystonia [14]. Or dystonia might appear as a symptom of s.c. hemithalamic syndrome (hemiparesis, hemidystonia, hemiataxia, hemihypesthesia, etc.). Nevertheless, one can consider a fresh focal brain lesion as a cause of dystonia practically anytime when the contralateral body parts are affected and even in the cases when the lesion localisation is rather atypical; particularly, the brainstem lesions can cause secondary dystonia when present at different levels. Usually the longer delay between the presence of focal brain lesion on imaging and onset of dystonia makes the diagnosis of secondary dystonia difficult, but in many cases of focal brain lesions, the dystonia manifests with a week- or month-long time gap [15].

According to Dressler, the focal brain lesions affecting together pyramidal and extrapyramidal structures cause the frequently so-called muscle hyperactivity in the paretic limb, which consists of different disorders including dystonia, rigidity, spasms and spastic response to the passive extension. Dystonia in this syndrome prevails, and it is often described as "spastic dystonia". The term "spastic condition" has been originally proposed by Dressler [5]; recently, we also introduced the term "spastic movement disorder" [16]; nevertheless, in both cases, the same syndrome is depicted in its description. In this disorder, the prevailing dystonia is present as tonic (with only rare fluctuations) and may be enhanced by the attempt to move the paretic limb. This type of secondary dystonia is probably the most common, due to the overall incidence of its most frequent cause, the stroke.

Secondary dystonia can appear in the perinatal encephalopathy (perinatal cerebral injury, cerebral palsy). It can be present in this disorder together with spasticity, so the differential diagnosis between two hypertonic disorders in these cases might be difficult. Pure secondary dystonia can manifest in cerebral palsy in the form of "athetoid cerebral palsy" or delayed-onset dystonia ([17–19]).

Diffuse brain disorder such as hypo- or hyperglycaemia and diffuse hypoxia may cause dystonia. In these cases, the diffuse disorder affects the whole brain, including the regions with a very high utilisation of oxygen and glucose [20]. Among them is the pallidum, and impact of these diffuse disorders on this structure may cause dystonia. The clinical presentation in these cases, however, is not monosymptomatic with dystonia, but rather complex, when other types of motor disorder (choreoathetosis, paresis, myoclonus) might be present. In some cases, the dystonia may have progressive course [21, 22].

Infections of the central nervous system can also trigger dystonia. Probably the most frequent movement disorder is dystonia in multiple sclerosis [23]; usually dystonia might be present in the Reye syndrome, subacute sclerosing panencephalitis, Creutzfeldt-Jakob disease and HIV-related encephalitis or in neurosyphilis. It also might be present as a post-infectious syndrome, for instance, following the Japanese encephalitis or tick-borne European encephalitis ([24–30]). Dystonia, together with chorea, can be a symptom of autoimmune systemic diseases [31, 32].

There was a long-term discussion whether the peripheral trauma can trigger dystonia [33–35]. More than 10 years ago, the term "reflex sympathetic dystrophy" was suggested for the complex disorder (dystonia, pain, amyotrophy) triggered by the peripheral limb trauma [36]. Nowadays, it has been replaced by the nosological category "regional pain syndrome" [37–39]. The dystonia itself in this syndrome might be progressive and devastating. Usually it does not respond to the botulinum toxin therapy, and in difficult cases, it deserves surgical treatment, including amputation [40–42].

Head trauma with the small, localised brain lesion can also be a cause of the secondary dystonia. This is usually caused by the contusion of thalamic region, and dystonia is present as hemidystonia together with other signs of hemithalamic syndrome [14, 43].

5.2 Dystonia-Plus Syndromes

Dystonia-plus syndromes represent a heterogeneous group of diseases, where dystonia is accompanied by other neurological features and gene mutations can be detected. At present, these disorders are categorized as dystonia-plus syndromes: dopa-responsive dystonia (DRD), myoclonus-dystonia, rapid-onset dystonia-parkinsonism and DYT-16 dystonia-parkinsonism. In this chapter, only the DRD and myoclonus-dystonia will be described; those syndromes connecting dystonia with parkinsonism are included in Chap. 6.

5.2.1 Dopa-Responsive Dystonia (DRD, DYT-5, DYT-14)

Dopa-responsive dystonia is the term with which Nygaard in 1988 tried to replace the previously used name for this disease, Segawa's syndrome [44]. For almost 10 years, these terms were used promisque, and the more recent literature uses practically only the term dopa-responsive dystonia. Nevertheless, Masaya Segawa was the first to have recognised and described this syndrome in Japanese children [45]. Nowadays, he has aimed to introduce the term "autosomal dominant GTP cyclohydrolase 1 deficiency".

Dopa-responsive dystonia is a rare disorder. The classical description of this type of dystonia is the "early-onset lower limb dystonia that can cause gait disorder, has a diurnal fluctuation pattern and promptly responds to L-DOPA treatment [44]".

Usually dystonia is present in the early course of the disease, and only later parkinsonism can develop; in adult patients, parkinsonism can be present together with dystonia in the early stage. Diurnal fluctuations are one of the pathognomonic features of DRD, when symptoms get worse in the afternoon or evening and improved in the morning, after sleeping.

DRD is an inherited disorder; however, it may be caused by various gene defects with different penetrance. All known mutations affect dopamine metabolism. The most frequent type of DRD, DYT-5, is caused by the guanosine triphosphate cyclohydrolase 1 gene defect on chromosome 14q22.1–q22.2 [46]. This type of DRD manifests in the childhood, most frequently in the age of 5–7 years, although adult-onset variants were also reported [47]. DYT-5 seems to have gender-different penetration, and thus females are more frequently affected than males. The typical initial clinical manifestation in childhood is the asymmetric dystonia affecting one of lower limbs, but the premotor initial symptoms might be nonspecific, bothering the patient with leg stiffness, cramps or tiredness on exercise. Bilateral, often symmetric, lower limb dystonia and parkinsonism with predominant bradykinesia develop later in disease course. The diagnosis is usually made on the basis of disease history, clinical picture and response to the L-DOPA challenge [48]. Differential diagnosis versus cerebral palsy might be an issue in atypical cases; nevertheless, the progressive course of the disease and its typical L-DOPA responsiveness are usual clues. L-DOPA treatment works well in DRD in low doses and usually without any sign of tachyphylaxis.

5.2.2 Myoclonus – Dystonia (DYT-11)

The disease with several name variations (myoclonus-dystonia, myoclonic dystonia, dystonia-myoclonus) is a rare autosomal dominant inherited disorder. The penetrance is incomplete and there is no gender difference present. The genetic basis of disease are the multiple mutations in the epsilon-sarcoglycan gene on the chromosome 7q21 [49]. The typical manifestation of this disease is formed by the dystonia with jerks of affected muscles, giving the clinical picture of myoclonus. Predominantly the axial muscles are affected, with some involvement of upper limbs; only rarely the manifestations of myoclonic dystonia on lower limbs were documented. There are several psychiatric disorders which may accompany motor symptoms of myoclonus-dystonia: anxiety, panic attacks, depression, obsessive-compulsive disorder and personality disorders [50]. Myoclonus-dystonia usually manifests in the late childhood or early adulthood; 80 % of affected individuals experienced the first symptoms of disease before 20 years of life. The course of myoclonus-dystonia is usually mild and does not affect significantly the quality of life; sometimes the spontaneous remissions have been reported. Myoclonus-dystonia may be alleviated by the consumption of low amount of alcohol; this may also serve in the differential diagnostic process [51]. Valproate may be helpful in the treatment; nevertheless, botulinum toxin injections are widely used for the successful treatment. In patients who do not respond to botulinum toxin injection, deep brain stimulation should be considered.

5.3 Heredodegenerative Dystonia

From the taxonomical point of view, the dystonia itself in heredodegenerative dystonias forms only a part of the clinical picture and always is accompanied by other neurological symptoms. It is also important to highlight that – in contrast to the dystonia-plus syndromes – in heredodegenerative dystonia, dystonia itself is never a dominant symptom of the disease. On the other hand, these disorders can start with dystonia as a sole symptom, which may be later accompanied and overcome by other symptoms. Current classification of heredodegenerative dystonia is done on the basis of underlying pathological processes (see Chap. 4).

5.3.1 Heredodegenerative Dystonia in the Mitochondriopathies

Mitochondriopathies are a group of disorders, which are inherited in a special way, since the gene defect is found in the mitochondrial DNA, which is only maternally inherited. In the mitochondriopathies, the defect of mitochondrial metabolic process affects mainly the tissues with the high level of energy consumption, i.e. the muscles and the brain. Probably the most frequent appearance of dystonia might be seen in the Leigh disease; nevertheless, it may be present in any of the mitochondrial diseases [52]. The muscle biopsy is a diagnostic clue in these diseases, and there is practically no cure for them. Dystonia may be temporarily alleviated by the botulinum toxin treatment.

5.3.2 Heredodegenerative Dystonia in the Inherited Metabolic Disorders

There are several conditions, in which dystonia may form an important part of the clinical picture: the Lesch-Nyhan syndrome (hyperuricaemia), Hartnup's disease, homocysteinaemia, methylmalonic acid disease and galactosaemia.

The *Lesch-Nyhan disease* is a rare disorder, which – due to the X-linked recessive inheritance – affects only males. The metabolic defect causes hyperuricaemia and secondary hyperuricosuria. Dystonia in the Lesch-Nyhan disease is usually localised in the head, forming a typical oro-mandibulo-lingual pattern. Self-mutilations in this body area due to dystonia are frequent in the Lesch-Nyhan disease [53]. Diagnosis is based on the evidence of hyperuricaemia or results of genetic examination.

Hartnup's disease is caused by the impaired absorption (in the intestine) and re-absorption (renal) of tryptophan and other neutral amino acids [54]. Dystonia might be a prominent feature; other symptoms include ataxia, nystagmus and tremor. Psychiatric disturbances, as hallucinations, are present in some cases. The symptoms of this inherited condition appear usually in the period of childhood and early adulthood [55]. The presence of increased amino acids level in the urine makes the diagnosis.

Homocysteinaemia is an autosomal recessive disorder of the homocysteine metabolism, in which serum levels of homocysteine are excessively high. The leading symptoms of this disease are bone deformities, and in laboratory tests homocysteinaemia (the normative data in laboratories may differ!) and homocystinuria are found. Dystonia may also be present [56].

Methylmalonemia or methylmalonic aciduria is also inherited in the autosomal recessive manner; it presents with the chronic metabolic acidemia and the presence of methylmalonic acid in the urine. Together with chronic renal failure and haematological abnormalities (thrombopenia, leucopenia), dystonia may be present [57].

Galactosaemia is also autosomal recessive inherited metabolic disease, in which the serum level of galactose is pathologically increased due to enzymatic disorder. The symptoms include hepatomegaly, chronic renal failure, cataract and the presence of neurological symptoms with prominent dystonia [58].

5.3.3 Heredodegenerative Dystonia Caused by the Storage Diseases

Accumulation of different toxic or noxious substances in body tissues can be caused by many gene defects, virtually all autosomal recessive. The disposal of these substances can be specific and may be limited to some body compartments, or it may be diffuse. These rather rare disorders appear in the childhood or late childhood, but manifestation in the adulthood may also occur (Wilson's disease). Somatic symptoms include usually hepatosplenomegaly and chronic renal failure, neurological picture may vary from case to case, and dystonia is very often present.

Wilson's disease, which has been known under the name "hepatolenticular degeneration" is caused by the numerous mutations of the locus localised on the short arm of chromosome 13 [59]. In Wilson's disease, the metabolism of copper is defected due to the lack of its transporting serum protein, ceruloplasmin. The copper is not excreted in the urine and accumulates in the body tissue, predominantly in the liver and brain. In the brain, the maximum deposits are found in the basal ganglia. This involvement is the cause of predominant presence of extrapyramidal symptoms in the clinical neurological picture, including dystonia. Besides dystonia, tremor, chorea, ataxia and epilepsy may be present. Together with neurological impairment, the psychiatric symptoms are often present in the form of depression or other psychosis. The diagnosis of Wilson's disease is based on the increased level of copper in the serum and urine and decreased level of ceruloplasmine ibidem. When this evidence is not sufficient, liver biopsy should be used to confirm the accumulation of copper in hepatocytes. Copper can accumulate also in the cornea and manifest as Keyser-Fleischer ring in practically all Wilson's patients. On magnetic resonance imaging, one can find typical "panda sign" in T2-weighted images in the horizontal projection of the brainstem. The diagnosis of Wilson's disease is not a difficult issue in the case that Wilson's disease is considered in the differential diagnostic process. Chelation therapy (penicillamine, zinc) can alleviate the neurological symptoms almost completely; however, its chronic use can cause renal failure [60].

GM1-gangliosidosis is one of the lysosomal storage diseases. The lack of the enzyme beta-galactosidase causes the block of lysosomal metabolism of lipids. As a consequence, the GM1-ganglioside and other oligosaccharides accumulate in the body including the brain. There are three types of GM1-gangliosidosis according to the age of manifestation of disease. In type 3, which usually appears in the early adulthood, neurological symptoms may predominate; they include dystonia and parkinsonism [61].

The *Niemann-Pick disease* is also an autosomal recessive storage disease. The lysosomal accumulation of sphingomyelin in the liver, spleen, brain and bone marrow occurs in the Niemann-Pick disease of A and B types; in the C type, cholesterol and glycolipids are stored in the lysosomes. Clinically apparent neurological symptoms are present only in the A and C types of the Niemann-Pick disease, and dystonia can be present only in the C type, usually together with epilepsy, dementia and vertical gaze palsy [62]. On somatic examination, hepatosplenomegaly is usually found. The disease manifests virtually in the childhood, and the diagnosis is made on the presence of hepatosplenomegaly and neurological symptoms; it can be confirmed by the bone marrow (sea-blue histiocytes) and genetic examination.

Metachromatic dystrophy is the disease in which sulphatides are accumulated in the lysosomes, then the myelin is destructed in both the peripheral and central nervous systems. Disease itself is caused by the impairment of the arylsulphatase A. The main neurological symptoms in metachromatic dystrophy are dementia and polyneuropathy, which are relatively frequently accompanied by dystonia [63]. Disease can manifest in both childhood and adulthood, and the course is often rapidly progressive. The diagnosis is made on the basis of conclusive brain MRI examination results and confirmed by the finding of sulphatides in the urine.

Other rare lysosomal storage diseases are the *neuronal ceroid lipofuscinosis* and *Gaucher's disease*; both disorders can manifest with dystonia [64]. Diagnosis is made on the results of biochemical examinations of concurrent enzymes, palmitoyl protein thioesterase and beta-glucocerebrosidase in the leucocytes.

Mitochondrial storage diseases are caused by the accumulation of the metabolites in mitochondria. They are virtually all rare, and the *glutaric acidemia type 1* is probably the most frequently diagnosed. The disease itself is caused by the lack of glutaryl-coenzyme A dehydrogenase, which is needed for the metabolism of amino acids tryptophan, hydroxylysine and lysine. As a result, hydroxyglutaric acid and glutamic acid accumulate in the brain and other tissues. In the brain, basal ganglia are usually most seriously affected [65]. Glutaric acidemia type 1 appears practically only in childhood and manifests with athetosis or dystonia, which may be accompanied by the spasticity and epilepsy. Diagnosis is made on the presence of neurological symptoms and course of the disease; it can be confirmed by the biochemical examination and presence of reduced glutaryl-coenzyme A hydroxylase in fibroblasts and also by the genetic testing. Glutaric acidemia type 1 is treatable by the low-lysine diet, which usually stops progression, but the remission never occurs.

5.3.4 Heredodegenerative Dystonia Caused by the Nucleotide Repeat Diseases

Mostly overseen, dystonia may be also present as a symptom of CAG repeat diseases. They are usually inherited in the autosomal dominant manner; the clear inheritance can in some generation disappear due to the anticipation effect.

Huntington's disease is probably the most frequent and most widely known example of CAG repeat diseases. It is caused by the CAG repeat expansion in the 4q21–22 (or 4p16.3), where the usual number is 29–35 repeats. Any number above that value means the manifestation of disease, and its age of manifestation and seriousness depends on the excessive repeats number. Huntington's disease manifests usually in the middle age; however, the onset before age of 30 years is not uncommon. In such a case, the clinical picture resembles young-onset Parkinson's disease. The term "Westphal's variant" is commonly used for this type of manifestation. Nevertheless, Huntington's disease much more frequently manifests with the slowly progressive chorea and cognitive disturbance, accompanied by the symptoms of psychosis and behavioural disorder. The pathological protein product is called huntingtin, and it is widely accumulated within the neurons, with predominance in the basal ganglia, namely, the striatum. Dystonia can manifest in cases of early-onset Huntington's disease (Westphal's variant); it is not rare in the later phases of the typical-onset disease [66]. The diagnosis of Huntington's disease is, in typical cases, made on the unique presence of progressive chorea, dementia, psychosis and behavioural disorder, accompanied by the typical MRI finding, the atrophy of the caudate head; it is confirmed by the genetic testing. Treatment of Huntington's disease is only symptomatic (dopamine depletors, atypical neuroleptics, tetrabenazine); botulinum toxin injections for dystonia did not show any considerable effect.

Spinocerebellar ataxias are a large group of autosomal dominant CAG repeat diseases; currently, 38 loci (SCA1–SCA38) have been described, which may manifest with the clinical picture of combined cerebellar and spinal cord syndrome, accompanied by the diverse mixture of symptoms originating in other systems. Dystonia may appear in the SCA1, SCA3, SCA6, SCA7, SCA14, SCA15 and SCA17, with SCA3 being the most frequently present ion (Machado-Joseph in the older terminology) together with lower-body parkinsonism [5].

Dentatorubral-pallidoluysian atrophy (*DRPLA*), a rare autosomal dominant CAG repeat disease, can even more rarely manifest with dystonia [67].

5.4 Dystonia as a Feature of Another Neurologic Diseases

The *Hallervorden-Spatz disease* (pantothenate kinase-associated neurodegeneration, neurodegeneration with brain iron accumulation) is a rare degenerative disorder caused by the PKAN gene defect localised in the locus 20p13. The disease manifests usually in the childhood; however, adult-onset cases exist. When present in its typical form, it manifests with oro-mandibulo-lingual and limb dystonia, sometimes also with torticollis [68]. Parkinsonism may complete the clinical picture. Both dystonia and parkinsonism may be accompanied by the retinitis pigmentosa. The diagnosis is made on the presence of described clinical picture, childhood onset and typical brain MRI finding: eye-of-the-tiger sign, caused by the iron accumulation in the pallidum. Only symptomatic treatment exists; dystonia was treated with botulinum toxin injections with some effect. Acanthocytosis is another possible feature of this disease, but it may exist also in its pure form, neuroacanthocytosis.

Neuroacanthocytosis is an autosomal recessive hereditary disorder, in which the structural proteins of erythrocytes are altered causing their conversion into acanthocytes. The focal or segmental facial, lingual and mandibular dystonia is accompanied by parkinsonism and polyneuropathy; epilepsia may be also present [69]. The disease manifests in the young age, and its progression lasts from 6 to 10 years.

Friedreich's ataxia is the most common hereditary ataxia, with a prevalence of 1 per 30,000. In about 98 %, it manifests due to homozygosity for a GAA trinucleotide repeat expansion in intron 1 of FXN; this triplet expansion causes deficiency of frataxin, a small nuclear genome-encoded mitochondrial protein. Low frataxin levels lead to insufficient biosynthesis of iron-sulphur clusters that are required for mitochondrial electron transport and assembly of functional aconitase and iron dysmetabolism of the entire cell. The disease affects multiple sites in the central and peripheral nervous systems as well as a number of other organ systems, resulting in multiple signs and symptoms. Onset of this autosomal recessive condition is usually in the first two decades of life. Major clinical features include progressive ataxia, absent lower limb reflexes, increased plantar reflexes, peripheral sensory neuropathy and also dystonia, usually torticollis or oromandibular or limb dystonia [70]. Visceral pathology includes the heart, resulting in cardiomyopathy, and the pancreas, resulting in diabetes mellitus.

Parkinson's disease The Lubag complex (dystonia-parkinsonism) or DYT-3 X-linked dystonia-parkinsonism is described in another chapter, as well as the DYT-12, or rapid-onset dystonia-parkinsonism, autosomal dominant hereditary neurodegenerative condition. Dystonia may occur also in the initial or very early phase of "sporadic" or "idiopathic" Parkinson's disease manifestation, usually in young patients, particularly female. The dystonia might be present in the form of torticollis, blepharospasm or task-specific hand dystonia. The painful foot dystonia is present particularly if Parkinson's disease manifest as a disorder with the onset in the young age (in these terms, it means between 40 and 50 years of life). A broad spectrum of dystonic symptoms is present in the late, complicated phase of Parkinson's disease with on-off fluctuations. It may be present either as painful "off" dystonia of the foot or as "start of dose", "peak of dose" or "end of dose" dystonia in the "on" phase, usually together with choreic movements [71]. It is important to stress here that anytime the dystonia is present in parkinsonism, the diagnosis of multiple system atrophy should be considered or reconsidered.

References

1. Fahn S. Concept and classification of dystonia. In: Fahn S, Marsden CD, Calne DB editors. Dystonia 2. Adv Neurol. 1988;50:1–8.
2. Fahn S, Marsden CD, Calne DB. Classification and investigations of dystonia. In: Marsden CD, Fahn S, editors. Movement disorders 2. London: Butterworths; 1987. p. 332–58.
3. Asmus F, Gasser T. Dystonia-plus syndromes. Eur J Neurol. 2010;17 Suppl 1:37–45.
4. Colosimo C, Berardelli A. Clinical phenomenology of dystonia. Int Rev Neurobiol. 2011;98:509–23.
5. Dressler D. Nonprimary dystonias. In: Weiner W, Tolosa E, editors. Handbook of clinical neurology, Hyperkinetic movement disorders, vol. 101. Amsterdam: Elsevier; 2011. p. 513–38.
6. Fahn S. Classification of movement disorders. Mov Disord. 2011;26:947–57.
7. Phukan J, Albanese A, Gasser T, Warner T. Primary dystonia and dystonia-plus syndromes: clinical characteristics, diagnosis and pathogenesis. Lancet Neurol. 2011;10:1074–85.
8. Schneider SA, Bhatia KP. Secondary dystonia – clinical clues and syndromic associations. Eur J Neurol. 2010;17 Suppl 1:52–7.
9. Alarcón F, Tolosa E, Munoz E. Focal limb dystonia in a patient with a cerebellar mass. Arch Neurol. 2001;58:1125–7.
10. Chuang C, Fahn S, Frucht SJ. The natural history and treatment of acquired hemidystonia: report of 33 cases and review of the literature. J Neurol Neurosurg Psychiatry. 2002;72:59–67.
11. Kostic V, Svetel M, Kacar A. Symptomatic dystonias associated with structural brain lesions: report of 16 cases. Can J Neurol Sci. 1996;23:53–6.
12. Marsden CD, Obeso J, Zarranz J, et al. The anatomical basis of symptomatic hemidystonia. Brain. 1985;108:463–83.
13. Schwartz M, De Deyn PP, Van den Kerchove M, Pickut BA. Cervical dystonia as a probable consequence of focal cerebral lesion. Mov Disord. 1995;10:797–8.
14. Lee MS, Rinne OJ, Ceballos – Baumann A, et al. Dystonia after head trauma. Neurology. 1994;44:1374–8.
15. Berardelli A. Symptomatic or secondary basal ganglia diseases and tardive dyskinesias. Curr Opin Neurol. 1995;8:320–2.
16. Rosales RL, Kanovsky P, Fernandez HH. What's the "catch" in upper-limb post-stroke spasticity: expanding the role of botulinum toxin applications. Parkinsonism Relat Disord. 2011;17 Suppl 1:3–10.
17. Fairhurst C. Cerebral palsy: the whys and hows. Arch Dis Child Educ Pract Ed. 2012;97:122–31.
18. Singer H. Movement disorders in children. In: Tolosa E, Jankovic J, editors. Parkinson's disease and movement disorders. Philadelphia: Lippincott, Williams and Wilkins; 1998. p. 739–54.
19. Nelson KB. The neurologically impaired child and alleged malpractice at birth. Neurol Clin. 1999;17:283–93.
20. Johnston MV, Hoon AH. Possible mechanisms in infants for selective basal ganglia damage from asphyxia, kernicterus or mitochondrial encephalopathies. J Child Neurol. 2000;15:588–91.
21. Venkatesan A, Frucht S. Movement disorders after resuscitation from cardiac arrest. Neurol Clin. 2006;24:123–32.
22. Kostic VS, Covickovic-Sternic N, Svetel-Stojanovic M, Filipovic S. Delayed-onset dystonia due to asphyxia in the perinatal period. Srp Arh Celok Lek. 1997;125:84–8.
23. Tranchant C, Bhatia KP, Marsden CD. Movement disorders in multiple sclerosis. Mov Disord. 1995;10:418–23.
24. Alarcón F, Duenas G, Cevallos N, Lees AJ. Movement disorders in 30 patients with tuberculous meningitis. Mov Disord. 2000;15:561–9.

25. Carroll E, Sanchez – Ramos J. Hyperkinetic movement disorders associated with HIV and other viral infections. Hand Clin Neurol. 2011;100:323–34.
26. Hur YJ, Hwang T. Secondary paroxysmal dyskinesia associated with 2009 H1N1 infection. Kor J Pediatr. 2013;56:42–4.
27. Misra UK, Kalita J. Overview: Japanese encephalitis. Prog Neurobiol. 2010;91:108–20.
28. Tong ML, Lin LR, Zhang HL, et al. Spectrum and characterization of movement disorders secondary to neurosyphilis. Parkinsonism Relat Disord. 2013;19:441–5.
29. Maltéte D, Guyant-Marechal L, Mihout B, Hannequin D. Movement disorders and Creutzfeld-Jakob disease: a review. Parkinsonism Relat Disord. 2006;12:65–71.
30. Netravathi M, Pal PK, Indira Devi B. A clinical profile of 103 patients with secondary movement disorders: correlation of etiology with phenomenology. Eur J Neurol. 2012;19:226–33.
31. Baizabal-Carvallo JF, Jankovic J. Movement disorders in autoimmune diseases. Mov Disord. 2012;27:935–46.
32. Janavz LJ, Aminoff MJ. Dystonia and chorea in acquired systemic disorders. J Neurol Neurosurg Psychiatry. 1998;65:436–45.
33. Jankovic J. Post-traumatic movement disorders: central and peripheral mechanisms. Neurology. 1994;44:2006–14.
34. Jankovic J. Can peripheral trauma induce dystonia and other movement disorders? Yes! Mov Disord. 2001;16:7–12.
35. Weiner W. Can peripheral trauma induce dystonia? No! Mov Disord. 2001;16:13–22.
36. Serrano-Dueñas M. Reflex sympathetic syndrome and peripheral dystonia. Mov Disord. 2003;18:1212–3.
37. Maihofner C, Seifert F, Markovic K. Complex regional pain syndromes: new pathophysiological concepts and therapies. Eur J Neurol. 2010;17:649–60.
38. Schott GD. Peripherally-triggered CRPS and dystonia. Pain. 2007;130:203–7.
39. van Hilten JJ, Geraedts EJ, Marinus J. Peripheral trauma and movement disorders. Parkinsonism Relat Disord. 2007;13 Suppl 3:S395–9.
40. Edwards MJ, Alonso-Canovas A, Schrag A, Bloem BR, Thompson PD, Bhatia K. Limb amputations in fixed dystonia: a form of body integrity disorder? Mov Disord. 2011;26:1410–4.
41. Hawley JS, Weiner WJ. Psychogenic dystonia and peripheral trauma. Neurology. 2011;77:496–502.
42. van Roijen DE, Geraedts EJ, Marinus J, Jankovic J, van Hilten JJ. Peripheral trauma and movement disorders: a systematic review of reported cases. J Neurol Neurosurg Psychiatry. 2011;82:892–8.
43. Pettigrew LC, Jankovic J. Hemidystonia: a report of 22 patients and a review of the literature. J Neurol Neurosurg Psychiatry. 1985;48:650–7.
44. Nygaard TG, Marsden CD, Duvoisin RC. Dopa-responsive dystonia. Adv Neurol. 1988;50:377–84.
45. Segawa M, Hosaka A, Miyagawa F, Nomura Y, Imai H. Hereditary progressive dystonia with marked diurnal fluctuation. Adv Neurol. 1976;14:215–33.
46. Ichinose H, Ohye T, Segawa M et al. GTC cyclohydrolase I gene in hereditary progressive dystonia with marked diurnal fluctuation. Neurosci Lett. 1995;196:5–8.
47. Segawa M, Nomura Y, Nishiyama N. Autosomal dominant guanosine triphosphate cyclohydrolase I deficiency (Segawa disease). Ann Neurol. 2003;54:S32–45.
48. Nygaard T. Dopa-responsive dystonia. Delineation of the clinical syndrome and clues to pathogenesis. Adv Neurol. 1993;60:577–85.
49. Zimprich A, Grabowski M, Asmus F, et al. Mutation in the gene encoding epsilon-sarcoglycan cause myoclonus-dystonia syndrome. Nat Genet. 2001;29:66–9.
50. Klein C. Myoclonus and myoclonus-dystonia. In: Pulst S, editor. Genetics of movement disorders. San Diego: Academic; 2002. p. 449–69.
51. Quinn N. Essential myoclonus and myoclonic dystonia. Mov Disord. 1996;11:119–24.
52. Finsterer J. Leigh and Leigh-like syndrome in children and adults. Pediatr Neurol. 2008;39:223–35.

53. Schneider SA, Aggarwal A, Bhatt M, et al. Severe tongue protrusion dystonia: clinical syndromes and possible treatment. Neurology. 2006;67:940–3.
54. Kraut JA, Sachs G. Hartnup disorder: unraveling the mystery. Trends Pharmacol Sci. 2005;26:53–5.
55. Darras BT, Ampola MG, Dietz WH, et al. Intermittent dystonia in Hartnup's disease. Pediatr Neurol. 1989;5:118–20.
56. Berardelli A, Thompson PD, Zaccagnini M, et al. Two sisters with generalised dystonia associated with homocystinuria. Mov Disord. 1991;6:163–5.
57. Wortmann SB, Rodenburg RJ, Jonckheere A, et al. Biochemical and genetic analysis of 3-methylglutaconic aciduria type IV: a diagnostic strategy. Brain. 2009;132:136–46.
58. Rubio-Augusti I, Carecchio M, Bhatia KP, et al. Movement disorders in adult patients with classical galactosemia. Mov Disord. 2013;28:804–10.
59. Ala A, Walker AP, Ashkan K, et al. Wilson's disease. Lancet. 2007;369:397–408.
60. Kitzberger R, Madl C, Ferenci P. Wilson disease. Metab Brain Dis. 2005;20:295–302.
61. Inui K, Namba R, Ihara Y, et al. A case of chronic GM1 gangliosidosis presenting as dystonia: clinical and biochemical studies. J Neurol. 1990;237:491–3.
62. Shulman LM, David NJ, Weiner WJ. Psychosis as the initial manifestation of adult-onset Niemann-Pick disease type C. Neurology. 1995;45:1739–43.
63. Baumann N, Turpin JC, Lefevre M, et al. Motor and psycho-cognitive clinical types in adult metachromatic leukodystrophy: genotype/phenotype relationships. J Physiol Paris. 2002;96:301–6.
64. Capablo JL, Franco R, de cabezon AS, et al. Neurologic improvements in a type 3 Gaucher disease patient treated with imiglucerase/miglustat combination. Epilepsia. 2007;48:1406–7.
65. Funk CB, Prasad AN, Frosk P, et al. Neuropathological, biochemical and molecular findings in a glutaric acidemia type 1 cohort. Brain. 2005;128:711–22.
66. Post B, van Belzen M, Marcelis C, Meijer FJ, Willemsen MA, van de Warrenburg BP. Early onset dystonia and parkinsonism with abnormal globus pallidus signal in MRI: a diagnostic challenge. Mov Disord. 2013;28:2035–6.
67. Hatano T, Okuma Y, Ilijima M, et al. Cervical dystonia in dentatorubral-pallidoluysian atrophy. Acta Neurol Scand. 2003;308:287–9.
68. Gregory A, Hayflick SJ. Neurodegeneration with brain iron accumulation. Folia Neuropathol. 2005;43:286–96.
69. Bader B, Walker RH, Vogel M, Prosiegel M, McIntosh J, Danek A. Tongue protrusion and feeding dystonia: a hallmark of chorea-acanthocytosis. Mov Disord. 2010;25:127–9.
70. Garcia-Ruiz PJ, Mayo D, Hernandez J, Cantarero S, Ayuso C. Movement disorders in hereditary ataxias. J Neurol Sci. 2002;202:59–64.
71. Tolosa E, Compta Y. Dystonia in Parkinson's disease. J Neurol. 2006;253:7–13.
72. Maltête D, Guyant-Maréchal L, Mihout B, Hannequin D. Movement disorders and Creutzfeldt-Jakob disease: a review. Parkinsonism Relat Disord. 2006;12:65–71.

Dystonia–Parkinsonism Syndromes 6

Ericka Samantha A. Teleg and Raymond L. Rosales

6.1 Introduction

Clinicians throughout history have painstakingly provided us with classification and nomenclature of the several movement disorders that obscure diagnosis. As we come to understand, several of these movement disorders overlap with one another as part of the course of the syndrome or illness.

This chapter attempts to help the clinician explore disorders with such overlap of dystonia and parkinsonism and offer a way to enumerate clinical features in the context of Parkinson's disease (PD) and its therapy-related disorders.

Parkinsonism has been classified aetiologically into four categories:

1. Primary parkinsonism: This includes Parkinson's disease as a progressive, neurodegenerative disorder which involves the pervasive loss of dopaminergic neurons not only within the *substantia nigra* but also in the brainstem nuclei such as *the raphe nucleus* and *locus ceruleus* and in the frontal regions of the brain. Moreover, accumulation of Lewy bodies within the neurons is also implicated in the disease process.
2. Secondary parkinsonism: This includes toxin-induced parkinsonism, as well as parkinsonism as an aftermath of encephalitis, and vascular parkinsonism.
3. Parkinson-plus syndromes: These include diseases that may imitate clinical manifestations of Parkinson's disease accompanied by other neurological signs.

E.S.A. Teleg, MD (✉) • R.L. Rosales, MD, PhD
Department of Neurology and Psychiatry,
The Royal and Pontifical University of Santo Tomas,
Manila, Philippines
e-mail: ericka_teleg@yahoo.com; rlrosales@mnl.ust.edu.ph

P. Kanovsky et al. (eds.), *Dystonia and Dystonic Syndromes*,
DOI 10.1007/978-3-7091-1516-9_6

Among these syndromes are multisystem atrophy (MSA), olivopontocerebellar atrophy, Shy–Drager syndrome and striatonigral degeneration.

4. Heredo-degenerative parkinsonism: Among others, this includes progressive diseases such as Wilson's disease, X-linked dystonia–parkinsonism (lubag), frontotemporal dementia (FTD), juvenile Huntington's disease and neuroacanthocytosis [1].

Dystonia can also exist in four subgroups: (1) pure dystonia; (2) dystonia-plus syndromes, which includes dystonia and other movement disorders; (3) secondary dystonias that are caused by infection, stroke, trauma, tumour and toxins; and (4) heredo-degenerative dystonias – these are dominated by other neurologic deficits as in X-linked dystonia–parkinsonism, Huntington's disease and Wilson's disease [1].

Based on the categories mentioned of both movement disorders, the overlap is recognized in the classification of heredo-degenerative dystonia–parkinsonism syndromes. Thus, this chapter will only limit its discussion on both heredo-degenerative dystonia–parkinsonism as both these movement disorders in overlap present a challenge to clinicians. The hereditary dystonia-plus (parkinsonism) syndromes (with a special account on 'lubag'), neurochemical disorders as above-mentioned and other characterized movement disorders in a mix with dystonia–parkinsonism will be dealt in brief. Included in this chapter will also be the Guadeloupean parkinsonism in view of coexisting neurological findings, dystonia included.

The clinician's acumen in ruling out Wilson's disease and the secondary dystonia–parkinsonism syndromes, best exemplified by neuroleptic exposure, will be an important step in the diagnostic approach. The latter and related topics with dystonia–parkinsonism, such as progressive supranuclear palsy, cortico-basal ganglionic syndrome, multisystem atrophy and post-traumatic, metabolic–toxic, vascular, infectious, autoimmune and paraneoplastic disorders will be discussed elsewhere in this book.

6.2 Parkinson's Disease and Dystonia

The clinician is challenged when confronted with dystonia, considering that it figures as a prominent manifestation in PD (see Table 6.1). A major factor in the development of dystonia in PD is the age of onset, in that approximately 53–60 % of PD

Table 6.1 Summary of apparent dystonic manifestations in PD

Focal dystonias
Cervical dystonia
Limb dystonia
Hemidystonia
Blepharospasm
Striatal foot/hand
Axial dystonia
Scoliosis, kyphoscoliosis
Camptocormia
Freezing gait

patients with onset before age 40 years develop dystonia. It has been clearly shown that dystonia is related to young-onset PD rather than late-onset PD.

Focal dystonia could be an initial symptom presenting as cervical dystonias, limb dystonias or hemidystonias as that seen in X-linked dystonia–parkinsonism of Panay (XDP or DYT3 dystonia) [2]. Increasingly recognized to also antedate PD classic features, blepharospasm may often be associated with apraxia of eyelid opening or syndrome of levator inhibition. Updating from our own series of followed up PD patients and our botulinum toxin type A (BTA) database, we found blepharospasm antedating PD diagnosis in 1:100 cases [3]. Axial dystonia and scoliosis and kyphoscoliosis are frequent concomitant manifestations of PD. The phenomenon may involve truncal flexion – *camptocormia* – wherein there is an apparent fixed deformity of the trunk most apparent during walking or when leaning against a wall. Largely debated, the side-leaning posture (*Pisa syndrome*) is quite different from the bent head and neck which are frequent PD features.

Freezing gait in PD or referred to as motor blocks has been gleaned to be mechanistically related to a disinhibited foot grasp rather than a lower limb dystonia. This has generally been attributed to premature co-contraction of the tibialis anterior and gastrocnemius muscles as reflected in EMG [14].

Dystonia preceding other PD symptoms involves the feet. The 'striatal' foot, characterized as that with a unilateral equinovarus dystonic posture of the foot and extension of the big toe, clearly mimics the *Babinski sign*. The hyperextension of the metacarpophalangeal joint and extension of the interphalangeal joints may be seen in more than 10 % of the untreated patients with advanced disease [14].

6.2.1 Parkinson's Disease (PD) Therapy and Dystonia

Clinically, the abnormal movements in PD therapy are varied and may present as chorea, choreoathetosis, ballism, myoclonus and akathisia. Dystonia can occur purely or in combination with chorea. Dystonia contributes to the disability when present than any other of the involved abnormal movements. Nevertheless, dystonia is a more challenging and disabling problem.

The most common presentation of dystonia in PD is in the context of the therapeutic strategies in the use of levodopa (see Table 6.2). Levodopa-related dystonias in PD include the pretreatment dystonia, peak-dose dystonia, diphasic dystonia, wearing off dystonia and early-morning dystonia. This classification includes the onset of dyskinesias in relation to levodopa dose [4]. Additionally, levodopa-induced

Table 6.2 Dystonic features in PD treatment

Occurrence	Dopamine levels
Pretreatment dystonia	Before exposure to levodopa
Peak-dose dystonia	High plasma dopamine levels
Diphasic dystonia	Low but falling or rising plasma dopamine levels
'Off' dystonia	Low/no plasma dopamine levels
Early-morning dystonia	Low plasma dopamine levels

dyskinesias usually involve the most affected side of the PD. It is more apparent in the lower limbs than the upper limbs. This may be explained by dopaminergic loss in the somatotopic area of the foot, the dorsolateral striatum [6].

Around 2.4 % in the PD population develops pretreatment dystonia (i.e. before exposure to levodopa) or dystonia induced by volitional activity and usually involving the most apparent parkinsonian side [7].

The most common type of levodopa-induced dyskinesia is the peak-dose type. This occurs during the said ‘on’ period, in which an overdose state occurs. An increased amount of levodopa is present in the system at this time [5]. Dyskinesias documented in early treatment are choreic movements to be followed by dystonia during continuing of treatment in later stages. The dystonia affects the lower limbs although involvement of the neck and trunk is apparent. This characteristic of dystonia that is observed appears with an onset of 30 min to 2 h upon levodopa intake and ends after 15 min to 2 h [4]. A therapeutic strategy includes adding higher doses of dopamine receptor agonists while reducing the dosage of levodopa. Although the major risk factor for peak-dose dyskinesia is the severity of the disease, peak-dose chorea and dystonia can also occur at subtherapeutic doses of levodopa. On the other hand, lowering the dose of levodopa can bring the patient into a parkinsonian state. The glutamate antagonist, amantadine, is a useful anti-dyskinetic agent that is dose dependent. Amantadine has a dopaminergic effect and it can be combined with smaller and more frequent doses of levodopa [5].

Diphasic dyskinesias, otherwise known as the D-I-D (dyskinesia–improvement–dyskinesia), occur when the plasma levels of levodopa are fluctuating (rise or fall). This is a phenomenon that may involve presence of a differential sensitivity of at least two dopamine receptors. The more sensitive one would respond to lower levels of levodopa to induce the dyskinetic state. The other receptor would be activated at higher levels and inhibit dyskinesia. This phenomenon poses a problem since a higher dose of levodopa would render the patient in a dyskinetic state, while in the other end of the spectrum, lowering the dosage will render the patient in a parkinsonian state [5]. Therapeutic strategy of this type of occurrence may still involve utilizing a long-acting dopamine agonist [5].

‘Off period’ dystonia occurs during the night and early morning, when patients are not medicated for an extended period of time. This phenomenon could present as a pure dystonia, occasionally combined with painful and disabling sustained leg spasms [6]. The therapeutic strategy of preventing ‘offs’ may be the effective way to control ‘off’ dystonia – such as the use of a controlled-release levodopa/dopa decarboxylase inhibitor preparation [5] or perhaps a combination with a COMT inhibitor in some patients.

Early-morning dystonia is seen in patients with PD upon awakening as the name implies. Dystonia appears before the first dose of levodopa is taken. Patients with this type have often been taking a higher dosage of levodopa for quite a long period of time. The unilateral or bilateral involvement of the lower limbs is common. In addition, the parkinsonian side does not seem to be always affected by this dyskinesia. Reducing or adding levodopa treatment does not appear to have any benefit at all [4]. Nevertheless, ‘rescue’ doses of apomorphine pen injections may likewise be offered to these 'off' periods with dystonia in PD, who may need immediate relief, especially those with painful dystonia. In the same token, botulinum neurotoxin-A (BNTA) injections may be tried in these patients who have repeated dystonic disabling episodes in the limbs, and for which approach, the effects may last 3–4 months, as it combines with usual PD therapy.

6.3 Inherited Diseases of Dystonia and Parkinsonism

6.3.1 Juvenile-Onset Parkinson's Disease and Dystonia

In view of the high occurrence of dystonia in young-onset (i.e. juvenile-onset PD) PD, perhaps even antedating the PD manifestations, recognition of these disorders as differential diagnoses is imperative. Several gene studies have identified mutations in the *parkin gene* responsible for a varied presentation of PD and dystonia.

The PARK1 juvenile parkinsonism has an autosomal-dominant type of inheritance. It includes an age range of 20–85 years old. Clinical manifestations include asymmetric resting tremor, rigidity and gait and postural abnormalities [7]. Apparently, no dystonic features are evident. Alpha-synuclein mutation in chromosome 4q21 is implicated.

The PARK2 juvenile parkinsonism encodes a gene mutation for *parkin*. It has an autosomal recessive pattern of inheritance, and as differentiated from PARK1, foot dystonia is evident in PARK2 (see Fig. 6.1).

Asymmetric parkinsonism, behavioural manifestations and early severe motor complications are among its other features. From the monogenic forms of parkinsonism, approximately 50 % of cases with autosomal recessive young-onset parkinsonism have *parkin* gene mutations [7].

PARK6 encodes for PINK1 (PTEN-induced kinase 1) gene. The mutation involves a common defective gene rivalling parkin, being a putative protective kinase with certain mitochondrial localization for mutated forms. A mitochondrial

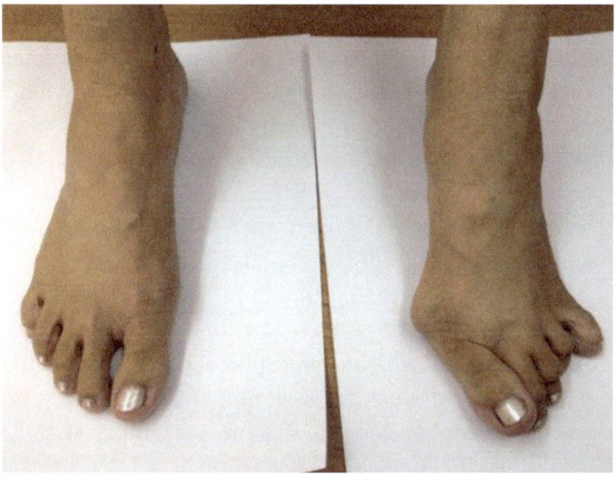

Fig. 6.1 Foot dystonia in a patient with autosomal recessive early-onset PD

dysfunction is implicated after a PINK1 gene mutation. Mode of transmission is autosomal recessive. It has been reported that dystonia at onset of disease is present. Age range involves 9–48 years old. Interestingly, the characteristic clinical features of PINK1-positive families include a mild, slow progression, except in a few cases, and there is good levodopa response [8, 10].

The gene DJ-1 mutation responsible for PARK7 familial early-onset PD is known to be abundant in neurons and astrocytes. DJ-1 is known to be a hydroperoxide-responsive protein that responds to oxidative stress insults produced by hydrogen peroxide. Clearly, mutation of this protein or its downregulation may indeed augment oxidative stress damage. Significant clinical manifestations include a mild progression of parkinsonism, sustained response to levodopa and psychiatric abnormalities. Dystonia does not appear to be evident [8]. Thus, the lack of dystonia might be a distinct clinical sign for differentiating this form from parkin cases. However, DJ-1 mutations may interact with parkin and alpha-synuclein during oxidative stress.

6.4 Heredo-degenerative Disorders

A *DYT* label is based upon a chromosomal location or a phenotype of interest. It represents a clinically heterogeneous group of disorders that include (A) primary dystonia, with dystonia as the only clinical manifestation (DYT1, DYT2, DYT4, DYT6, DYT7, DYT13, DYT17 and DYT21); (B) dystonia-plus syndromes, which include dystonia in the presence of other clinical features such as parkinsonism or myoclonus (DYT3, DYT5/DYT14, DYT11, DYT12, DYT15 and DYT16); and (C) paroxysmal forms of dystonia (DYT8, DYT9, DYT10, DYT18, DYT19 and DYT20) [9].

DYT disorders (as the list expands and enunciated elsewhere in this book) of relevance herein under are those that include the dystonia and parkinsonism.

6.4.1 DYT3 or X-Linked Dystonia–Parkinsonism (XDP)

Otherwise known by its native name, *lubag*, DYT3 dystonia or X-linked dystonia–parkinsonism (XDP) was first identified in Panay Island of the Philippines, where the prevalence is 5.74 per 100,000. Cases worldwide have been linked to Filipino ancestry, with the mean age of onset to be 34.8 years. The mode of inheritance appears to be an X-linked recessive pattern. Although described anecdotally by her predecessors, it was Lillian V. Lee, neurologist from the Philippines, who first introduced in detail the XDP phenomenology in the world literature and who, to date, is deeply involved in registry, research and therapeutic trials of the disorder.

The abnormal protein product in XDP remains to be determined. Two schools of thought have been developed as to the genetics of XDP. In the first notion, a family linkage analysis with polymorphic DNA markers that span the X-chromosome, clearly made possible the assignment of the XDP gene to subchromosomal locations. Mapping of the XDP gene was done with the use of a yeast artificial chromosome (YAC) constructed from Xq13.1 and other markers to pinpoint the gene location. Furthermore, sequencing by Nolte et al. delineated an XDP disease gene that described five disease-specific nucleotide changes (DSCs) in multiple transcript system (MTS) and 48-bp (base-pair) deletion unique to XDP. These DSCs influence splicing of transcripts that clearly contribute to the disease [22].

In the second notion, another group, reported a disease-specific short interspersed nuclear element, tandem repeats and SVA retrotransposon in intron 32 of TAF1. This significantly reduced repression of TAF1, and the D2 (dopamine) receptor genes were found in the caudate nucleus of his studied patients. MTS as previously described was not present, leading to deduce that it might be in the TAF1 gene. The impairment of TAF1 expression might be brought about by the SVA retrotransposon insertion [22]. To date, active research is ongoing to further narrow down and pinpoint the exact protein gene aetiology of this interesting, indigenous and disabling disease.

Distribution of focal dystonia in XDP is apparent in the following descending order of regions: lower limbs (31 %) > craniofacial area (28 %) > neck and shoulder (23 %) > upper extremities (14 %) and trunk (2.6 %) (see Fig. 6.2). The parkinsonian symptoms that become apparent later on in the progression of the disease are tremors, bradykinesia, masked facies, hypomimia and drooling [13].

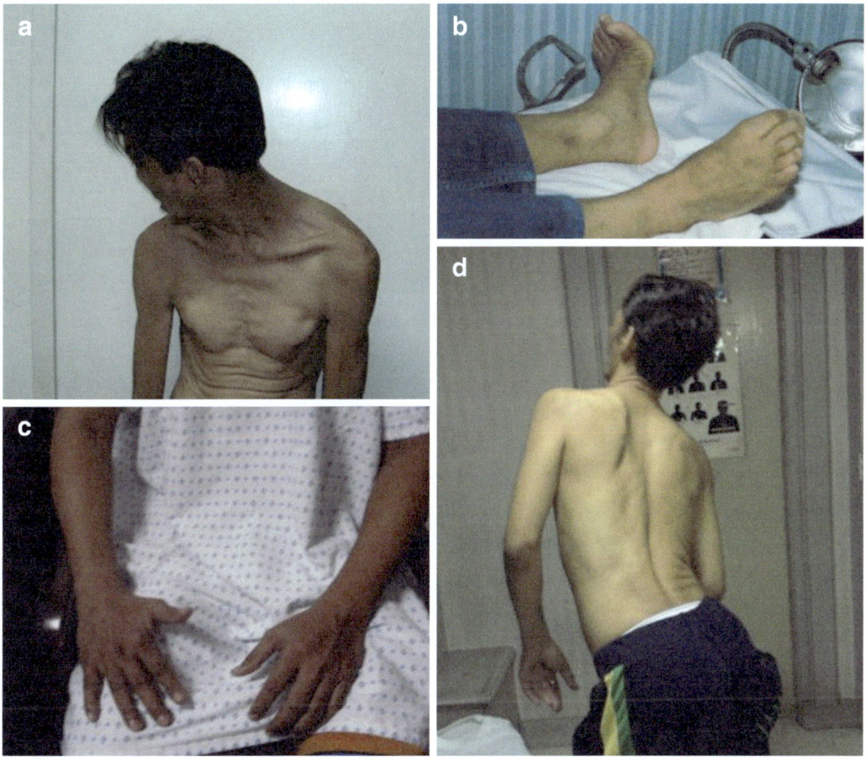

Fig. 6.2 (**a**) Cervical dystonia in XDP; (**b**) lower limb dystonia in XDP; (**c**) upper limb dystonia in XDP; (**d**) truncal dystonia in XDP

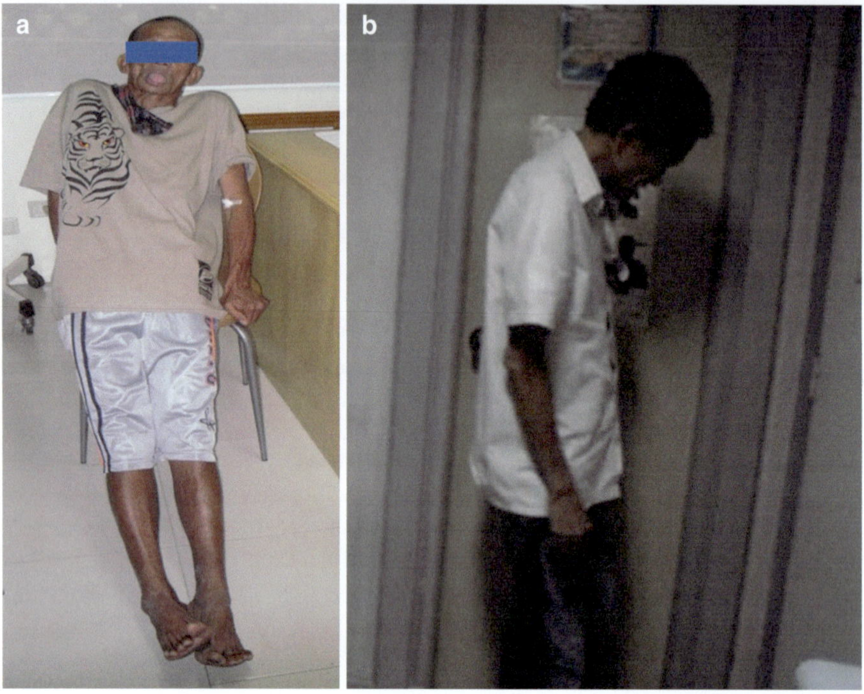

Fig. 6.3 (**a**) Generalized dystonia with parkinsonism in XDP; (**b**) *bent head* in parkinsonism

The clinical course of XDP can very well explain the occurrence of parkinsonism and dystonic features. It was initially included among the primary dystonias before the parkinsonism features were entirely recognized. Focal dystonia appears in the third or fourth decade of the illness and becomes generalized from 2 to 5 years, and by the 7th to the 10th year, dystonia and parkinsonism coexist (see Fig. 6.3). Later on, in the 15th year of the illness, a parkinsonian state predominates [13] and in fact a clinical staging has been proposed in this regard [12].

Medical treatment of XDP remains inadequate, considering how therapies (i.e. anticholinergics, benzodiazepines, baclofen, levodopa and D2 blocking agents, among others) are unable to alter the disease course nor provide significant symptomatic relief of the disabling focal or generalized manifestations. Perhaps approaching temporary symptom relief from severe muscle spasms and pain will be BNTA [20]. This chemodenervating agent may improve certain quality of life parameters like feeding, speech, neck and limb postures as well as writing and ambulation. On the other hand, as the expanding list of XDP cases significantly responding to deep brain stimulation (i.e. GPi-DBS) is promising [21], a larger cohort of cases observed over time will eventually provide us the jury on this form of functional surgery. Family counselling among affected families is also highly recommended.

6.5 Neurochemical Dystonia–Parkinsonism

Certain disorders are derived from neurochemical abnormalities. The ones that would clinically present with dystonia–parkinsonism among these neurochemical disorders are *dopa-responsive dystonia (DYT5), rapid-onset dystonia (DYT12) and early-onset dystonia (DYT16)*. GM1 Gangliosidosis (Type 3, adult form) having a combination of generalized dystonia, parkinsonism and dementia will be discussed elsewhere (Table 6.3).

6.5.1 DYT5 or Dopa-Responsive Dystonia (DRD)

The DYT5 otherwise named as dopa-responsive dystonia occurs in childhood between the age of 5 and 6 years old. It is an important differential diagnosis during assessment of cerebral palsy in children. Other than significant response to levodopa therapy, one feature that makes this type of dystonia–parkinsonism unique is the diurnal occurrence of the symptoms. The children appear to have no dystonic movements and posturing in the morning but may be severely affected in the afternoon or evening [11]. Thus, worsening occurs as the day progresses and improvement is observed after sleep [13].

Dystonia is more apparent in the lower limbs than the upper limbs. The adult-onset oromandibular dystonia is observed also in a number of patients. Spastic paraplegia resulting in abnormal gait with a tendency to walk on the toes is common [11, 13].

Parkinsonism features observed are bradykinesia, hypomimia, flexed posture and postural instability.

The recognition of autosomal-dominant DYT5 pattern when done can lead to an easy treatment. The cause is by mutations in the gene encoding GTP cyclohydrolase I (GCH1) on chromosome 14q22. It is known that GCH1 is important in the conversion of GTP to tetrahydrobiopterin (BH4), the cofactor for tyrosine hydroxylase (TH). This is the rate-limiting enzyme for dopamine synthesis [13]. Thus, DYT5 is very much responsive to levodopa therapy.

6.5.2 DYT12 or Rapid-Onset Dystonia–Parkinsonism (RDP)

This is an autosomal-dominant movement disorder that involves mutations in the ATP1A3 gene encoding the α3 subunit of the Na+/K+-ATPase on chromosome 19q12-13. The Na+/K+-ATPase is known to maintain the plasma membrane electrochemical gradient.

Clinical features are characterized by an abrupt onset of persistent dystonia and parkinsonism evident during adolescence and early adulthood with age range of 15–45 years old.

Symptoms begin suddenly or insidiously but with rapid progression. After worsening, symptoms tend to stabilize. There is usually a rostro-caudal (face>arm, leg) distribution of affected body parts. Prominent bulbar manifestations such as dysphagia and dysarthria are common. The symptoms can be aggravated by stress or fatigue.

Dystonic features are characterized as focal limb posturing, with waxy effortful rapid successive movements. Parkinsonism symptoms that may coexist include bradykinesia and postural instability.

Dopaminergic therapy is not yielding in this disorder [11, 13].

6.5.3 DYT16: Early-Onset Dystonia

This autosomal recessive early-onset dystonia has its occurrence in age ranging from 2 to 18 years old. Leg pain and gait abnormalities herald this disease. Mutation in the PRKRA gene whose function remains to be elucidated is coded in chromosome 2q31.

The clinical phenomenology is followed by a combination of focal and generalized dystonia involving the larynx, cervical spine, limb and truncal (with opisthotonic posturing) regions. In addition, facial grimacing and orofacial dystonia are notable features.

Reported in a minority of patients, a parkinsonian manifestation by way of bradykinesia (usually in the setting of severe generalized dystonia) may occur but is unresponsive to levodopa therapy [15].

6.6 Dystonia–Parkinsonism in Other Characterized Syndromes

6.6.1 Guadeloupean Parkinsonism

This type of parkinsonism hails from the Caribbean island of Guadeloupe. Patients labelled with this type develop parkinsonism similar to that of progressive supranuclear palsy (PSP) syndrome. The combination of parkinsonism, dementia and motor neuron disease was first noted among native inhabitants of Guam [16]. Three parkinsonian phenotypes, with equal frequencies, were distinguished: *Parkinson's disease*, *a progressive supranuclear palsy (PSP)-like syndrome designated as Gd-PSP and a parkinsonism–dementia syndrome designated as Gd-PDC*.

Initially constituted on the basis of the presence of impaired vertical gaze plus other oculomotor signs and presence of dementia were the Gd-PSP and Gd-PDC groups, respectively. However, analysis showed that, aside from the oculomotor signs, there were few statistically significant clinical differences between these two groups. Gd-PSP and Gd-PDC patients had the classic parkinsonian symptoms such as truncal rigidity, akinesia and tremor, but these symptoms were not responsive to levodopa, and they only rarely developed levodopa-induced dyskinesias. They also had early postural instability, frequent falls and autonomic dysfunction and cognitive impairment.

Furthermore, aside from the oculomotor signs, the Gd-PSP patients differed markedly from patients with classical PSP.

Compared to idiopathic PD where 6 % will have dystonia, about 8 % in Gd-PSP and 5 % in Gd-PDC may develop dystonia. The dystonia among these types of patients is the retrocollis type. Blepharospasm was also observed in some patients [17].

6.7 Spinocerebellar Ataxia Syndromes

Parkinsonism, postural tremor and dystonia appear to be common in many of these hereditary ataxias. Other movement disorders that are infrequent are chorea and myoclonus as well [18].

6.7.1 Spinocerebellar Ataxia 2 (SCA 2)

This type of ataxia has an autosomal-dominant pattern of inheritance characterized by features such as early saccade slowing, early hyporeflexia, severe tremor of postural or intention type and early myoclonus. The postural tremor is noted with other parkinsonism features such as bradykinesia and rigidity.

The females are most affected with a CAG expansion and mutational repeats identified in chromosome 12q. Mutation lies in the polyglutamine domain of the protein ataxin-2 which can be present in expansion in families over generations.

Movement disorders include cervical dystonia, hemidystonia and hand cramp. The presence of these dystonic features may suggest a dysfunctional pontocerebellar pathway, important in hereditary ataxias [19].

6.7.2 Spinocerebellar Ataxia 3 (SCA 3)

Spinocerebellar ataxia 3 or Machado–Joseph disease (MJD) is characterized primarily by cerebellar ataxia and pyramidal signs variably associated with a dystonic–rigid extrapyramidal syndrome. The generalized dystonia appears to mask the ataxia that is observed. Significantly, other clinical manifestations include a dysfunction of motor coordination that can affect gaze, speech, gait and balance. Others include external progressive ophthalmoplegia and fasciculation-like movements of facial and lingual muscles [19].

Notably, postural tremor associated with focal dystonia in the upper limb can be present [19].

Pathophysiology is associated with a CAG repeat expansion in the coding region of MJD1 gene that codes ataxin-3, a polyubiquitin-binding protein in chromosome 14q32.1. This protein is related to a ubiquitin-mediated proteolysis event. As a result, expanded trinucleotide repeats of range 56–96 explain the CAG repeat expansion.

MJD1, together with mutant ataxin-3, aggregates into intranuclear inclusions that bring about neurodegeneration in axons within fibre tracts of the corpus callosum, nigrostriatal tracts and the olivocerebellar fibres.

6.8 Huntington's Disease

Huntington's disease (HD) is a progressive degenerative disorder that manifests a CAG trinucleotide expansion in the gene encoding for a protein known as *huntingtin* or IT-5 in chromosome 4. When excessive CAG triplet repeats are present, excessive *huntingtin* protein is produced. The age of onset is usually between 30 and 40 years old.

Clinical manifestations begin with cognitive and emotional difficulties such as irritability, depression, clumsiness, speech difficulty, memory loss, lack of motivation, paranoia, intellectual decline, sleep disturbance, hallucinations and sexual problems. When movement disorders appear, these include parkinsonism, chorea, myoclonus and dystonia. Parkinsonism features in particular include bradykinesia, while dystonia may appear at the early course of the disease. Other presenting symptoms include chorea, unsteadiness and gait difficulty. Thus, the symptoms altogether can be classified into the motor, cognitive and psychiatric spectrum of HD. Parkinsonism in HD may not respond to dopaminergic therapy, save perhaps for the late-onset levodopa-responsive HD [14].

This present list of dystonia–parkinsonism syndromes is far from complete, because more disorders are being described with this overlap, Fahr's disease, brain iron accumulation and ataxia–telangiectasia included.

In summary, dystonia in parkinsonian disorders is a well-recognized entity. For the most part, the diagnosis of a parkinsonian syndrome is largely clinical. This makes the classification of patients challenging. Symptoms can also overlap between different parkinsonian disorders. Complicating this picture is that dystonia can occur in primary parkinsonian conditions and parkinsonism can occur in primary dystonic conditions. Thus, the clinician may face the diagnostic challenge by approaching each patient from the phenomenology of dystonia and parkinsonism in a mix with other distinct neurological features (not only movement disorders), nevertheless also associated with non-neurological features. A good way to start the phenomenological approach is to consider the dominant movement disorder(s) and the regions/sides most involved. In the end, other than neuroimaging, genetic studies guided by the clinical clues will zero in on the diagnosis. Needless to say, these clinical clues include the patient's sex, movement disorder, age of onset, ethnicity, genogram and response to levodopa therapy.

Please refer to the table below for a summary of the disorders described (Table 6.3).

Acknowledgements Photos have been taken from Dr. Raymond L. Rosales patient archive. Consent from patients and their caregivers has been properly obtained for this purpose.

Table 6.3 Summary of parkinsonism and dystonia features

Disease	Parkinson feature	Dystonia feature	Clinical and therapeutic clues
PD Parkinson's disease PD therapy related (see Table 6.2)	Asymmetric tremor, rigidity and bradykinesia	Cervical, limb, truncal, striatal foot/hand and blepharospasm	Good response to levodopa PD treatment related to diurnal patterns
Juvenile-onset PD and dystonia	Asymmetric resting tremor, rigidity and gait	Focal dystonia – usually lower limbs	Good response to levodopa Behavioural symptoms apparent
DYT3/XDP	Late-onset tremors, rigidity, bradykinesia	Early-onset focal to generalized dystonia	Moderate response to levodopa Moderate response to levodopa; Filipino ancestry, males
DYT5	Hypomimia flexed posture	"diurnal focal lower limb dystonia"	Good response to levodopa
DYT12	Bradykinesia	Mainly limb dystonia rostro-cauda of abrupt onset	Poor response to levodopa Bulbar symptoms
DYT16	Bradykinesia	Focal to generalized truncal posturing	Poor or no response to levodopa
Guadeloupean	Truncal rigidity, akinesia and tremor	Focal, largely cervical	Poor response to levodopa
SCA 2	Tremor – postural or intention type, bradykinesia and rigidity	Cervical dystonia	Moderate response to levodopa; Early hyporeflexia
SCA 3	Dystonic–rigid syndrome; postural tremor	Focal to generalized dystonia	With moderate response to levodopa Gaze, speech, gait affected Fasciculation-like movements of facial and lingual muscles
Huntington's disease	Bradykinesia Rigidity	Dystonic features: focal limb dystonia, foot inversion, internal rotation of shoulder, sustained clenching of fist	With moderate response to levodopa Psychiatric and cognitive symptoms
Wilson's disease	Tremors, rigidity, hypomimia	Focal or generalized – segmental distribution	Responsive to levodopa to some extent Behavioral changes and eye Kayser-Fleischer rings
GM1 Gangliosidosis (adult form, type 3)	Parkinsonian feature: akinetic-rigid parkinsonism	Dystonia feature: focal/generalized-largely facial, truncal and limb	Clinical and therapeutic clues: demenia, cherry red macular spot, dysmorphic feature, skeletal dysplasia

References

1. Frucht S, Fahn S. Movement disorder emergencies, diagnosis and treatment. Totowa: Humana Press; 2005. p. 2–4.
2. Rosales RL, Ng A, Delgado Delos Santos M, Go C, Joya-Tanglao M, Teleg R, Dantes M. Botulinum toxin-A targeted for focal/multifocal dystonias – an open-label application in X-linked dystonia-parkinsonism. Phil J Neurol. 2012;16(1):37–45.
3. Rosales RL. Dystonia, spasticity and botulinum toxin therapy: rationale, evidences and clinical context. In: Rosales RL, editor. Dystonia: the many facets. Croatia: Intech Open Access Publishers; 2012. p. 83–98.
4. Lugo R, Fernandez H. Dystonia in parkinsonian syndromes, Dystonia: the many facets. Croatia: Intech Open Access Publishers; 2012. p. 65–75.
5. Fahn S, Jankovic J. Chapter 6 medical treatment of parkinson disease. In: Principles and practice of movement disorders. Philadelphia: Churchill Livingstone Elsevier; 2011. p. 153–4.
6. Thanvi L. Levodopa-induced dyskinesia in PD: clinical features, pathogenesis, prevention and treatment. Postgrad Med J. 2007;83(980):384–8.
7. Bhidayasiri R, Pulst S. Gene table: juvenile parkinsonism. Eur J Pediatr Neurol. 2009;13:290–2.
8. Hatano Y, Akasawa K, Li S, et al. Novel PINK1 mutations in early-onset parkinsonism. Ann Neurol. 2004;56:424–7.
9. Chen S-F, Shen Y-C, et al. Dystonia and genetics. In: Rosales RL, editor. Dystonia: the many facets. Croatia: Intech Open Access Publishers; 2012. p. 11–26.
10. Rosales RL. X-linked dystonia parkinsonism: clinical phenotype, genetics and therapeutics. J Mov Disord Kor. 2010;2:32–8.
11. Fahn S, Jankovic J. Chapter 13. Dystonia, phenomenology, classification, etiology, pathology, genetics. In: Principles and practice of movement disorders. Philadelphia: Churchill Livingstone Elsevier; 2011. p. 325.
12. Lee LV, Rivera, Teleg R, Dantes M, Pasco PM, Jamora RD, Arancillo J, Villareal-Jordan R, Rosales RL, Demaisip C, Maranon E, Peralta O, Monding M, Sarcia S. The Unique Phenomenology of Sex-Linked Dystonia Parkinsonism (XDP, DYT3, "Lubag"). International J Neurol Sci. 2011;121:3–11.
13. Bressman SB. Genetics and dystonia. Parkinsonism Relat Disord. 2007;13:S347–S35.
14. Jankovic J, Tinter R. Dystonia and parkinsonism. Parkinsonism Relat Disord. 2001;8: 109–21.
15. Carmargos, et al. DYT16, a novel young-onset dystonia-parkinsonism disorder: identification of a mutation in the stress-response protein: PRKRA. Lancet Neurol. 2008;7(3):207–15.
16. Fahn S, Jankovic J. Chapter 10 parkinsonism-plus syndromes and secondary parkinsonian disorders. In: Principles and practice of movement disorders. Philadelphia: Churchill Livingstone Elsevier; 2011.
17. Caparros-Lefebvre D, Sergeant N, Lees A, et al. Guadeloupean parkinsonism – a cluster of progressive supranuclear palsy-like tauopathy. Brain J. 2002;125:801–11.
18. Garcia Ruiz PJ, Mayo D, Hernandez J, et al. Movement disorders in hereditary ataxias. J Neurol Sci. 2002;202:59–64.
19. Boesch SM, Mu¨ller J, et al. Cervical dystonia in spinocerebellar ataxia type 2: clinical and polymyographic findings. J Neurol Neurosurg Psychiatry. 2007;78(5):520–2.
20. Rosales RL, Delos Santos MM, Ng AR, Teleg R, Dantes M, Lee LV, Fernandez HH. The broadening application of chemodenervation in X-linked dystonia-parkinsonism (part I): muscle afferent block versus botulinum toxin-A in cervical and limb dystonias. Int J Neurosci. 2011;121:35–43.
21. Aguilar J, Vesagas TS, Jamora RD, Teleg R, Ledesma L, Rosales RL, Fernandez HH, Lee LV. The promise of deep brain stimulation in X-linked dystonia parkinsonism. Int J Neurosci. 2011;121:57–63.
22. Pasco P, Ison C, Munoz E, Magpusao N, Cheng A, Tan K, Lo R, Teleg R, Dantes M, Borres R, Maranon E, Demaisip C, Reyes M, Lee LV. Understanding XDP through imaging, pathology and genetics. Int J Neurosci. 2011;121:1–2.

7 Dystonia in Multiple System Atrophy, Progressive Supranuclear Palsy, and Corticobasal Degeneration

Luca Marsili and Carlo Colosimo

Dystonia associated with parkinsonism may occur in various clinical settings. Rare hereditary conditions involving the basal ganglia, such as Wilson's disease, Huntington's disease, neurodegeneration with brain iron accumulation type 1 disease (NBIA1), and familial basal ganglia calcifications, are frequently characterized by both these symptoms [1–3]. Generalized dystonia with parkinsonian features is also a classical presentation of dopa-responsive dystonia [4]. As regards more common sporadic neurodegenerative disorders, dystonia is frequently observed in Parkinson's disease (PD) patients experiencing motor complications, though it may also be seen, usually in the form of foot or hand posturing, in early untreated patients. Clinicopathological studies have also revealed that dystonia is frequently present in all atypical parkinsonian syndromes (APS), such as multiple system atrophy (MSA), progressive supranuclear palsy (PSP), and corticobasal degeneration (CBD). The phenomenology and evolution over time of dystonia in APS varies [2, 5–7].

The recognition of specific postural abnormalities in APS might be of diagnostic value as such abnormalities are often mistakenly attributed to rheumatic or orthopedic problems when present at disease onset, particularly if the typical parkinsonian signs are not evident. In addition to the deformity it causes, dystonia may reduce dexterity and interfere with activities of daily living and gait, increase the likelihood of falls, and induce discomfort and pain. Consequently, the presence of dystonia may exacerbate overall functional disability, which is already high in patients affected by APS [8, 9] (Table 7.1).

L. Marsili • C. Colosimo (✉)
Department of Neurology and Psychiatry, Sapienza University of Rome, Viale dell'Università, 30, 00185 Rome, Italy
e-mail: carlo.colosimo@uniroma1.it

P. Kanovsky et al. (eds.), *Dystonia and Dystonic Syndromes*,
DOI 10.1007/978-3-7091-1516-9_7

Table 7.1 Main types of dystonia and relative frequencies in atypical parkinsonism

Type of dystonia	MSA	PSP	CBD
Blepharospasm	–	++	±
Laryngeal stridor	++	–	–
Craniocervical	++[a]	±	±
Antecollis	++	±	–
Retrocollis	–	+	±
Limb dystonia	±	++	++

(++) very common, (+) common, (±) uncommon, (–) occasional
[a]This is frequently drug induced

7.1 Multiple System Atrophy

The lack of detailed studies on dystonia in pathologically proven MSA means it is difficult to define its prevalence, nature, and extent in such patients: while the first review of the literature on this topic, conducted by Rivest and coworkers, concluded that dystonia is relatively rare in pathologically confirmed MSA [2], a subsequent prospective clinical study suggested that dystonia is common in untreated MSA-P patients, it being observed in 42 % of patients [6]. According to Boesch and coworkers (24 cases) and Kollensperger and coworkers (100 cases), the distribution and prevalence of dystonia in MSA is as follows: craniocervical dystonia (often drug induced) 58.3 %, Pisa syndrome 42.1 %, antecollis 25–36.8 %, diurnal laryngeal stridor 22.8 %, and limb dystonia 20.8 % [6, 10]. Blepharospasm has also been reported in MSA patients, though much less frequently than in PSP [5].

7.1.1 Pisa Syndrome

Pisa syndrome was originally described as a rare subacute dystonic reaction that appears 3–10 days after the start of neuroleptic treatment and is characterized by an abnormal tonic lateral flexion of the trunk (including the head and neck) associated with backward rotation [11]. This is typically a mobile picture, resolving in the supine position. A similar picture may occur as a form of tardive dystonia [12]. Pisa syndrome has more recently been described in PD [9, 13, 14], and, more frequently, in MSA [15]; it is consequently now considered to be one of the most common red flags of MSA [10]. Pisa syndrome may also occur in Alzheimer's disease treated with cholinesterase inhibitors [16] or as an idiopathic phenomenon, possibly representing a form of idiopathic axial dystonia. Pisa syndrome is different from scoliosis, the latter being defined as a lateral curve of the spine, usually combined with a rotation of the vertebrae [17]. The onset of Pisa syndrome may be either chronic or acute [14]. When the deformity progresses, patients may develop pain [18] as well as postural instability, which increases the likelihood of falling. It is not yet entirely clear whether parkinsonian patients with lateral trunk flexion lean toward or away from their more severely affected side. Most investigators have found that patients tend to lean away from the more affected side [18], while some have found no association between the direction of the curve and the laterality of PD [19]. Pisa syndrome may be secondary not only to dopamine-blocking agents but also to dopaminergic antiparkinsonian medications [20]. In the latter case, Pisa syndrome may occur following any changes (initiation, increase, or decrease) in dopaminergic medication. Trunk deviation tends to disappear following revision of the medication change [20], though it may become irreversible in some cases. Other non-dopaminergic therapies (lithium carbonate, valproic acid, antidepressants, and cholinesterase inhibitors) may also trigger Pisa syndrome [13, 16, 21, 22]. Finally, Pisa syndrome has been reported to develop in patients after pallidotomy, at varying time intervals following surgery [23, 24].

There are as yet no widely accepted diagnostic criteria for Pisa syndrome, though Bonanni and colleagues [25] have proposed the following definition for lateral axial dystonia: more than 15° lateral flexion of the trunk, increasing during walking, not present when supine, and in the absence of any mechanical restriction to trunk movement (i.e., degenerative spinal disease), with continuous electromyographic activity in the lumbar paraspinal muscles ipsilateral to the bending side. Another definition of Pisa syndrome is that proposed by Doherty and coworkers: a pronounced (at least 10°) lateral flexion, which can be alleviated completely by passive mobilization or by lying supine. This definition differs from the previous one insofar as the mechanism might not be exclusively dystonic and electrophysiological studies are not required to define the clinical syndrome [9]. In MSA, progressive striatal degeneration may be the underlying pathological substrate of this unusual dystonic picture [26].

7.1.2 Antecollis

Antecollis refers to a marked forward flexion (more than 45°) of the head and neck. Antecollis may be part of the so-called stooped parkinsonian posture, though some patients display disproportionate antecollis, that is, neck flexion that is much more pronounced than would be expected when the trunk and limbs are flexed [24, 27]. The term "dropped head syndrome" is instead used to describe marked neck flexion, though only in neuromuscular disorders such as myasthenia gravis, polymyositis, and motor neuron disease [28], in which neck extension weakness is often observed. In PD, antecollis may also represent a form of off-dystonia or might develop as dyskinesias secondary to increases in dopaminergic medications; as a consequence, any fluctuations in antecollis and its relationship with medication times should always be investigated. Several case reports have suggested that antecollis might be induced or aggravated by dopamine agonist therapy [29–32] or amantadine [33]. These medications should be suspended if there is a temporal association with the onset of the syndrome, even though the antecollis may not be reversible [30, 31]. Antecollis, which is, unlike the idiopathic form of spasmodic torticollis, characterized by an almost fixed neck, is considered to be a common feature of MSA [27]. Its prevalence in MSA is estimated to range from 37 % [10] to 42.1 % [8], which is much higher than that reported in PD (5.8 %) [29, 34, 35]. Moreover, antecollis is more frequent in women, in akinetic-rigid patients [34], and in subjects of Japanese ethnicity [32]. Although this symptom may appear before other motor features [34], it usually emerges several years after disease onset. Patients may complain of neck pain or develop problems secondary to neck flexion (difficulty in swallowing, excessive drooling, or visual limitation). In the early stages, hypertrophy and spasms might be visible in various anterior and posterior neck muscles, whereas contraction of the posterior neck muscles, particularly of the splenius capitis and trapezius, becomes prominent later [24]. Most studies report normal strength upon testing of neck extension [24, 34], while some note prominent contractions, which limit voluntary neck extension, in the sternocleidomastoid muscles [29]. In antecollis, muscles that originate at the transverse processes of the cervical vertebrae (scalenus scapulae and longus colli muscles) are likely to be affected on both sides, though they cannot easily be assessed without laboratory testing (e.g., EMG). Unlike idiopathic cervical dystonia, there is no "geste antagoniste" that can improve abnormal posture [6]. In the differential diagnosis of antecollis, imaging is useful to rule out cervical spine pathology (though it should be borne in mind that long-term antecollis can damage the cervical spine).

7.1.3 Laryngeal Stridor

Respiratory dysfunction characterized by deep involuntary sighs or inspiratory stridor, both diurnally and nocturnally, is frequent in MSA but much rarer in patients with PD [10], where it has been described as a dyskinetic side effect of L-dopa treatment [36]. Inspiratory stridor, particularly at night, is commonly attributed to vocal cord paralysis [37], though it might also reflect dystonia of the vocal cords [38]. Stridor, which has been documented in 9–34 % of MSA patients, may develop at any time during the natural history of the disease and represents a negative prognostic factor [39–41].

7.1.4 Limb Dystonia

Limb dystonia in MSA includes dystonic arm posturing, task-specific dystonia (such as writer's cramp), or equinovarus foot posturing [6]. Limb dystonia is most likely to occur in MSA subjects who are levodopa naïve [6], as previous reports on early PD also suggest [42]. Boesch and coworkers noted that limb dystonia disappeared after therapy in all levodopa-responsive patients with MSA, thereby indicating that dystonia in levodopa-naïve patients with MSA predominantly reflects early nigrostriatal dysfunction due to reduced endogenous dopaminergic stimulation rather than to postsynaptic striatal pathology [6, 43].

7.2 Progressive Supranuclear Palsy

As in MSA, the assessment of the frequency and nature of dystonic manifestations in PSP must rely on clinicopathological observations, which are unfortunately scarce in many cases [2] owing, among other things, to the frequent misdiagnosis with PD in the early stages of the disease. According to a study by Barclay and Lang (83 cases), the distribution and prevalence of dystonia in PSP is as follows: limb dystonia 27 %, blepharospasm 24 %, and nuchal dystonia 17 % [5].

7.2.1 Limb Dystonia

Limb involvement is the most common dystonic manifestation of PSP patients (27 %), while hemidystonia is considered the most frequent distribution (11 %) [5, 44]. The so-called "pointing gun" posture, with extension of the index finger and thumb, is typically described in dystonic PSP patients. Hemidystonia may in some patients be a very early feature of this disease that leads to a mistaken diagnosis of CBD because of the pronounced unilateral dystonia. However, the presence of vertical supranuclear gaze palsy upon presentation and the absence of cortical sensory loss and stimulus-sensitive myoclonus should point to a diagnosis of PSP. Moreover, significant limb apraxia is usually absent in PSP, though this symptom may be difficult to assess owing to the pronounced disability caused by dystonia, bradykinesia, and rigidity. Single limb involvement is described in 10 % of PSP cases, with the arm being the limb affected most often. Arm dystonia may also be the initial manifestation of PSP, occurring before the onset of the typical eye movement abnormalities [5, 45]. Limb dystonia as a presenting feature was reported in up to 50 % of the patients studied by Rafal and Friedman and in 27 % of the patients studied by Barclay and Lang [5, 44]. Dopaminergic therapy may cause involuntary choreic movements of the dystonic limb that usually cease following discontinuation of the drug [5].

7.2.2 Blepharospasm

A well-known dystonic manifestation of PSP is blepharospasm [46]. In some clinical-based series, the presence of blepharospasm has been reported to range from 8 to 26 % of cases [47, 48]. Blepharospasm in PSP patients can occur either spontaneously or reflexively in response to facial stimuli or voluntary facial movements (reflex blepharospasm). Up to 50 % of the PSP patients with blepharospasm studied by Barclay and Lang also displayed apraxia of eyelid opening (AEO) [5]. AEO has been described as a nonparalytic motor abnormality characterized by difficulty in opening the eyes at will in the absence of a visible contraction of the orbicularis oculi muscle [49, 50]. Some studies have highlighted the presence of an isolated contraction of the pretarsal component of the orbicularis oculi muscle in several cases of AEO, which suggests that this phenomenon may be a variant of dystonic blepharospasm [51]. Golbe and coworkers noted that the duration and severity of parkinsonian features in patients with eyelid function disturbances were not significantly different from that in patients without this feature [47]. Blepharospasm is often a late manifestation of the disease, there being few cases in which this symptom occurs as an early manifestation, whereas apraxia of eyelid opening or closing frequently appears in the middle stages of the disease [52]. Levodopa therapy may worsen this dystonic phenomenon, and levodopa withdrawal leads to its resolution. Histologically, the distribution of degeneration in classic PSP includes involvement of several midbrain nuclei. As midbrain lesions have been known to result in blepharospasm [53], it is not surprising that this sign occurs in a consistent number of cases of PSP, as well as in a certain number of MSA patients.

7.2.3 Retrocollis

Retrocollis is an abnormal neck posture, with the head held in extension. It may occur as a subtype of primary cervical dystonia [54] or in patients exposed to neuroleptics [9]. Retrocollis associated with axial rigidity is most typically seen in patients with the classic variant of PSP (Richardson's syndrome). Since retrocollis is very rarely seen in PD, this feature should be regarded as being highly suggestive of PSP, with a reported prevalence of around 17 % [5]. Some authors do not consider this neck posture to be dystonic; indeed, retrocollis in PSP is characterized by recurrent local pain with extensor posturing of the neck, though not usually associated with involuntary dystonic neck movements; in addition, it is not associated with muscle hypertrophy nor can it be relieved by antagonistic gestures, which are typical features of classic cervical dystonia [5]. For these reasons, several authors consider that a more appropriate term is "nuchal rigidity in extension" and that the term "axial dystonia" should only be used in the rare cases in which it is associated with torticollis [55].

7.3 Corticobasal Degeneration

Dystonia is considered one of the classical features of CBD and is reported in up to 83 % of clinically diagnosed CBD patients [56–58]. In a recent review by Stamelou and coworkers [7], 404 pathologically proven cases of CBD were reviewed to assess the frequency and the clinical characteristics of dystonia in this pathology. In this series, 37.5 % of the CBD cases had dystonia, while the remaining 62.5 % did not. Age at onset and disease duration were not significantly different in CBD patients with dystonia and those without dystonia. A similar, or even more frequent, association between dystonia and CBD (60–83 % of cases) has been reported in other studies [57, 58], probably because the majority of these studies included mixed populations of clinically and pathologically proven cases [7]. Since the pathological substrates that may underlie corticobasal syndrome (CBS) include a wide range of pathologies, such as PSP, Alzheimer's dementia (AD), Pick's disease, frontotemporal lobar degeneration with ubiquitin- and TDP-43-positive inclusions, Lewy body disease, frontotemporal lobar degeneration with fused-in sarcoma-positive inclusions, and Creutzfeldt-Jakob disease [59, 60], the clinical diagnosis of CBS has a low specificity for CBD. CBS is the most frequent phenotype in CBD with dystonia, whereas FTD is more often associated with CBD without dystonia. Dystonia in pathologically proven CBD may develop either in the early stages (within 2 years of disease onset) or late stages. The majority of early-onset dystonic CBD patients display a CBS phenotype, while the remainder have a PSP phenotype. The majority of patients with late-onset dystonic CBD display a "dementia" phenotype (FTD, AD, or progressive nonfluent aphasia) [7]. Initial distribution of dystonia is typically in the upper limbs, which are usually affected most, followed by cervical dystonia and blepharospasm, with few cases presenting lower limb dystonia (which is associated with CBS and is usually levodopa induced and reversible) [61, 62]. Upper limb dystonia is more often associated with the CBS phenotype, while cervical dystonia and blepharospasm are more common in other phenotypes, particularly in PSP [7]. Dystonia in CBS associated with CBD appears to occur early in the course of the disease; it is initially most commonly focal (upper limb), and it then progresses to hemidystonia and may subsequently also affect the other side but rarely begins in the leg. Cervical dystonia and blepharospasm are rarely isolated. The typical clinical distribution of dystonia in CBD is adduction and flexion of the arm, forearm, wrist, and metacarpophalangeal joints, extension of the interphalangeal joints, as well as internal rotation and flexion of the hip, flexion of the knee, and inversion of the foot [7, 57]. As might be expected, the most frequent distributions of dystonia in CBD with the classic PSP phenotype are blepharospasm and axial dystonia. The pathophysiology of dystonia in CBD is largely unknown. Neurophysiological studies based on transcranial magnetic stimulation have revealed enhanced cortical excitability in CBS [63], which may result from the loss of inhibitory input from the somatosensory to the motor cortex because of parietal atrophy or the existence of an alternative hyperexcitable thalamocortical pathway [64], possibly following the loss of inhibitory neurons in the cortex or thalamus, the morphological changes in somatosensory cortices, or the existence of alternative cortical-subcortical pathways [64–66].

7.4 Therapy

Dystonic features associated with atypical parkinsonism may be treated using either pharmacological or non-pharmacological approaches. A reduction in or withdrawal of levodopa therapy should be considered if a close relationship between the onset of the involuntary movements/contractions and dopaminergic therapy is documented. Botulinum toxin therapy may be recommended, in selected cases, as symptomatic therapy for both motor and nonmotor symptoms. Motor symptoms that may benefit from the injection of botulinum toxin are painful foot dystonia, laryngeal and cervical dystonia, cranial dystonia (including trismus), upper limb contractures, as well as freezing of gait. Nonmotor symptoms that may be treated with botulinum toxin include sialorrhea and urinary symptoms (bladder hyperreflexia or associated prostatic hyperplasia) [67, 68]. Non-pharmacological approaches include exercise programs that focus on the coordination of muscle activity while maintaining posture and movement, which may help to perform skills and activities such as grasping, rolling over in bed, and preserving balance [69]. Moreover, changes in posture can be corrected by applying verbal or visual feedback, for example, by using a mirror [52, 70].

As regards more specific situations, postural deformities such as Pisa syndrome can be treated by means of both medical and surgical strategies. Drug treatment approaches include anticholinergics and clozapine [25], whose efficacy is however limited. When Bonanni and colleagues performed a blinded crossover trial of botulinum toxin injected into the lumbar paraspinal muscles compared with placebo in nine patients with Pisa syndrome, the majority of the patients reported an improvement in their posture and opted to continue receiving botulinum toxin treatment at the end of the study [25]. These results have not been replicated in other centers. In another study, subthalamic DBS was applied to ten parkinsonian patients with Pisa syndrome, though the results were inconclusive [71].

With regard to antecollis, some patients have reported improved head position following treatment with levodopa, though this improvement has not been reported consistently in the literature [6, 24, 34]. Muscle relaxants such as clonazepam may also be helpful [34]. Botulinum toxin therapy is usually attempted if there are active dystonic spasms upon examination, though the benefit of this therapy has only been reported in patients treated with a bilateral injection in the levator scapulae muscle [24]. Botulinum toxin treatment of prevertebral deep neck flexors (longus colli and longus capitis) has proved beneficial in selected cases, though CT guidance is required [72]. As other neck muscles, such as the scalene (anterior and medius) and submental groups, might contribute to antecollis, the approach to each patient should probably be tailored according to the examination findings. The frequent development of dysphagia is, however, a common limiting factor in the use of this treatment for antecollis. Intensive physiotherapy and the use of neck collars might be beneficial, though there is as yet no evidence to support this hypothesis. A limited number of reports on the management of severe antecollis by means of surgical fusion [73] and DBS [35] exist, though these should be reserved as possible third-line options after oral treatment and local botulinum toxin injections.

The lack of a consistent approach to treatment in PSP and the complex symptomatology make it difficult to accurately assess the response of dystonia to medical management. While patients with limb dystonia or blepharospasm may display a beneficial response to tricyclic agents, only those who receive botulinum toxin injections for blepharospasm consistently experience a good response to such injections [5]. Local injections of botulinum toxin may also prove effective in the treatment of severe retrocollis associated with specific disabilities and in particular with pain.

As regards the treatment of dystonia in CBD, the beneficial effect of levodopa on dystonia, which has been studied in a limited number of cases, is either partial or absent in the majority of cases [57, 74], though there have been sporadic reports of a moderate improvement in the symptoms [75]. Anticholinergics, amantadine, propranolol, primidone, bromocriptine, amitriptyline, levetiracetam, and valproate have provided little or no benefit, though it should be borne in mind that none of these agents was tested in a controlled clinical study [57, 74, 76]. No definite data are available on the efficacy of botulinum toxin in pathologically proven cases [74]. Botulinum toxin may be beneficial in the treatment of CBD-associated limb dystonia [57, 58, 66] and could therefore be offered to such patients to relieve symptoms.

References

1. Adams RD, Victor M. Principles of neurology. 2nd ed. New York: Mc Graw-Hill; 1981. p. 690.
2. Rivest J, Quinn N, Marsden CD. Dystonia in Parkinson's disease, multiple system atrophy and progressive supranuclear palsy. Neurology. 1990;40:1571–8.
3. Shoulson I. Huntington's disease. A decade of progress. Neurol Clin. 1984;2:515–26. Review.
4. Segawa M. Segawa disease (hereditary progressive dystonia with marked diurnal fluctuation-HPD) and abnormalities in pteridin metabolism. Rinsho Shinkeigaku. 1996;36:1322–3. Japanese.
5. Barclay CL, Lang AE. Dystonia in progressive supranuclear palsy. J Neurol Neurosurg Psychiatry. 1997;62:352–6.
6. Boesch SM, Wenning GK, Ransmayr G, et al. Dystonia in multiple system atrophy. J Neurol Neurosurg Psychiatry. 2002;72:300–3.
7. Stamelou M, Alonso-Canovas A, Bhatia KP. Dystonia in corticobasal degeneration: a review of the literature on 404 pathologically proven cases. Mov Disord. 2012;27:696–702.
8. Ashour R, Jankovic J. Joint and skeletal deformities in Parkinson's disease, multiple system atrophy, and progressive supranuclear palsy. Mov Disord. 2006;21:1856–63.
9. Doherty KM, van de Warrenburg BP, Peralta MC, et al. Postural deformities in Parkinson's disease. Lancet Neurol. 2011;10:538–49.
10. Kollensperger M, Geser F, Seppi K, et al. Red flags for multiple system atrophy. Mov Disord. 2008;23:1093–9.
11. Ekbom K, Lindholm H, Ljungberg L. New dystonic syndrome associated with butyrophenone therapy. Z Neurol. 1972;202:94–103.
12. Suzuki T, Koizumi J, Moroji T, et al. Clinical characteristics of the Pisa syndrome. Acta Psychiatr Scand. 1990;82:454–7.
13. Villarejo A, Camacho A, Garcia-Ramos R, et al. Cholinergic-dopaminergic imbalance in Pisa syndrome. Clin Neuropharmacol. 2003;26:119–21.
14. Yokochi F. Lateral flexion in Parkinson's disease and Pisa syndrome. J Neurol. 2006;253:17–20.
15. Colosimo C. Pisa syndrome in a patient with multiple system atrophy. Mov Disord. 1998;13:607–9.
16. Vanacore N, Suzzareddu G, Maggini M, et al. Pisa syndrome in a cohort of Alzheimer's disease patients. Acta Neurol Scand. 2005;111:199–201.
17. Schwab FJ, Smith VA, Biserni M, et al. Adult scoliosis: a quantitative radiographic and clinical analysis. Spine. 2002;27:387–92.
18. Di Matteo A, Fasano A, Squintani G, et al. Lateral trunk flexion in Parkinson's disease: EMG features disclose two different underlying pathophysiological mechanisms. J Neurol. 2011;258:740–45.
19. Baik JS, Kim JY, Park J, et al. Scoliosis in patients with Parkinson's disease. J Clin Neurol. 2009;5:91–4.
20. Cannas A, Solla P, Floris G, et al. Reversible Pisa syndrome in patients with Parkinson's disease on dopaminergic therapy. J Neurol. 2009;256:390–485.
21. Salazar Z, Tschopp L, Calandra C, et al. Pisa syndrome and parkinsonism secondary to valproic acid in Huntington's disease. Mov Disord. 2008;23:2430–1.

22. Suzuki T, Matsuzaka H. Drug-induced Pisa syndrome (pleurothotonus): epidemiology and management. CNS Drugs. 2002;16:165–74.
23. Spanaki C, Zafeiris S, Plaitakis A. Levodopa-aggravated lateral flexion of the neck and trunk as a delayed phenomenon of unilateral pallidotomy. Mov Disord. 2010;25:655–6.
24. van de Warrenburg BP, Bhatia KP, Quinn NP. Pisa syndrome after unilateral pallidotomy in Parkinson's disease: an unrecognised, delayed adverse event? J Neurol Neurosurg Psychiatry. 2007;78:329–30.
25. Bonanni L, Thomas A, Varanese S, et al. Botulinum toxin treatment of lateral axial dystonia in Parkinsonism. Mov Disord. 2007;22:2097–103.
26. Hozumi I, Piao YS, Inuzuka T, et al. Marked asymmetry of putaminal pathology in an MSA-P patient with Pisa syndrome. Mov Disord. 2004;19:470–2.
27. Quinn N. Disproportionate antecollis in multiple system atrophy. Lancet. 1989;333:844.
28. Lange D, Fetell M, Lovelace R, et al. The floppy head syndrome. Annal Neurol. 1986;20:133.
29. Fujimoto K. Dropped head in Parkinson's disease. J Neurol. 2006;253:21–6.
30. Suzuki M, Hirai T, Ito Y, et al. Pramipexole-induced antecollis in Parkinson's disease. J Neurol Sci. 2008;264:195–7.
31. Taguchi Y, Takashima S, Tanaka K. Pramipexole-induced dropped head syndrome in Parkinson's disease. Intern Med. 2008;47:2011–2.
32. Uzawa A, Mori M, Kojima S, et al. Dopamine agonist-induced antecollis in Parkinson's disease. Mov Disord. 2009;24:2408–11.
33. Kataoka H, Ueno S. Dropped head associated with amantadine in Parkinson disease. Clin Neuropharmacol. 2011;34:48–9.
34. Kashihara K, Ohno M, Tomita S. Dropped head syndrome in Parkinson's disease. Mov Disord. 2006;21:1213–6.
35. Yamada H, Katayama Y, Yamamoto T. Parkinson's disease and dropped head. Rinsho Shinkeigaku. 2003;43:955.
36. Corbin DO, Williams AC. Stridor during dystonic phases of Parkinson's disease. J Neurol Neurosurg Psychiatry. 1987;50:821–2.
37. Hughes RG, Gibbin KP, Lowe J. Vocal fold abductor paralysis as a solitary and fatal manifestation of multiple system atrophy. J Laryngol Otol. 1998;112:177–8.
38. Merlo IM, Occhini A, Pacchetti C, et al. Not paralysis, but dystonia causes stridor in multiple system atrophy. Neurology. 2002;58:649–52.
39. Wenning GK, Ben Shlomo Y, Magalhaes M, et al. Clinical features and natural history of multiple system atrophy. An analysis of 100 cases. Brain. 1994;117:835–45.
40. Wenning GK, Ben Shlomo Y, Magalhaes M, et al. Clinicopathological study of 35 cases of multiple system atrophy. J Neurol Neurosurg Psychiatry. 1995;58:160–6.
41. Wenning GK, Tison F, Ben Shlomo Y, et al. Multiple system atrophy: a review of 203 pathologically proven cases. Mov Disord. 1997;12:133–47.
42. Poewe W, Lees AJ, Steiger D. Foot dystonia in Parkinson's disease: clinical phenomenology and neuropharmacology. Adv Neurol. 1987;45:357–60.
43. Poewe W, Wenning GK. The natural history of Parkinson's disease. Ann Neurol. 1998;44:1–9.
44. Rafal RD, Friedman JH. Limb dystonia in progressive supranuclear palsy. Neurology. 1987;37:1545–9.
45. Oide T, Ohara S, Yazawa M, et al. Progressive supranuclear palsy with asymmetric tau pathology presenting with unilateral limb dystonia. Acta Neuropathol. 2002;104:209–14.
46. Pfaffenbach DD, Layton DD, Kearns TP. Ocular manifestations in progressive supranuclear palsy. Am J Ophthalmol. 1972;74:1179–84.
47. Golbe LI, Davis PH, Lepore FE. Eyelid movement abnormalities in progressive supranuclear palsy. Mov Disord. 1989;4:297–302.
48. Jankovic J, Friedman DI, Pirozzolo FJ, et al. In: Streifler MB, Korczyn AD, Melamed E, Youdim MBH, editors. Progressive supranuclear palsy: motor, neurobehavioural and neuro-ophthalmological findings, vol. 53. New York: Raven; 1990. p. 293–304. Advances in neurology.
49. Boghen D. Apraxia of lid opening: a review. Neurology. 1997;48:1491–4.
50. Lamberti P, De Mari M, Zenzola A, et al. Frequency of apraxia of eyelid opening in the general population and in patients with extrapyramidal disorders. Neurol Sci. 2002;23:81–2.

51. Krack P, Marion MH. "Apraxia of lid opening", a focal eyelid dystonia: clinical study of 32 patients. Mov Disord. 1994;9:610–5.
52. Colosimo C, Riley DE, Wenning GK. Handbook of atypical parkinsonism. Cambridge: Cambridge University Press; 2011. Published by Cambridge University press.
53. Jankovic J, Patel SC. Blepharospasm associated with brain-stem lesions. Neurology. 1983;33:1237–40.
54. Papapetropoulos S, Baez S, Zitser J, et al. Retrocollis: classification, clinical phenotype, treatment outcomes and risk factors. Eur Neurol. 2008;59:71–5.
55. Constatinidis J, Tissot R, De Ajuriaguerra J. Dystonie oculo-facio-cervicale ou paralysie supranucleire progressive de Steele-Richardson-Olszewski. Pseudo-paralysie du regard, troubles visuo-spatiaux, pseudo-demence, alterations neuronales. Rev Neurol. 1970;122: 249–62.
56. Halpern C, McMillan C, Moore, et al. Calculation impairment in neurodegenerative diseases. J Neurol Sci. 2003;208:31–8.
57. Kompoliti K, Goetz CG, Boeve BF, et al. Clinical presentation and pharmacological therapy in corticobasal degeneration. Arch Neurol. 1998;55:957–61.
58. Vanek Z, Jankovic J. Dystonia in corticobasal degeneration. Mov Disord. 2001;16:252–7.
59. Boeve BF. The multiple phenotypes of corticobasal syndrome and corticobasal degeneration: implications for further study. J Mol Neurosci. 2011;45:350–3.
60. Rebeiz JJ, Kolodny EH, Richardson EP. Corticodentatonigral degeneration with neuronal achromasia. Arch Neurol. 1968;18:20–33.
61. Kouri N, Whitwell JL, Josephs KA, et al. Corticobasal degeneration: a pathologically distinct 4R tauopathy. Nat Rev Neurol. 2011;7:263–72.
62. Ling H, O'sullivan SS, Holton JL, et al. Does corticobasal degeneration exist? A clinicopathological re-evaluation. Brain. 2010;133:2045–57.
63. Huang KJ, Lu MK, Kao A, et al. Clinical, imaging and electrophysiological studies of corticobasal degeneration. Acta Neurol Taiwan. 2007;16:13–21.
64. Thompson PD, Day BL, Rothwell JC, et al. The myoclonus in corticobasal degeneration. Evidence for two forms of cortical reflex myoclonus. Brain. 1994;117:1197–207.
65. Frasson E, Bertolasi L, Bertasi V, et al. Paired transcranial magnetic stimulation for the early diagnosis of corticobasal degeneration. Clin Neurophysiol. 2003;114:272–8.
66. Muller J, Wenning GK, Wissel J, et al. Botulinum toxin treatment in atypical parkinsonian disorders associated with disabling focal dystonia. J Neurol. 2002;249:300–4.
67. Giannantoni A, Rossi A, Mearini E, et al. Botulinum toxin A for overactive bladder and detrusor muscle overactivity in patients with Parkinson's disease and multiple system atrophy. J Urol. 2009;182:1453–7.
68. Seppi K, Weintraub D, Coelho M, et al. The movement disorder society evidence-based medicine review update: treatments for the non-motor symptoms of Parkinson's disease. Mov Disord. 2011;26:42–80.
69. Stallibrass C, Sissons P, Chalmers C. Randomized controlled trial of the Alexander technique for idiopathic Parkinson's disease. Clin Rehabil. 2002;16:695–708.
70. Formisano R, Pratesi L, Modarelli FT, et al. Rehabilitation and Parkinson's disease. Scand J Rehabil Med. 1992;24:157–60.
71. Umemura A, Oka Y, Ohkita K, et al. Effect of subthalamic deep brain stimulation on postural abnormality in Parkinson disease. J Neurosurg. 2010;112:1283–8.
72. Herting B, Wunderlich S, Glockler T, et al. Computed tomographically-controlled injection of botulinum toxin into the longus colli muscle in severe anterocollis. Mov Disord. 2004;19:588–90.
73. Pereira EAC, Wilson-MacDonald J, Green AL, et al. Posterior occipitocervical instrumented fusion for dropped head syndrome after deep brain stimulation. J Clin Neurosci. 2010;17:541–2.
74. Gibb WR, Luthert PJ, Marsden CD. Corticobasal degeneration. Brain. 1989;112:1171–92.
75. Boeve BF, Maraganore DM, Parisi JE, et al. Pathologic heterogeneity in clinically diagnosed corticobasal degeneration. Neurology. 1999;53:795–800.
76. Tiwari D, Amar K. A case of corticobasal degeneration presenting with alien limb syndrome. Age Ageing. 2008;37:600–1.

8 Functional (Psychogenic) Dystonia

Antonella Macerollo and Mark J. Edwards

Functional (psychogenic) neurological disorders are common throughout neurological practice, and functional movement disorders (FMDs) form a considerable proportion of these cases. In this chapter, we will consider the entity of functional (psychogenic) dystonia.

8.1 Terminology

A number of different words can be used to describe the broad topic of this chapter. The word used for centuries – hysteria – has now almost completely been abandoned given its connotations that the source of the problem is in the uterus and its lay use as an insulting term [1, 2]. However, one of the most eminent movement disorder specialists of recent times, David Marsden, asserted that neurologists "have clung on to hysteria because its modern roots sprang from neurology" [3]. The term "psychogenic" is the one in most common usage among movement disorder physicians and is one of a number of terms (conversion disorder, somatization, psychosomatic) that suggest a primary (even sole) role for psychological factors in the genesis

A. Macerollo • M.J. Edwards (✉)
Sobell Department of Motor Neuroscience and Movement Disorders,
Institute of Neurology, University College London, Queen Square, London, UK
e-mail: m.j.edwards@ucl.ac.uk

P. Kanovsky et al. (eds.), *Dystonia and Dystonic Syndromes*,
DOI 10.1007/978-3-7091-1516-9_8

of these movement disorders [4]. However, epidemiological studies fail to find the expected high rates of psychological trauma or distress in these patients [5]. In a survey of members of the movement disorder society reported by Espay et al., only 18 % of specialists used the presence of psychological disorder to make the diagnosis, and, interestingly, psychiatrists often sent patients to neurologists clarifying the diagnosis [6].

In place of these terms and other terms that define the disorder by saying what it is not (nonorganic, medically unexplained), it has recently been suggested that the historical term "functional" is the most appropriate [7]. It is certainly a term which is more acceptable to patients than many of the alternatives [8]. There are arguments to support the use of other terms, but while acknowledging these, we will use the term "functional" in this chapter [9]. In doing so, we seek to define these disorders by positive clinical characteristics as movement disorders that are significantly altered by distraction of attention or nonphysiological maneuvers (including dramatic placebo effects) and which are incongruent with movement disorders known to be caused by neurological disease.

8.2 Epidemiology

Functional neurological symptoms are very common, accounting for 16 % of the newly referred outpatients attending neurology clinics [10], and are the second most common issue for neurological outpatient consultation. The most common functional neurological symptoms are nonepileptic attacks and functional weakness [11]. Functional movement disorders (FMDs) constitute about 15 % of those with functional neurological symptoms. In movement disorder clinics, FMDs account for about 2–5 % of patients, but in tertiary movement disorder clinics, such patients account for 20–30 % of cases [12, 13].

The most common FMDs are tremor and dystonia that together account for about 70 % of patients, followed by myoclonus and gait disturbances. Functional parkinsonism, tics, and chorea are rare [12]. Women are more often affected than men, and the mean age at onset is from 37 to 50 years depending on the study selected [12, 13]. Age at onset for patients with functional dystonia is younger than the average for FMD in general: Schrag et al. found an age at onset ranging from 16.6 to 42.8 [14]. FMDs are also seen in children and elderly patients [15, 16]. Functional dystonia, however, seems to be very rare in those over 60 [16].

FMDs are characterized by the level of disability and impairment in the quality of life equivalent to that reported by patients affected by Parkinson's disease [17]. Consequentially, considerable health and social care costs are associated with this group of patients. Indeed, Carson et al. demonstrated that, in a large cohort of patients with functional neurological disorders (1,144 cases), 50 % of cases had stopped working for health reasons and 27 % were receiving disability-related financial benefits. We can estimate that the economic impact is similar for patients with FMD [18].

8.3 Historical Development of the Concept of "Psychogenic" Dystonia

One of the oldest descriptions of cases of "psychogenic" dystonia comes from Allen in his *Observations on the Motor Phenomena of Hysteria*. Historical descriptions of cases of proposed psychogenic dystonia (until at least the 1980s) are complicated by the fact that many forms of dystonia that are now recognized to be organic were classified as psychogenic. However, cases III and VI by Allen are emblematic descriptions, respectively, of functional foot dystonia and hand dystonia. Clinically, he describes these conditions as "tonic paralysis" and "tonic paralysis with contracture" highlighting the fixity of the postures in these cases that over the time can cause permanent contractures [19].

Fahn and Williams proposed diagnostic criteria for psychogenic dystonia and analyzed the clinical characteristics of 21 patients categorized as having either documented ($n=17$) or clinically established ($n=4$) psychogenic dystonia. Seven of these individuals had paroxysmal dystonia. All these groups of patients presented inconsistent or incongruous dystonic movements. Other important clues commonly exhibited were give-way weakness, the complaint of pain and tenderness to touch, "false" sensory findings, multiple somatizations, and inconsistent slowness of movements. The patients with psychogenic paroxysmal dystonia also had nonepileptic attacks (defined so as patients had abnormal movement associated with apparent loss of consciousness). They highlighted the presence of dystonia at rest in their patients, contrasting with the triggering/worsening of organic dystonia by action. The most common site of onset was in the foot, and they highlighted the role of injury and immobilization (casting) as triggers to onset. In several patients, there was a spread of involvement to other parts of the body. The importance of this paper can be judged by the fact that the diagnostic criteria proposed by them were then adopted as criteria for psychogenic movement disorders in general [20].

Sa et al. described 16 patients with "post-traumatic torticollis," highlighting the issue of peripheral trauma in triggering the onset of dystonia in these patients. These patients had fixed abnormal postures of the neck and lacked sensory tricks or response to botulinum toxin injections [21]. The role of peripheral trauma had previously been highlighted in other papers, and then as of now, it was an issue of some controversy with some suggesting that peripheral trauma could cause dystonia via an "organic" route. Authors also highlighted the issue of severe pain, often identical to that seen in "reflex sympathetic dystrophy," now termed complex regional pain syndrome type I (CRPSI) as an accompanying feature. Several case series reported dystonic posture following peripheral nerve injury [22–24] or surgery [13, 25, 26]. Marsden and Schott described cases of fixed dystonia with CRPS following peripheral trauma, and Bhatia et al. described a series of such cases, calling them "causalgia-dystonia." These patients had vasomotor, sudomotor, and trophic changes and typically had a poor response to botulinum toxin and were very disabled. In line with other FMDs, young women predominated in this group [27].

Van Hilten et al. labeled this condition "tonic dystonia of chronic regional pain," highlighting the presence of severe pain associated to fixed postures similar to that reported by patients affected by complex regional pain syndrome type I (CRPSI) and highlighted an organic explanation for such symptoms. In this series, 26 patients with CRPSI progressed toward a multifocal or generalized tonic dystonia. The abnormal posture began distally, involved mainly flexor muscles, and was associated with sensory (hyperesthesia, hyperalgesia) and autonomic symptoms (sudomotor activity, swelling, discolored skin, decreased skin temperature). They reported an association with the HLA-DR13 haplotype in these patients, proposing it to be a susceptibility factor for this distinct phenotype of CRPS associated with tonic dystonia [28].

In the same period, Verdugo et al. argued in favor of a functional explanation of post-traumatic dystonia. In this series, no structural central or peripheral nervous system anomaly was present in any CRPSI patient displaying abnormal postures. Furthermore, they found that every patient presented at least one additional functional motor sign [29].

The largest clinical series of patients is that reported by Schrag et al. They described clinical features of fixed dystonia in 103 patients (84 % female) [13].The fixed postures involved predominantly the limbs (90 %) and rarely the neck (6 %) or jaw (4 %), and in 61 %, the spread occurred to other body parts. Schrag et al. confirmed the high frequency of the temporal association between fixed postures and physical trauma. Indeed, 64 % presented with dystonia after a minor peripheral injury. Many had associated pain and 20 % of patients fulfilled the criteria for CRPSI. In this series, the triggering injury reported was mainly injuries of soft tissues, but limb overuse, fractures, and consequent immobilization in a plaster cast were also reported as precipitating events. In line with previous studies, dystonia developed subacutely over days or weeks and showed typical features of functional dystonia, as emphasized by Verdugo et al. [29].

8.4 Clinical Features of Functional Dystonia

In recent times, there has been a shift in diagnostic criteria for functional neurological disorders in general with an emphasis placed on positive features in the history and examination and a reduction in emphasis on concomitant psychological distress, which itself is present in a great many patients with neurological disease. This process is reflected in newer diagnostic criteria for psychogenic movement disorders, in attempts at formalizing electrophysiological test batteries, and in the latest criteria for conversion disorder reported in DSM-5 [30–33].

The mainstay of positive diagnosis in patients with functional motor symptoms such as weakness and tremor is the demonstration that normal movement can occur when attention is distracted. Thus, Hoover's sign of functional weakness [34] and various techniques used to assess functional tremor (tapping tasks, ballistic movement tasks) [35] all rely on accessing movement in an unattended fashion and demonstrating a significant change in function compared to when the patient's attention is not distracted.

Table 8.1 Diagnostic criteria for FMD [20, 30]

Fahn-Williams criteria	
Category	Criteria
Documented	Persistent relief by psychotherapy, suggestion of placebo response, which may be helped by physiotherapy, or absence of abnormal postures when patient is unobserved
Clinically established	The posture is inconsistent over time, or it is incongruent with classical dystonia (e.g., impossibility to move limbs on request and great resistance during passive movements), plus one of the following manifestations: other psychogenic signs, multiple somatizations, or an obvious psychiatric disorder
Probable	The dystonia is incongruent or inconsistent with typical organic dystonia, or there are psychogenic signs or multiple somatizations
Possible	Evidence of an emotional disturbance
Gupta-Lang criteria	
Clinically definite	Documented category
	Clinically established category
	Incongruent and/or inconsistent movement disorders (even without the additional presence of psychogenic signs, multiple somatizations, or psychiatric disturbance)
Laboratory-supported definite	Electrophysiological tests proving an FMD (primary evidence of premovement potentials before jerks or data from tremor studies)

The discussion of the historical development of the concept of functional dystonia has revealed that the most common manifestation is with fixed abnormal postures. These postures are often not changed a great deal by distraction. This could of course be because they are not an FMD as some have argued, but an alternative explanation is that maintenance of such fixed postures requires very little attentional focus as opposed to functional tremor or weakness. The key diagnostic points are therefore historical features, pattern recognition, and the search for additional functional signs and symptoms (Table 8.1).

Patients with functional fixed dystonia typically present acutely or subacutely after painful injury to the limb/body part where the abnormal posture arises. The posture may emerge after a period of casting of the limb. Postures tend to be fixed and cannot be moved even passively from their position. There is often a paradoxical relaxation of muscles at rest (even though the posture is maintained) and a very strong contraction of muscles as soon as an attempt is made to move the limb. This is often despite the fact that voluntary contraction of the same muscles to command is weak or nonexistent. The distribution of the dystonia is unusual for the age at onset with multifocal/generalized dystonia common despite onset in adult life. There is typically an absence of sensory gestes or task specificity. The typical pattern of hand functional dystonia results in flexion of the fingers at the metacarpophalangeal or interphalangeal joints. Often the last three fingers are more affected than the second one, with the thumb least affected or not affected. Involvement of the lower extremities results in plantar flexion and inversion of the foot and sometimes additional clawing, curling, or scissoring of toes. Consequent to prolonged maintenance of tonic postures, patients may have muscle atrophy and joint contractures [13, 36].

Fixed dystonia is often associated with other functional movement disorders or other functional neurological symptoms, including tremor and nonepileptic attacks. With distraction, one can sometimes appreciate a sudden giving way of the abnormal fixed posture. There is an absence of "hard" neurological signs such as pathologically brisk reflexes and extensor plantar responses.

Schrag et al. reported higher rates of affective disorder, dissociative symptoms, and somatization disorder in fixed dystonia than in organic dystonia patients [13]. An association between fixed dystonia with or without clinical features of CRPS and joint hypermobility syndrome has recently been proposed [37]. Joint hypermobility syndrome is characterized by joint laxity which causes articular dislocations, subluxations, and arthralgia [38]. In these patients, the clinical pattern often involves urinary dysfunction and postural orthostatic tachycardia syndrome [39].

There are other patterns of functional dystonia, in particular fixed postures affecting the face ("the smirk") [40] and paroxysmal attacks of dystonic posturing. These latter attacks are phenomenologically different from the known patterns of paroxysmal dystonia occurring as part of paroxysmal kinesigenic dyskinesia, paroxysmal non-kinesigenic dyskinesia, and paroxysmal exercise-induced dystonia. They often have multiple triggers, have highly variable attack duration, and are complicated by other symptoms not seen in organic paroxysmal attacks including very severe pain and apparent alteration of conscious level.

8.5 Pathophysiology and Neurobiology of Fixed Dystonia

Historically, concepts of dissociation, conversion, and somatization formed the basis of the diagnosis and pathophysiological explanation of functional neurological symptoms including functional dystonia. Such concepts, while accepting the reality and involuntary nature of symptoms, suggest a key role for psychological trauma in the genesis of such symptoms. In turn, this suggests that the mainstay of treatment should be uncovering and treatment of this psychological trauma.

As mentioned above, the evidence in favor of psychological trauma playing a key role in the pathophysiology of functional neurological symptoms, including functional dystonia, is weak. In addition, purely psychologically based explanations for symptoms leave a large "explanatory gap": how does one go from a psychological trauma to physical symptoms which are experienced as involuntary, and how can this process be implemented within the brain?

A more neurobiologically focused model of the genesis of functional neurological symptoms in general has highlighted three key processes: abnormal self-directed attention (self-monitoring), abnormal beliefs/expectations, and abnormal sense of agency [41]. The first key feature that distinguishes clearly patients with FMD from those with "organic" movements is the abnormal self-directed attention, an essential factor in manifesting symptoms. When attention is distracted, there is typically improvement of functional movements. Conversely, symptoms get worse during physical examination, during which attention is drawn toward the body. An experimental exploration of the key role of attention toward FMD was conducted by Parees et al. [42], where patients with FMD showed the impaired performance in movements that were highly predicted (a condition in which there is opportunity for attention toward movement production) compared to healthy subjects.

The nature of attention in fixed dystonia has been added recently by a study of regional cerebral blood flow by positron emission tomography [43]. They observed prefrontal activation in functional dystonia as well as in organic forms; it could represent abnormal movement-related attention. An activation of the basal ganglia and cerebellum has been seen only in functional patients, and it could explain additional problems with self-directed attention/self-monitoring or could reflect a contribution from connected limbic structures.

The second key concept in the neurobiological model of FMD is the presence of abnormal (high-level) beliefs related to symptoms. This does not just mean consciously reportable beliefs about symptoms but involves abnormal predictions and expectations in the setting of active inference in the brain, where the brain is proposed to actively predict and seek to explain sensory input on the basis of past experience. Expectations or prior beliefs play an important role in altering sensory experience. Voon et al., through a number of functional imaging studies in FMD, proposed a model of generation of functional symptoms. Emotional arousal events might be a trigger for the development of movements controlled by the supplementary motor area, functionally disconnected from top-down control by the prefrontal cortex [44].

The third pathophysiological aspect is the abnormal sense of agency exhibited by this group of patients. They have movements that have characteristics one would associate with voluntary movement, for example, the requirement for attention for movement abnormalities to manifest, but yet the movements are not associated with a sense that they are under self-control. There is evidence that patients with FMD and motor conversion disorders experience a lack of sense of agency for movement in general, supporting their own self-report that they cannot voluntarily control the abnormal movements they produce [45].

In the realm of functional dystonia itself, several studies have been conducted with the aim of comparing organic and functional dystonia. Starting from the knowledge that organic dystonia is characterized by abnormal cortical and spinal inhibition [46], Espay et al. examined in the affected limb cortical and spinal inhibitory circuits and cortical activity associated with voluntary movements in functional and organic dystonia patients. Surprisingly, they found that these patients had a similar reduction of short- and long-interval intracortical inhibition (SICI and LICI), as well as the same increase of cortical silent period [47]. Consistent with these findings, Avanzino et al., testing both affected and unaffected sides, observed that SICI and silent period were reduced bilaterally to a similar extent in patients with primary dystonia and those with fixed dystonia [48]. While these findings could be explained by there being a similar endophenotypic trait to reduced cortical inhibition in patients with organic and functional dystonia, an alternative explanation is that there are confounding factors that influenced the results in those with functional dystonia. Reduced SICI and silent period are not specific markers of dystonia. They are abnormal in a number of neurological disorders but are also influenced in healthy subjects by attention toward the limb, muscle contraction itself, and psychological state.

Quartarone et al. found that organic dystonia patients only exhibited an abnormal (excessive) response to a motor cortical plasticity protocol, in comparison to a normal response in patients with functional dystonia [49, 50].

In a recent study, Katschnig and associates explored the pathophysiology of fixed dystonia applying two experimental techniques: mental rotation of body parts and somatosensory temporal discrimination. They observed that patients with fixed dystonia were slower at performing a mental rotation of a body part displayed in different orientations on a screen than healthy controls, showing a similar level of impairment to patients with organic dystonia. Furthermore, only patients with organic dystonia had abnormalities of temporal discrimination [51]. However, a recent study has found abnormal temporal discrimination in functional dystonia. There are, however, clearly possible influences from abnormal attentional focus in such paradigms which are based on self-report [52].

The impairment in the mental rotation task supports the concept that patients with fixed dystonia may have a distortion of body image, since the ability to perform mental rotation of corporal objects is linked to the concept of "body schema" [53]. It has been proposed that aberrant proprioceptive input due to the prolonged abnormal limb posture could produce secondary changes in central body image. Body image disturbance might contribute to pain and other features, in particular driving some patients to seek the amputation of the limb affected, as experienced in our clinical practice [54]. This phenomenon is also seen in patients with "apotemnophilia" or "body integrity identity disorder" [55]. This fits with previous studies regarding amputation sought by patients with CRPSI and is evidence that at least a proportion of these patients reported fixed dystonia [56, 57]. Therefore, fixed dystonia with or without CRPSI could be considered as a type of body integrity identity disorder, in which a peripheral painful stimulus might be the trigger. Further support for the notion of preexisting disturbed central body schema is the almost uniform lack of success of such amputations with the development of chronic pain and functional movement disorder in the stump and spread of the movement disorder to other limbs [58, 59].

Another important area of interest is the pathophysiological link between peripheral trauma and dystonia. Peripheral mechanisms, such as sensitization of peripheral nociceptors or ectopic or ephaptic transmission of nerve impulses, have been suggested by Jankovic et al. [60]. The spread of symptoms in different body parts, even those not involved in physical injury, seen in fixed dystonia cannot be explained by peripheral mechanisms. Impairment of neuronal circuits at the spinal and brainstem level and central synaptic reorganization have been suggested by van Hilten [28, 61].

8.6 Prognosis and Approach to Treatment of Fixed Dystonia

Evidence regarding the most useful approach to treatment is still controversial, but early intervention seems to be the best way in preventing the development of chronic symptoms and irreversible contractures [13]. Delay in management tends to lead to an unfavorable long-term outcome. The first important step in a successful treatment approach is the optimal communication of the diagnosis with rational explanation of physical symptoms, highlighting the positive way in which the clinician reached the diagnosis.

As reported by Schrag et al. [13], pharmacological and surgical treatments are often unsuccessful. Therefore, an important part of treatment is reduction and/or withdrawal of unnecessary medications and avoidance of unnecessary invasive procedures. Antidepressants and anxiolytics may be helpful in those with anxiety and depression [62]. Two studies have investigated the efficacy of intrathecal baclofen in patients with fixed dystonia associated with CRPSI. Initially, a placebo-controlled study done by van Hilten et al showed impressive response to this drug with marked improvement of the postures, especially of the upper limbs, and of the pain and sensory symptoms [63]. The limitation of this study was the small number of patients and the systematic unblinding of subjects that is likely to have occurred given the prominent systemic effects of intrathecal baclofen. A follow-up study of a larger group of patients found beneficial effects in many patients but a high rate of long-term complications [64].

Cognitive behavioral therapy (CBT) may lead to partial or complete remission of functional symptoms [65, 66]. CBT is directed toward how thoughts, beliefs, and expectations influence the occurrence of unexplained symptoms and to identify behaviors that may be contributing to maintenance of symptoms.

Several studies provide evidence that physical rehabilitation is useful to treat functional motor symptoms [67–69]. Such an approach can be combined usefully with non-pharmacological approaches to pain management, including desensitization, graded mobilization, and the use of specialist pain management CBT techniques.

One approach to the management of severely affected patients is inpatient multidisciplinary rehabilitation [13], providing both physical and psychological treatments in a more intensive fashion than can be achieved as an outpatient.

In addition, placebo interventions can have dramatic effects in patients with functional dystonia, but the long-term benefit of such approaches is not known. In this regard, dramatic immediate responses have been reported to botulinum toxin injections [70, 71] and transcranial magnetic stimulation [72].

Prognosis from published studies is generally poor [73, 74]. One long-term follow-up study has reported improvement in less than 25 % of patients, remission in only 6 %, and progressive worsening in one third. In this study, patients with CRPS and fixed dystonia had poorer outcome [75].

Conclusion

Functional dystonia can be an extremely disabling condition. Future studies are required to develop more objective diagnostic criteria to distinguish functional dystonia from other forms. However, with appropriate management, improvement and even resolution of symptoms can occur. There is a clear need for studies that can delineate the most effective treatment strategies and to build these into clear management pathways for these patients.

References

1. Charcot JM. Clinical lectures on diseases of the nervous system, vol. 3. London: The New Sydenham Society; 1889. p. 12.
2. Head H. The diagnosis of hysteria. BMJ. 1922;1:827–9.
3. Marsden CD. Hysteria–a neurologist's view. Psychol Med. 1986;16:277–88.
4. Stone J, Edwards MJ. How "psychogenic" are psychogenic movement disorders? Mov Disord. 2011;26:1787–8.
5. Kranick S, Ekanayake V, Martinez V, Ameli R, Hallett M, Voon V. Psychopathology and psychogenic movement disorders. Mov Disord. 2011;26:1844–50.
6. Espay AJ, Goldenhar LM, Voon V, Schrag A, Burton N, Lang AE. Opinions and clinical practices related to diagnosing and managing patients with psychogenic movement disorders: an international survey of movement disorder society members. Mov Disord. 2009;24: 1366–74.
7. Edwards MJ, Stone J, Lang AE. From psychogenic movement disorder to functional movement disorder: it's time to change the name. Mov Disord. 2014;29:849–52.
8. Stone J, Wojcik W, Durrance D, Carson A, Lewis S, MacKenzie L, Warlow CP, Sharpe M. What should we say to patients with symptoms unexplained by disease? The "number needed to offend". BMJ. 2002;325:1449–50.
9. Trimble MR. Functional diseases. Br Med J (Clin Res Ed). 1982;285:1768–70.
10. Stone J, Carson A, Duncan R, Roberts R, Warlow C, Hibberd C, Coleman R, Cull R, Murray G, Pelosi A, Cavanagh J, Matthews K, Goldbeck R, Smyth R, Walker J, Sharpe M. Who is referred to neurology clinics? – The diagnoses made in 3781 new patients. Clin Neurol Neurosurg. 2010;112:747–51.
11. Stone J. Functional neurological symptoms. Clin Med. 2013;13:80–3.
12. Factor SA, Podskalny GD, Molho ES. Psychogenic movement disorders: frequency, clinical profile, and characteristics. J Neurol Neurosurg Psychiatry. 1995;59:406–12.
13. Williams DT, Ford B, Fahn S. Phenomenology and psychopathology related to psychogenic movement disorders. Adv Neurol. 1995;65:231–57.
14. Schrag A, Trimble M, Quinn N, Bhatia K. The syndrome of fixed dystonia: an evaluation of 103 patients. Brain. 2004;127:2360–72.
15. Schwingenschuh P, Pont-Sunyer C, Surtees R, Edwards MJ, Bhatia KP. Psychogenic movement disorders in children: a report of 15 cases and a review of the literature. Mov Disord. 2008;23:1882–8.
16. Batla A, Stamelou M, Edwards MJ, Pareés I, Saifee TA, Fox Z, Bhatia KP. Functional movement disorders are not uncommon in the elderly. Mov Disord. 2013;28:540–3.
17. Anderson KE, Gruber-Baldini AL, Vaughan CG, Reich SG, Fishman PS, Weiner WJ, Shulman LM. Impact of psychogenic movement disorders versus Parkinson's on disability, quality of life, and psychopathology. Mov Disord. 2007;22:2204–9.
18. Carson A, Stone J, Hibberd C, Murray G, Duncan R, Coleman R, Warlow C, Roberts R, Pelosi A, Cavanagh J, Matthews K, Goldbeck R, Hansen C, Sharpe M. Disability, distress and unemployment in neurology outpatients with symptoms 'unexplained by organic disease'. J Neurol Neurosurg Psychiatry. 2011;82:810–3.

19. Allen IM. Observations on the motor phenomena of hysteria. J Neurol Psychopathol. 1935;16:1–25.
20. Fahn S, Williams DT. Psychogenic dystonia. Adv Neurol. 1988;50:431–55.
21. Sa DS, Mailis-Gagnon A, Nicholson K, Lang AE. Posttraumatic painful torticollis. Mov Disord. 2003;18:1482–91.
22. Marsden CD, Obeso JA, Traub MM, Rothwell JC, Kranz H, La Cruz F. Muscle spasms associated with Sudeck's atrophy after injury. Br Med J (Clin Res Ed). 1984;288:173–6.
23. Schott GD. Induction of involuntary movements by peripheral trauma: an analogy with causalgia. Lancet. 1986;2:712–6.
24. Schott GD. The relationship of peripheral trauma and pain to dystonia. J Neurol Neurosurg Psychiatry. 1985;48:698–701.
25. Jankovic J, Van der Linden C. Dystonia and tremor induced by peripheral trauma: predisposing factors. J Neurol Neurosurg Psychiatry. 1988;51:1512–9.
26. van Rooijen DE, Geraedts EJ, Marinus J, Jankovic J, van Hilten JJ. Peripheral trauma and movement disorders: a systematic review of reported cases. J Neurol Neurosurg Psychiatry. 2011;82:892–8.
27. Bhatia KP, Bhatt MH, Marsden CD. The causalgia-dystonia syndrome. Brain. 1993;116:843–51.
28. Van Hilten JJ, Van de Beek WJ, Roep BO. Multifocal or generalized tonic dystonia of complex regional pain syndrome: a distinct clinical entity associated with HLA-DR13. Ann Neurol. 2000;48:113–6.
29. Verdugo RJ, Ochoa JL. Abnormal movements in complex regional pain syndrome: assessment of their nature. Muscle Nerve. 2000;23:198–205.
30. Gupta A, Lang AE. Psychogenic movement disorders. Curr Opin Neurol. 2009;22:430–6.
31. Schwingenschuh P, Katschnig P, Seiler S, Saifee TA, Aguirregomozcorta M, Cordivari C, Schmidt R, Rothwell JC, Bhatia KP, Edwards MJ. Moving toward "laboratory-supported" criteria for psychogenic tremor. Mov Disord. 2011;26:2509–15.
32. Stone J, LaFrance Jr WC, Levenson JL, Sharpe M. Issues for DSM-5: conversion disorder. Am J Psychiatry. 2010;167:626–7.
33. Stone J, LaFrance Jr WC, Brown R, Spiegel D, Levenson JL, Sharpe M. Conversion disorder: current problems and potential solutions for DSM-5. J Psychosom Res. 2011;71:369–76.
34. Ziv I, Djaldetti R, Zoldan Y, Avraham M, Melamed E. Diagnosis of "non-organic" limb paresis by a novel objective motor assessment: the quantitative Hoover's test. J Neurol. 1998;245:797–802.
35. Roper LS, Saifee TA, Parees I, Rickards H, Edwards MJ. How to use the entrainment test in the diagnosis of functional tremor. Pract Neurol. 2013;13(6):396–8.
36. Lang AE. Psychogenic dystonia: a review of 18 cases. Can J Neurol Sci. 1995;22:136–43.
37. Kassavetis P, Batla A, Pareés I, Saifee TA, Schrag A, Cordivari C, Bhatia KP, Edwards MJ. Joint hypermobility syndrome: a risk factor for fixed dystonia? Mov Disord. 2012;27: 1070.
38. Grahame R. Joint hypermobility syndrome pain. Curr Pain Headache Rep. 2009;13:427–33.
39. Gazit Y, Nahir AM, Grahame R, Jacob G. Dysautonomia in the joint hypermobility syndrome. Am J Med. 2003;115:33–40.
40. Fasano A, Valadas A, Bhatia KP, Prashanth LK, Lang AE, Munhoz RP, Morgante F, Tarsy D, Duker AP, Girlanda P, Bentivoglio AR, Espay AJ. Psychogenic facial movement disorders: clinical features and associated conditions. Mov Disord. 2012;27:1544–51.
41. Edwards MJ, Fotopoulou A, Pareés I. Neurobiology of functional (psychogenic) movement disorders. Curr Opin Neurol. 2013;26:442–7.
42. Pareés I, Kassavetis P, Saifee TA, Sadnicka A, Davare M, Bhatia KP, Rothwell JC, Bestmann S, Edwards MJ. Failure of explicit movement control in patients with functional motor symptoms. Mov Disord. 2013;28:517–23.
43. Schrag AE, Mehta AR, Bhatia KP, Brown RJ, Frackowiak RS, Trimble MR, Ward NS, Rowe JB. The functional neuroimaging correlates of psychogenic versus organic dystonia. Brain. 2013;136:770–81.

44. Voon V, Brezing C, Gallea C, Hallett M. Aberrant supplementary motor complex and limbic activity during motor preparation in motor conversion disorder. Mov Disord. 2011;26:2396–403.
45. Kranick SM, Moore JW, Yusuf N, Martinez VT, Lafaver K, Edwards MJ, Mehta AR, Collins P, Harrison NA, Haggard P, Hallett M, Voon V. Action-effect binding is decreased in motor conversion disorder: implications for sense of agency. Mov Disord. 2013. doi:10.1002/mds.25408.
46. Hallett M. Dystonia: abnormal movements result from loss of inhibition. Adv Neurol. 2004;94:1–9.
47. Espay AJ, Morgante F, Purzner J, Gunraj CA, Lang AE, Chen R. Cortical and spinal abnormalities in psychogenic dystonia. Ann Neurol. 2006;59:825–34.
48. Avanzino L, Martino D, van de Warrenburg BP, Schneider SA, Abbruzzese G, Defazio G, Schrag A, Bhatia KP, Rothwell JC. Cortical excitability is abnormal in patients with the "fixed dystonia" syndrome. Mov Disord. 2008;23:646–52.
49. Quartarone A, Rizzo V, Terranova C, Morgante F, Schneider S, Ibrahim N, Girlanda P, Bhatia KP, Rothwell JC. Abnormal sensorimotor plasticity in organic but not in psychogenic dystonia. Brain. 2009;132:2871–7.
50. Rosenkranz K. Plasticity and intracortical inhibition in dystonia–methodological reconsiderations. Brain. 2010;133:e146.
51. Katschnig P, Edwards MJ, Schwingenschuh P, Aguirregomozcorta M, Kägi G, Rothwell JC, Bhatia KP. Mental rotation of body parts and sensory temporal discrimination in fixed dystonia. Mov Disord. 2010;25:1061–7.
52. Morgante F, Tinazzi M, Squintani G, Martino D, Defazio G, Romito L, Albanese A, Di Matteo A, Quartarone A, Girlanda P, Fiorio M, Berardelli A. Abnormal tactile temporal discrimination in psychogenic dystonia. Neurology. 2011;77:1191–7.
53. Berlucchi G, Aglioti SM. The body in the brain revisited. Exp Brain Res. 2010;200:25–35.
54. Edwards MJ, Alonso-Canovas A, Schrag A, Bloem BR, Thompson PD, Bhatia K. Limb amputations in fixed dystonia: a form of body integrity identity disorder? Mov Disord. 2011;26:1410–4.
55. First MB. Desire for amputation of a limb: paraphilia, psychosis, or a new type of identity disorder. Psychol Med. 2005;35:919–28.
56. Dielissen PW, Claassen AT, Veldman PH, Goris RJ. Amputation for reflex sympathetic dystrophy. J Bone Joint Surg Br. 1995;77:270–3.
57. Szeinberg-Arazi D, Heim M, Nadvorna H, Ner IZ, Szeinberg A, Azaria M. A functional and psychosocial assessment of patients with post-Sudeck atrophy amputation. Arch Phys Med Rehabil. 1993;74:416–8.
58. Steiner JC, DeJesus PV, Mancall EL. Painful jumping amputation stumps: pathophysiology of a "sore circuit". Trans Am Neurol Assoc. 1974;99:253–5.
59. Zadikoff C, Mailis-Gagnon A, Lang AE. A case of a psychogenic "jumpy stump". J Neurol Neurosurg Psychiatry. 2006;77:1101.
60. Jankovic J. Post-traumatic movement disorders: central and peripheral mechanisms. Neurology. 1994;44:2006–14.
61. Hawley JS, Weiner WJ. Psychogenic dystonia and peripheral trauma. Neurology. 2011;77:496–502.
62. Voon V, Lang AE. Antidepressant treatment outcomes of psychogenic movement disorder. J Clin Psychiatry. 2005;66:1529–34.
63. van Hilten BJ, van de Beek WJ, Hoff JI, Voormolen JH, Delhaas EM. Intrathecal baclofen for the treatment of dystonia in patients with reflex sympathetic dystrophy. N Engl J Med. 2000;343:625–30.
64. van Rijn MA, Munts AG, Marinus J, Voormolen JH, de Boer KS, Teepe-Twiss IM, van Dasselaar NT, Delhaas EM, van Hilten JJ. Intrathecal baclofen for dystonia of complex regional pain syndrome. Pain. 2009;143:41–7.
65. Hinson VK, Weinstein S, Bernard B, Leurgans SE, Goetz CG. Single-blind clinical trial of psychotherapy for treatment of psychogenic movement disorders. Parkinsonism Relat Disord. 2006;12:177–80.

66. Sharpe M, Walker J, Williams C, Stone J, Cavanagh J, Murray G, Butcher I, Duncan R, Smith S, Carson A. Guided self-help for functional (psychogenic) symptoms: a randomized controlled efficacy trial. Neurology. 2011;77:564–72.
67. Czarnecki K, Thompson JM, Seime R, Geda YE, Duffy JR, Ahlskog JE. Functional movement disorders: successful treatment with a physical therapy rehabilitation protocol. Parkinsonism Relat Disord. 2012;18:247–51.
68. Dallocchio C, Arbasino C, Klersy C, Marchioni E. The effects of physical activity on psychogenic movement disorders. Mov Disord. 2010;25(4):421–5.
69. Edwards MJ, Stone J, Nielsen G. Physiotherapists and patients with functional (psychogenic) motor symptoms: a survey of attitudes and interest. J Neurol Neurosurg Psychiatry. 2012;83:655–68.
70. Edwards MJ, Bhatia KP, Cordivari C. Immediate response to botulinum toxin injections in patients with fixed dystonia. Mov Disord. 2011;26:917–8.
71. Shamy MC. The treatment of psychogenic movement disorders with suggestion is ethically justified. Mov Disord. 2010;25:260–4.
72. Pollak TA, Nicholson TR, Edwards MJ, David AS. A systematic review of transcranial magnetic stimulation in the treatment of functional (conversion) neurological symptoms. J Neurol Neurosurg Psychiatry. 2014;85(2):191–7.
73. Feinstein A, Stergiopoulos V, Fine J, Lang AE. Psychiatric outcome in patients with a psychogenic movement disorder: a prospective study. Neuropsychiatry Neuropsychol Behav Neurol. 2001;14:169–76.
74. Thomas M, Vuong KD, Jankovic J. Long-term prognosis of patients with psychogenic movement disorders. Parkinsonism Relat Disord. 2006;12:382–7.
75. Ibrahim NM, Martino D, van de Warrenburg BP, Quinn NP, Bhatia KP, Brown RJ, Trimble M, Schrag A. The prognosis of fixed dystonia: a follow-up study. Parkinsonism Relat Disord. 2009;15:592–7.

Part II

Treatment of Dystonia

9 Traditional Pharmacotherapeutic Approaches to Dystonia: Mode of Action of Currently Used Drugs for the Treatment of Dystonia

Tarannum S. Khan and Hubert H. Fernandez

Dystonia is a movement disorder characterized by involuntary muscle contractions of opposing muscle groups, resulting in twisting postures and abnormal positions. As a phenomenology, it is classified under hyperkinetic movement disorders [1]. This chapter discusses the traditional pharmacological management of various types of focal and generalized dystonias. Despite the recent advances in the genetic etiologies of dystonia, the treatments predominantly still remain symptomatic at present. Current treatments can be generally subdivided into oral medications, injections or chemodenervation in the affected muscles, and surgical approaches. The chemodenervation treatment of dystonia and surgical treatments will be discussed elsewhere in the book. In this chapter, we will focus on the traditional pharmacotherapeutic options and their evidence-based support [2, 3].

As with any disease, correct diagnosis of dystonia is critical before deciding the treatment modality. In addition, differentiating and categorizing dystonias also help in deciding the mode of treatment [3]. Traditionally, dystonia is classified based on age of onset, etiology (primary versus secondary), and distribution of body regions affected [4, 5]. Patients can be categorized as having focal, segmental, or generalized symptoms. Focal dystonia involves a single body region; segmental dystonia involves contiguous body regions; and generalized dystonia is the involvement of multiple areas including the legs. As stated above, the classification makes the management decisions easier. Some modalities of treatments are more effective in certain types of dystonias. For example, pharmacological treatment in the form of medications is usually the first course of action in the early childhood-onset

T.S. Khan, MD (✉)
Department of Neurology, Movement Disorders Program, Cleveland Clinic Florida, Weston, FL, USA
e-mail: khant@ccf.org

H.H. Fernandez, MD
Movement Disorders Program, Center for Neurological Restoration, Cleveland Clinic, Cleveland, OH, USA

P. Kanovsky et al. (eds.), *Dystonia and Dystonic Syndromes*,
DOI 10.1007/978-3-7091-1516-9_9

dystonia. The first medication to try in childhood-onset dystonia is usually levodopa, although only 5 % of childhood dystonias are dopa-responsive dystonias (DRD). With increasing experience and understanding of the use of chemodenervation in dystonia, botulinum toxin injections are now the treatment of choice for most focal and segmental dystonias which are discussed in the later chapters in this book.

Finally, several medications have been used over the years for the treatment of dystonia. These include dopaminergic agents, antidopaminergic agents, anticholinergic drugs, and muscle relaxants. We will discuss these and a few others in detail in this chapter.

9.1 Dopaminergic Medications

9.1.1 DRD

The first medication to be tried in childhood-onset or adolescent-onset segmental or generalized dystonia should almost always be levodopa. This practice is due to excellent response to levodopa in DRD, although it represents a small number of 5 % of the total childhood-onset dystonias. Most patients show dramatic improvement with a small dose of levodopa. These patients typically present in early childhood or early adulthood with leg dystonia, parkinsonism, diurnal fluctuation of symptoms, and gait difficulties. Most cases of DRD are autosomal dominant, caused by mutations in GCH1 gene (GTP cyclohydrolase 1) located on chromosome 14q. This gene is responsible for the production of tetrahydrobiopterin which is essential in dopamine synthesis. The diurnal variation seen in DRD is due to depletion of low levels of tetrahydrobiopterin in affected patients which does not last all day, thus resulting in reduced dopamine synthesis and the symptoms. Rarely, it can be autosomal recessive, due to abnormalities in tyrosine hydroxylase and sepiapterin reductase which then affect the synthesis of dopamine [6, 7].

Levodopa supplementation is very effective in DRD. It is given in combination with the decarboxylase inhibitor carbidopa, which prevents peripheral conversion of levodopa into dopamine, thereby minimizing peripheral side effects (mainly nausea) and allowing more levodopa to cross the blood-brain barrier. Usually small doses are quite effective starting with 25/100 mg of carbidopa/levodopa, one-half to one tablet two to three times daily. A clear beneficial effect can often be seen in patients within a few days of the start of the therapy. While most patients usually respond to a few hundred milligrams of levodopa, some may need a high dose of levodopa up to 1,000 mg daily. The delayed levodopa-related side effects of dyskinesia and motor fluctuations seen in Parkinson's disease are usually not observed in DRD patients. The reason is that in DRD patients, the dopaminergic neurons in the substantia nigra remain intact (as opposed to neurodegenerative dopaminergic neuron loss in Parkinson's disease patients). The defect in DRD is in the dopamine synthesis process, thereby resulting in a metabolic deficiency of dopamine [8, 9]. The usual side effects of levodopa are nausea, lightheadedness, dizziness, hallucinations, and confusion. There have been some reports of DRD being responsive to tetrahydrobiopterin; however, it has not been found to be consistently effective.

9.2 Dopaminergic Treatment in Other Dystonias

There is no evidence-based data supporting the use of levodopa or dopamine agonists in other primary dystonias. But there are reports of up to 25 % improvement in various types of adult-onset primary and secondary dystonias [10].

Bromocriptine has been studied in high dosages in uncontrolled studies for both focal and generalized dystonias with good results [11–13].

9.3 Anticholinergic Drugs

These drugs block the action of acetylcholine on the central muscarinic receptors. Trihexyphenidyl is perhaps the most commonly used drug in this category. It is useful mainly for segmental and generalized dystonia. Two studies, both of which were done in the 1980s, have concluded the superiority of trihexyphenidyl over placebo in childhood-onset primary or secondary dystonias [7, 8]. It is often not helpful and less tolerated in adult-onset focal dystonia [9]. In a randomized, double-blind, placebo-controlled trial of segmental and generalized dystonia in young patients, trihexyphenidyl was found to be very effective [14]. In this study, the mean age was 19 years (range 9–32), and mean dose was 30 mg/day. The best results were seen if treatment was started within the first 5 years of symptom onset [15]. However, the tolerability and efficacy differed among patients [16]. Based on evidence from studies, trihexyphenidyl is highly recommended for childhood- and young-onset non-dopamine-responsive dystonias; however, the evidence for adult-onset dystonia remains weak. Because of its potential side effects, it is best started low with a gradual taper up. It can be started with one-half of a 2 mg tab twice a day for a week, then increasing gradually to 8–20 mg daily over the next 4–8 weeks. It can be dosed three times a day. The maximum dose can be anywhere from 60 to 100 mg daily [10]. The common side effects of trihexyphenidyl are confusion, hallucination, and drowsiness in addition to dry mouth, blurred vision, and urinary retention, and it may be very difficult to use in higher dosages in older individuals (especially older than 60 years) due to these side effects.

9.4 Antidopaminergic Drugs

The treatment of dystonia with antidopaminergic medications seems counterintuitive. Dopamine-blocking agents also known as neuroleptics are known to cause acute dystonic reaction and tardive dystonia [16]; however, the same dystonia-causing medications have been used to treat dystonia [16, 19]. Clozapine has been studied in trials with variable benefits and is not used commonly because of the potential for agranulocytosis, thereby requiring cumbersome blood monitoring [20–23]. Other medications which have shown some benefit are haloperidol, pimozide, and risperidone.

9.4.1 Tetrabenazine

Tetrabenazine works presynaptically to reduce the overall supply of monoamines including dopamine. Tetrabenazine is specifically useful in tardive dystonia [11]. There was marked improvement with up to 89 % improvement in symptoms in one study [18]. It has been shown useful for generalized and focal dystonia in a small, placebo-controlled, double-blind trial in 1982 [17]. The results have been reconfirmed in other open-label and retrospective data collections [18]. The dosage can be 12.5–25 mg daily with slow titration upward up to 150–200 mg daily as tolerated. The common side effects are drowsiness, parkinsonism, anxiety, restlessness, and insomnia.

9.4.2 Baclofen

Baclofen works as an agonist at the presynaptic gamma-aminobutyric acid receptors (GABAb). It was shown effective in two retrospective studies for the treatment of dystonia [24, 25]. Among 31 children with idiopathic dystonia, 30 % showed great improvement of symptoms. DYT1 dystonia patients have also been shown to have benefited from oral baclofen [26]. Variable symptomatic improvement has been seen in adults with focal dystonias. The effective dosage in one trial has been found to be 40–180 mg [30]. The initial dose can be 5–10 mg two to three times a day with gradual increment of 10–20 mg/day every 3–5 days up to usually 80 mg. The common side effects are drowsiness, dizziness, slurred speech, ataxia, and hypotonia.

Intrathecal baclofen (ITB) has been shown repeatedly to be effective in the treatment of dystonia [27–29]. A "trial run" of ITB can be quite beneficial before the actual final implant due to the reports of placebo effect which according to some reports can last up to 2 days [30]. It is thought to be especially effective in spastic dystonia or dystonia affecting predominantly the trunk and legs [29]. It should also be considered in patients who have failed anticholinergic treatments [31]. In one study of 86 patients, the mean dose of ITB used was 395 μg/day (micrograms/day) with initiation of the ITB at 200 μg/day. The positive symptomatic benefits were seen in 85 % of patients [44, 45]. The side effects can be similar to oral baclofen and some catheter-related issues of catheter breakage or disconnection [32]. Other muscle relaxants like tizanidine (Zanaflex), cyclobenzaprine (Flexeril), metaxalone (Skelaxin), and others can be used. However, not much data are available for their positive benefits in dystonia.

9.4.3 Benzodiazepines

The mode of action of benzodiazepines is through GABAa receptor complex stimulation. Clonazepam, lorazepam, and diazepam have been studied in different trials. Clonazepam is the most commonly used drug. Clonazepam was found to improve symptoms of blepharospasm and dystonic choreoathetosis [33]. In an analysis of open-label trials, clonazepam was seen as beneficial in 23 % of blepharospasm and 21 % of cervical dystonia patients [34].

9.4.4 Anticonvulsants

Valproic acid, carbamazepine, and phenytoin have shown variable beneficial effects in the treatment of dystonia in small trials [35, 36]. Levetiracetam showed improvement of dystonic symptoms in a couple of trials only to be rejected in two other trials looking for beneficial effects in cervical, segmental, and generalized dystonia [37–40].

9.4.5 Other Drugs

Mexiletine (lidocaine) was studied in oral and intravenous form in nine spasmodic torticollis patients [41, 42]. Riluzole was studied in a small open-label trial and was found effective in botulinum toxin-resistant patients [43]. Lithium has no clear convincing data.

9.5 Summary

In summary (Fig. 9.1), in childhood- and young-onset dystonia, with patients <28 years of age, and non-DYT1 dystonia, levodopa should be tried in small dosage of up to 25/100 mg carbidopa/levodopa three times a day. If no clear benefit is observed, as is the case of non-DRD dystonias, the next medication that can be tried in young patients is trihexyphenidyl starting at low dose and gradually increasing up to 60–70 mg daily in divided dosages as tolerated. In most other generalized dystonia patients, the "rule of ABC" comes in handy: A, Artane (trihexyphenidyl); B, baclofen; and C, clonazepam. These and other medications can be tried in various combinations as tolerated. Polypharmacy often is the rule for pharmacotherapy of dystonia. For tardive dystonia, tetrabenazine works very well and should be tried starting with a 12.5–25 mg daily dose increasing to 100 mg daily. For status dystonicus, anticholinergics, baclofen, and even pimozide can be considered. All the other drugs are investigational. For most focal dystonias, the first line of treatment these days is botulinum toxin followed with medications for refractory or difficult to treat cases (Table 9.1a–e). The side effects, mode of action, dosage, etc., are summarized in Table 9.2.

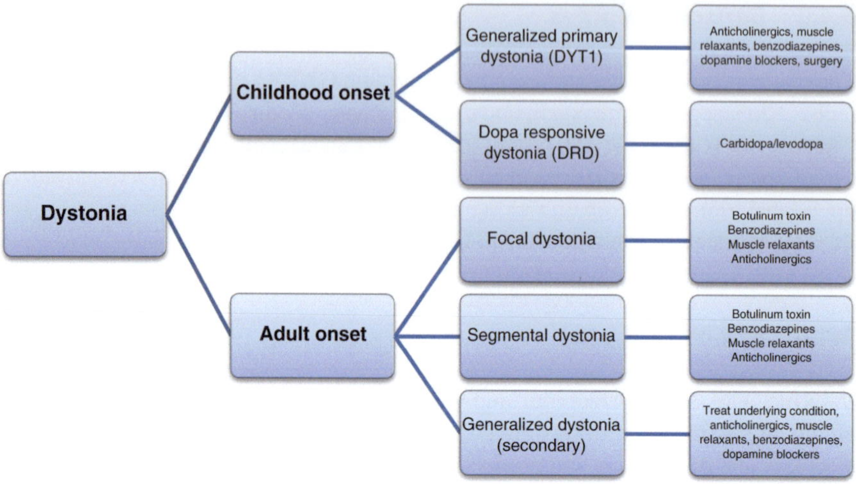

Fig. 9.1 General schema of treatment

Table 9.1 Treatment of choice in order of preference in focal dystonia

(a) Blepharospasm
1. Botulinum toxin
2. Benzodiazepines
3. Anticholinergics
4. Levodopa
5. Muscle relaxants
6. Surgical treatment
(b) Oromandibular dystonia
1. Botulinum toxin
2. Anticholinergics
3. Muscle relaxants
4. Benzodiazepines
(c) Cervical dystonia
1. Botulinum toxin
2. Anticholinergics
3. Muscle relaxants
4. Benzodiazepines
5. Deep brain stimulation
(d) Task-specific dystonia
1. Botulinum toxin
2. Anticholinergics
3. Muscle relaxants
4. Benzodiazepines
(e) Tardive dystonia
1. Tetrabenazine
2. Botulinum toxin

Table 9.2 Summary of medications

Medication	Indication	Dose	Mode of action	Side effects
Levodopa	DRD (should be tried in all early-onset dystonia patients <28 years of age)	25/100 mg carbidopa/levodopa ½–1 tab bid–tid, max dose: 1,000 mg	Replenishes dopamine	Agitation, anxiety, drowsiness, dizziness, faintness, fatigue
Trihexyphenidyl	Childhood-onset generalized and segmental dystonia, adult-onset focal and segmental dystonia	2 mg tabs, ½ tab 2–3 times a day; gradual increment of 1–2 mg/day every 5–7 days up to 20–30 mg/day, max dose: 60–100 mg	Blockage of action of acetylcholine on the central muscarinic receptors	Blurred vision, dry mouth, urinary retention, and cognitive slowing and memory loss. Poorly tolerated in patients >70 years of age
Baclofen	Generalized, segmental, focal	10 mg bid–tid with gradual increment of 10–20 mg/day every 3–5 days, max dose: 80 mg	Presynaptic GABAb agonists	Drowsiness, dizziness, slurred speech, ataxia, hypotonia
Intrathecal baclofen	Spastic painful generalized, failed anticholinergic treatment	100–200 μg/day	Presynaptic GABAb agonists	Drowsiness, dizziness, slurred speech, ataxia, hypotonia, catheter breakage, and disconnection
Clonazepam	Generalized, segmental, and focal	0.5–1 mg, bid or tid as tolerated with gradual increment, max dose: 6–8 mg as tolerated/day	Acts on benzodiazepine receptor in the brain facilitating GABAergic inhibition of neuronal firing	Agitation, anxiety, drowsiness, dizziness, confusion, slurred speech
Tetrabenazine	Tardive dyskinesia	12.5–25 mg daily with slow titration upward up to 150–200 mg daily as tolerated	Works presynaptically to reduce overall supply of monoamines including dopamine	Drowsiness, parkinsonism, anxiety, restlessness, and insomnia

References

1. Bressman SB. Dystonia update. Clin Neuropharmacol. 2000;23:239–51.
2. Jankovic J. Treatment of dystonia. Lancet Neurol. 2006;5:864–72.
3. Albanese A, Bhatia KP, et al. A systematic review on the diagnosis and treatment of primary (idiopathic) dystonia and dystonia plus syndromes: report of an EFNS/MDS-ES Task force. Eur J Neurol. 2006;13:433–44.
4. Fahn S, Bressman S, Marsden CD. Classification of dystonia. Adv Neurol. 1998;78:1–10.
5. Fahn S, Marsden CD, et al. Classification and investigation of dystonia. Movement disorders 2. London: Butterworths; 1987. p. 332–58.
6. Steinberger D, Korinthenber R, et al. Dopa responsive dystonia: mutation analysis of GCH1 and analysis of therapeutic doses of L-dopa. Neurology. 2000;55:1735–7.
7. Schneider SA, Mohire MD, et al. Familial dopa responsive cervical dystonia. Neurology. 2006;66:599–601.
8. Nygaard TG, Marsden CD, Fahn S. Dopa responsive dystonia: long term treatment response and prognosis. Neurology. 1991;41:174–81.
9. Nutt JG, Nygaard TG. Response to levodopa treatment in dopa-responsive dystonia. Arch Neurol. 2001;58:905–10.
10. Jankovic J. Dystonia: medical therapy and botulinum toxin. Adv Neurol. 2004;94:275–86.
11. Lees A, Shaw KM, Stern GM. Letter: bromocriptine and spasmodic torticollis. Br Med J. 1976;1:1343.
12. Stahl SM, Berger PA. Bromocriptine, physostigmine, and neurotransmitter mechanisms in the dystonias. Neurology. 1982;32:889–92.
13. Obeso JA, Luquin MR. Bromocriptine and lisuride in dystonias. Neurology. 1984;34:135–6.
14. Burke RE, Fahn S, Marsden CD. Torsion dystonia: a double blind, prospective trial of high dosage trihexyphenidyl. Neurology. 1986;36:160–4.
15. Fahn S. High dosage anticholinergic therapy in dystonia. Neurology. 1983;33:1255–61.
16. Marsden CD, Marion MH, Quinn N. The treatment of severe dystonia in children and adults. J Neurol Neurosurg Psychiatry. 1984;47:1163–73.
17. Jankovic J. Treatment of hyperkinetic movement disorders with tetrabenazine: a double blind crossover study. Ann Neurol. 1982;11:41–7.
18. Jancovic J, Beach J. Long term effects of tetrabenazine in hyperkinetic movement disorders. Neurology. 1997;48:358–62.
19. Marsden CD, Quinn NP. The dystonias. Neurological disorders affecting 20,000 people in Britain. Br Med J. 1990;300:139–44.
20. Karp BI, Goldstein SR, Chen R, et al. An open trial of clozapine for dystonia. Mov Disord. 1999;14:652–7.
21. Lang AE. Dopamine agonists and antagonists in the treatment of idiopathic dystonia. Adv Neurol. 1988;50:561–70.
22. Balash Y, Giladi N. Efficacy of pharmacological treatment of dystonia: evidence based review including meta-analysis of the effect of botulinum toxin and other cure options. Eur J Neurol. 2004;11:361–70.
23. Burbaud P, Giehl D, et al. A pilot trial of clozapine in the treatment of cervical dystonia. J Neurol. 1998;245:329–31.
24. Greene P. Baclofen in the treatment of dystonia. Clin Neuropharmacol. 1992;15:276–88.
25. Greene PE, Fahn S. Baclofen in the treatment of idiopathic dystonia in children. Mov Disord. 1992;7:48–52.
26. Anca MH, Falic Zaccai T, et al. Natural history of Oppenheim's dystonia (DYT1) in Israel. J Child Neurol. 2003;18:325–30.
27. Narayan RK, Loubser PG, Jankovic J, et al. Intrathecal baclofen for intractable axial dystonia. Neurology. 1991;41:1141–2.
28. Penn RD, Gianino, York MM. Intrathecal baclofen for motor disorders. Mov Disord. 1995;10:657–75.

29. Hou JG, Ondo W, Jankovic J. Intrathecal baclofen for dystonia. Mov Disord. 2001;16:1201–2.
30. Green P. Baclofen in the treatment of dystonia. Clin Neuropharmacol. 1992;15:276–88.
31. Ford B, Greene PE, Louis ED, et al. Intrathecal baclofen in the treatment of patients with dystonia. Arch Neurol. 1996;53:1241–6.
32. Albright AL, Barry MJ, Shafron DH, et al. Intrathecal baclofen for generalized dystonia. Dev Med Child Neurol. 2001;43:652–7.
33. Jankovic J, Ford J. Blepharospasm and orofacial cervical dystonia: clinical and pharmacological findings in 100 patients. Ann Neurol. 1983;13:402–11.
34. Greene P, Shale H, Fahn S. Analysis of open label trials in torsion dystonia using high dosage of anticholinergics and other drugs. Mov Disord. 1988;3:46–60.
35. Snoeck JW, van Weerden TW, et al. Meige syndrome: double blind crossover study of sodium valproate. J Neurol Neurosurg Psychiatry. 1987;50:1522–5.
36. Lotze T, Jancovic J. Paroxysmal kinesigenic dyskinesias. Semin Pediatr Neurol. 2003;10:68–79.
37. Zesiewicz TA, Louis ED, Sullivan KL, et al. Substantial improvement in a Meige's syndrome patient with levetiracetam treatment. Mov Disord. 2004;19:1518–21.
38. Sullivan KL, Hauser RA, Louis ED, et al. Levetiracetam for the treatment of generalized dystonia. Parkinsonism Relat Disord. 2005;11:460–71.
39. Tarsy D, Ryan RK, Ro SI. An open label trial of levetiracetam for treatment of cervical dystonia. Mov Disord. 2006;21:734–5.
40. Hering S, Wenning GK, Seppi K, et al. An open label trial of levetiracetam for segmental and generalized dystonia. Mov Disord. 2007;22:1649–51.
41. Ohara S, Hayashi R, et al. Mexiletine in the treatment of spasmodic torticollis. Mov Disord. 1998;13:934–40.
42. Lucetti C, Nuti A, et al. Mexiletine in the treatment of torticollis and generalized dystonia. Clin Neuropharmacol. 2000;23:186–9.
43. Muller J, Wenning GK, et al. Riluzole therapy in cervical dystonia. Mov Disord. 2002;17:198–200.
44. Alright AL, Barry MJ, et al. Intrathecal baclofen for generalized dystonia. Dev Med Child Neurol. 2001;43:652–7.
45. Walker RH, Danishi FO, et al. Intrathecal baclofen for dystonia: benefits and complications during six years of experience. Mov Disord. 2000;15:1242–7.

Botulinum Toxin Therapy of Dystonia

10

Dirk Dressler and Petr Kanovsky

10.1 History

Botulinum toxin (BT) is infamous as the compound with the highest toxic potency of any natural or man-made substance causing the clinical syndrome of botulism in man and animals. At the end of the 1970s, this perception began to change, when BT was first used by Alan B. Scott to treat strabismus in children [1, 2]. It soon became clear that this had established a completely novel therapeutic principle which could be used in various muscle hyperactivity syndromes. Subsequently BT was used in blepharospasm, hemifacial spasm and cervical dystonia, thus reaching neurology [3, 4]. Here its use exploded and soon numerous other medical specialties became involved. With its use in crocodile tears, pioneered by Manuel Meyer in Zurich, exocrine glands became a second principal target tissue for BT. Recently, BT's use for treatment of pain syndromes was established [5]. For most of its indications, BT therapy is the therapy of choice [6]. For some, it has revolutionised therapy altogether. This, together with its exploding use in cosmetics, has generated an industry with annual sales in excess of three billion US dollars. BT's use in dystonia, however, is still one of the most important indications for BT, both with respect to the amount of BT used and with respect to the therapeutic impact generated.

D. Dressler, MD, PhD (✉)
Movement Disorders Section, Department of Neurology, Hannover Medical School, Carl-Neuberg-Str. 1, 30625 Hannover, Germany
e-mail: dressler.dirk@mh-hannover.de

P. Kanovsky
Department of Neurology, Palacky University Medical School, Olomouc, Czech Republic
e-mail: petr.kanovsky@fnol.cz

P. Kanovsky et al. (eds.), *Dystonia and Dystonic Syndromes*,
DOI 10.1007/978-3-7091-1516-9_10

10.2 Pharmacology of Botulinum Toxin Drugs

10.2.1 Structure

As shown in Fig. 10.1, BT drugs consist of the BT component and of excipients. The BT component is formed by botulinum neurotoxin (BNT) and by non-toxic proteins also known as complexing proteins. BNT consists of a heavy amino acid chain with a molecular weight of 100 kD and a light amino acid chain with a molecular weight of 50 kD. BNT and complexing proteins form superstructures of different sizes [7]. In Xeomin®, the complexing proteins are removed during the manufacturing process [8].

10.2.2 Mode of Action

When a BT drug is injected into a target tissue, it is bound to glycoprotein structures located on the cholinergic nerve terminal. Subsequently, BNT's light chain is internalised by using synaptic vesicle proteins and syntagmins of the neuron's acetylcholine vesicle-recycling mechanism [9–12]. Intracellularly, BT cleaves different proteins of the acetylcholine transport protein cascade (**s**oluble **N**-ethylmaleimide-sensitive fusion *a*ttachment protein *re*ceptor or SNARE proteins) transporting the acetylcholine vesicle towards the synaptic cleft [13]. Different BT types target different SNARE proteins. BT interrupts the synaptic transmission only temporarily. Structural neuronal changes or functional neuronal impairment other than the synaptic blockade itself cannot be detected. Recently, we therefore suggested classifying BT not as a neurotoxin but as a temporary neuromodulator [14]. BT can block cholinergic neuromuscular transmission but also the cholinergic autonomic

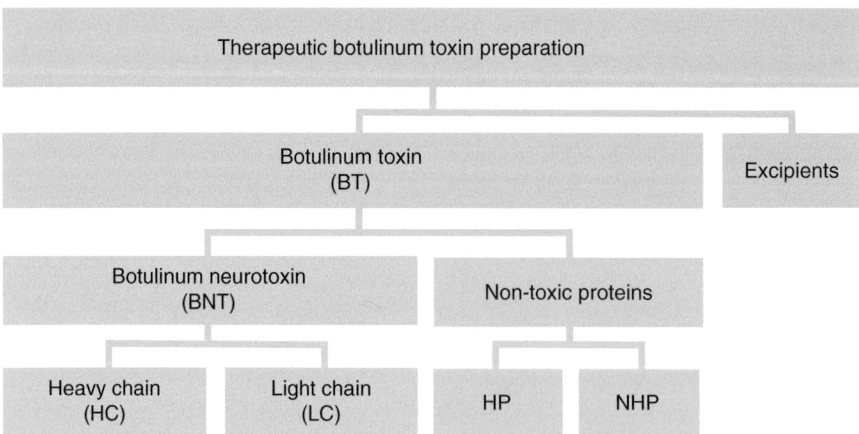

Fig. 10.1 Contents of therapeutic botulinum toxin preparations. *HP* haemagglutinating protein, *NHP* non-haemagglutinating protein

innervation of the sweat glands, the tear glands, the salivary glands and the smooth muscles depending on the target tissue selected.

Apart from a direct action upon the striated muscle, BT can act upon the muscle spindle organ reducing its centripetal information traffic [15–17]. Whether this BT-induced blockade of the muscle spindle's centripetal information is relevant to BT's therapeutic action remains unclear [18, 19].

Although BT can produce numerous indirect central nervous system effects, direct central nervous system effects beyond the alpha motoneuron are a controversial issue [20, 21].

In addition to the blockade of the acetylcholine secretion, animal experiments indicate BT-induced blockade of transmitters involved in pain perception, pain transmission and pain processing [22–28]. This may be the basis of BT's action on pain syndromes.

10.2.3 Time Course of Action

After intramuscular injection, initial BT effects can be detected within 2–3 days depending on the detection method used. BT reaches its maximal effect after about 2 weeks, maintains it and then gradually starts to decline after 2–2.5 months. BT injections into glandular tissue can exert prolonged effects of up to 6 or 9 months. BT's action follows a dose-effect correlation [29]. An additional dose-duration correlation can also be assumed. Both correlations are valid only within certain limits. Although there may be interindividual variability of the duration of action, there is remarkable intraindividual reproducibility of the duration of action throughout years of continued treatment [30, 31].

10.2.4 Botulinum Toxin Drugs

BT drugs are either based upon BT type A, such as Botox® (onabotulinumtoxinA, Allergan Inc, Irvine, CA, USA), Dysport® (abobotulinumtoxinA, Ipsen Ltd, Slough, Berks, UK) and Xeomin® (incobotulinumtoxinA, Merz Pharmaceuticals, Frankfurt/M, Germany), or upon BT type B, such as NeuroBloc®/MYOBLOC® (rimabotulinumtoxinB, Solstice Neurosciences Inc, Malvern, PA, USA). Additional BT drugs are distributed in fringe markets (Fig. 10.2). As shown in Table 10.1, the properties of BT type A drugs are similar. Those of the BT type B drug are considerably different. Differences include antigenicity, adverse effect profiles, adverse effect frequency, injection site pain, storage conditions, mode of preparation, pharmacological stability and potency labelling. Although all BT drugs are based upon BT, they are not generics. They can be compared in principle and they can be interchanged in an ongoing therapy, but comparisons and interchanges have to be performed with caution. Current understanding suggests that Botox® and Xeomin® have identical therapeutic effects and adverse effect profiles [32, 33]. Their potency labelling is identical, making exchanges between both products particularly easy [34].

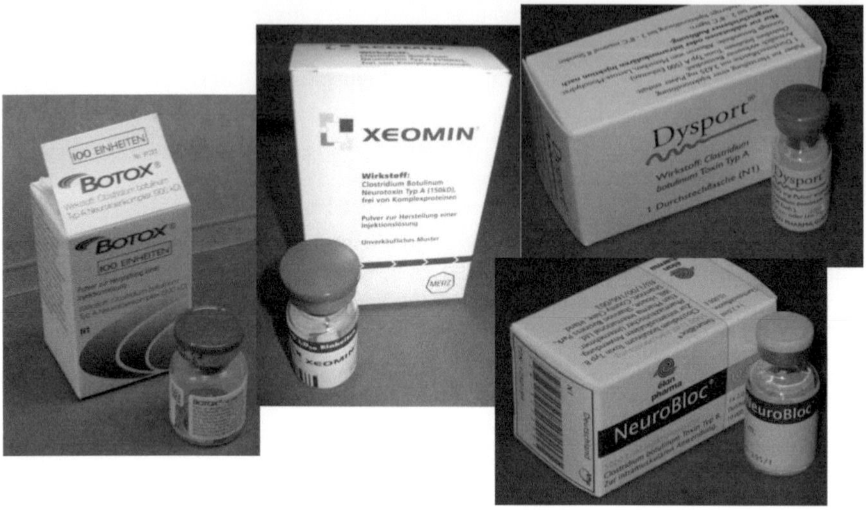

Fig. 10.2 Some of the currently available botulinum toxin drugs

With an improved specific biological activity, Xeomin® should have a reduced antigenicity as compared to conventional BT drugs [8].

Dysport® also has an identical therapeutic profile as Botox®. Its potency labelling is still a matter of debate. Currently, a conversion ratio between the mouse units measured by Allergan for Botox® and the mouse units measured by Ipsen for Dysport® of 1:2.5 must be assumed. NeuroBloc®/MYOBLOC®'s potency labelling can be compared to that of Botox® by using a conversion ratio 1:40. Its disadvantageous adverse effect profile and its problematic antigenicity will be discussed below. It can only be recommended for those dystonia patients with complete secondary therapy failure due to formation of antibodies against BT type A.

10.2.5 Antigenicity

Since BT consists of foreign proteins, antibodies can be formed against BNT and the non-toxic proteins. Antibodies formed against BNT (BNT-AB) block BT's biological activity and, thus, produce antibody-induced therapy failure (ABTF). BNT-AB are therefore called blocking or neutralising antibodies. BNT-AB and BT are in a functional balance [35]. The therapeutic relevance of BNT-AB titres has to be considered. Antibodies formed against the non-toxic proteins do not interfere with BT's biological activity and are called non-neutralising antibodies.

Risk factors for ABTF include the amount of BT applied at each injection series, the interinjection interval [36] and the immunological quality of the BT drug as described by their specific biological activity. The specific biological activity varies between different therapeutic preparations as shown in Table 10.1 [37–40]. A

Table 10.1 Properties of different botulinum toxin drugs

	Botox®	Dysport®	Xeomin®	NeuroBloc® MYOBLOC®
Manufacturer	Allergan Inc	Ipsen Ltd	Merz Pharmaceuticals	Solstice Neurosciences Inc
Pharmaceutical preparation	Powder	Powder	Powder	Ready-to-use solution 5,000MU-E/ml
Storage conditions	Below 8 °C	Below 8 °C	Below 25 °C	Below 8 °C
Shelf life	36 months	24 months	36 months	24 months
Botulinum toxin type	A	A	A	B
SNARE target	SNAP25	SNAP25	SNAP25	VAMP
pH value of the reconstituted preparation	7.4	7.4	7.4	5.6
Stabilisation	Vacuum drying	Freeze-drying (lyophilisate)	Vacuum drying	pH reduction
Excipients	Human serum albumin 500 ug/vial Sodium chloride 900 ug/vial Buffer system	Human serum albumin 125 ug/vial Lactose 2,500 ug/vial Buffer system	Human serum albumin 1 mg/vial Sucrose 4.7 mg/vial Buffer system	Human serum albumin 0.5 mg/ml Disodium succinate 0.01 M Sodium chloride 0.1 M H_2O hydrochloric acid
Biological activity	100 MU-A/vial	500 MU-I/vial	100 MU-M/vial	1.0/2.5/10.0 kMU-E/vial
Biological activity in relation to Botox®	1	1/3	1	1/40
Specific biological activity	60 MU-EV/ngBNT	100 MU-EV/ngBNT	167 MU-EV/ngBNT	5 MU-EV/ngBNT

BNT botulinum neurotoxin, *MU-A* mouse unit in the Allergan mouse lethality assay, *MU-E* mouse unit in the Solstice mouse lethality assay, *MU-I* mouse unit in the Ipsen mouse lethality assay, *MU-M* mouse unit in the Merz mouse lethality assay, *MU-EV* approximate equivalence mouse unit, 1MU-EV = 1MU-A = 1MU-M = 3MU-I = 40MU-E

probable risk factor for ABTF is the reactivity of the immune system of the individual patient [41, 42]. Potential risk factors include the target tissue; the type of injection, i.e. intradermal, intramuscular or intraglandular injections; and female sex. Cumulative dose, treatment time and patient age have been excluded as risk factors. Recent analysis revealed that ABTF usually develops within the first 2–3 years of BT therapy [43]. After a treatment time of more than 5 years, ABTF becomes rare.

10.2.6 Safety Aspects and Adverse Effects

Based upon a broad therapeutic window and strictly local effects avoiding contact with excretion organs, BT excels with a remarkably advantageous adverse effect profile. Adverse effects can be classified as obligate, local or systemic. BT adverse effects occur in a typical time window after BT application. They usually start after 1 week and last for 1–2 weeks. Severity and duration of adverse effects depend on the BT dose applied, the target tissue and the adjacent tissues. Systemic spread of BT becomes clinically relevant only when BT doses applied are very high. BT transport through the blood-brain barrier is excluded by BT's molecular size. The use of BT during pregnancy is contraindicated as a precautionary measure until further experience is gained. Extremely rarely, BT applications may trigger acute autoimmune reactions with brachial plexopathies or with dermatomyositis [44]. Caution is required when using BT in patients with pre-existing pareses, as in amyotrophic lateral sclerosis, myopathies and motor polyneuropathies, or in patients with impaired neuromuscular transmission, such as myasthenia gravis and Lambert-Eaton syndrome [45, 46]. Warnings not to use BT in patients receiving aminoglycoside antibiotics seem theoretical. With large numbers of patients being treated over prolonged periods of time, long-term experience is ample but does not indicate additional adverse effects [47].

Based upon a conversion factor of 1:1, the adverse effect profiles of Xeomin® and of Botox® are identical [32, 33]. The adverse effect profile of BT type B drugs is substantially different from the adverse effect profile of therapeutic BT type A drugs. Whereas even low and intermediate BT-B doses frequently produce systemic autonomic adverse effects, frequency of motor adverse effects is similar after BT type B and BT type A [48]. Whereas BT type A has a relatively strong effect on the motor system and a relatively weak effect on the autonomic nervous system, this correlation is reversed in BT type B [49]. Because of its systemic autonomic adverse effects, BT type B should be used with caution in patients with pre-existing autonomic dysfunction or in connection with anticholinergics.

10.2.7 Therapeutic Profile

With the features described above, BT's therapeutic profile can be summarised as shown in Table 10.2. BT can be used in muscles, exocrine glands and structures associated with pain. BT's therapeutic effect follows a time course, which is remarkably reproducible in the individual patient, but shows some interindividual variability. When BT is injected into muscle tissue, it produces a peripheral paresis. This paresis manifests clinically after a few days, reaches its maximum after 1–2 weeks, is usually stable for 6–12 weeks and then gradually resolves over several weeks. The extent of the therapeutic effect is well controllable by the amount of BT applied. Adverse effects are benign and fully reversible. With BT type B, systemic anticholinergic adverse effects and ABTF are frequent.

Table 10.2 Therapeutic profile of botulinum toxin drugs

Target tissues	Striate and smooth muscles ('neuromuscular junction') Exocrine glands Structures involved in generation, perception or transmission of pain
Therapeutic effect	Localised Predictable time course Fully reversible Extent well controllable Duration controllable within certain limits
Adverse effects	Fully reversible Obligate adverse effects manageable Local adverse effects few Systemic adverse effects BT-A: extremely rare BT-B: frequent anticholinergic adverse effects

Table 10.3 The multilayer concept of comprehensive dystonia therapy

Layer	Description	Modalities
1	Antidystonic treatment	Botulinum toxin therapy Oral drugs Intrathecal baclofen Surgical interventions
2	Adjuvant drugs	Analgesics Anxiolytics
3	Adjuvant treatment	Physiotherapy
4	Adjuvant measures	Social support Providing information Patient support groups

10.3 Basic Principles of Botulinum Toxin Therapy

10.3.1 The Multilayer Concept of Dystonia Treatment

Dystonia is a chronic condition for which no causal therapy exists. In most cases, dystonia treatment therefore has to be multimodal and long term. Co-ordinating the efforts of interdisciplinary treatment teams over prolonged periods of time may be supported by special working groups such as IAB (Interdisziplinärer Arbeitskreis Bewegungsstörungen) [50].

Treatment of dystonia can best be described as a multilayer concept (Table 10.3). The basal layer includes antidystonic therapies. Here, BT therapy is pivotal. It is entirely symptomatic. Nevertheless, its use may change the long-term perspective of patients affected. In children, symptomatic suppression of dystonic muscle activity may enable certain motor developments otherwise prevented. In adults, it can avoid the development of complications otherwise potentially dominating the clinical picture. The use of BT therapy early in dystonia's course therefore seems advisable. BT's therapeutic effect may involve central mechanisms. Its predominant effects, however, are peripheral. Usually, BT therapy should be complemented by physiotherapy where stretching exercises that can increase and maintain joint mobility, training of antagonistic muscles that can improve functional capabilities and retraining the impaired patient's self-perception in space are the main tasks [51]. BT therapy can easily be combined with other antidystonic treatment options, including oral antidystonic drugs, intrathecal baclofen and peripheral or central surgery including deep brain stimulation.

The second layer includes adjuvant drugs such as analgesics and anxiolytics. They can become necessary when the effects of the antidystonic drugs are unsatisfactory. Relaxation exercises can break the vicious circle of motor induction in dystonia.

The third layer consists of adjuvant measures including adequate information for the patient and his/her family about the BT therapy as well as about the treated condition. Sharing a solid information base is a prerequisite for a stable patient-physician relationship, which seems particularly important since dystonia is a chronic condition. Patient support groups have a role to play here. Social support gives advice to the patient about available social benefits.

10.3.2 Dosing

BT therapy is a local therapy. Dosing, therefore, depends on two elements, i.e. the number of dystonic muscles requiring treatment (target muscles) and the degree of their dystonic involvement. Given the enormous variability of dystonia with respect to localisation and intensity, communication of total doses per patient is only relevant with respect to systemic adverse effects. Recently, it has been demonstrated that Xeomin® in total doses of up to 840 MU is free of motor or autonomic systemic adverse effects [33]. Indirect evidence suggests the similar safety margins for Botox® [33]. Frequently this dose range is not exploited. Within the individual target muscles, broad consensus emerged amongst physicians over the last years about appropriate dose ranges. Within these dose ranges, the individual dose has to be selected according to the degree of dystonic involvement. Additionally, some general dose modifiers (Table 10.4) should be used to adapt BT doses to the individual strength of the patient's target muscles. When neuromuscular transmission is impaired or when underlying paresis is present, BT dosages may need to be reduced.

10.3.3 Planning of Botulinum Toxin Therapy

Planning of BT therapy is based upon identification of target muscles and their degree of dystonic involvement. This information may be obtained clinically by

Table 10.4 Factors predicting the individual strength of the patient's target muscles. These factors, when present, modify the botulinum toxin dose within the recommended dose ranges

Sex	Male	Dose increase
	Female	Dose reduction
Age	Young age	Dose increase
	Old age	Dose reduction
Body weight	High	Dose increase
	Low	Dose reduction
Proportional muscle mass	High	Dose increase
	Low	Dose reduction

analysing the dystonic movements or postures. Based upon physiological activation patterns, dystonic muscles can be deduced. However, with unphysiological co-activation of antagonistic muscles as a key feature of dystonia, this clinical approach has limitations. An additional problem is compensatory muscle activity and preventive postures may be difficult to identify. The clinical approach does not allow determination of the degree of dystonic involvement. A superior way of planning of BT therapy is based upon electromyography [52]. Needle electromyography allows precise target muscle identification. It also allows comparison between maximal voluntary muscle activation and dystonic muscle activity for calculation of the dystonic involvement. Still, identification of compensatory muscle activation may be a problem. However, recordings of prolonged periods of electromyographic activity often allow distinction between dystonic and compensatory muscle activity. Target muscle selection by estimation of muscle hypertrophy using tomographic imaging techniques is inadequate, since muscle hypertrophy is not a necessary feature of dystonia and precise measurement of hypertrophy would require muscle relaxation which dystonia patients usually cannot maintain. At the end of the planning process stands the injection scheme describing the target muscles with their respective BT doses. This injection scheme is a prediction of the patient's therapeutic response. Subsequent injection series will usually improve the injection scheme. Recommendations not to exceed certain initial total doses and to very gradually build them up are counterproductive and risk the patient's compliance. Once the optimal injection scheme is established, it should be adhered to unless there is evidence of major changes of the treated symptomatology.

10.3.4 Intramuscular Placement of Botulinum Toxin Drugs

Placement of the BT drug into the target muscle can be done clinically by palpation (ideally under voluntary activation of the target muscle), by using surface landmarks or by back-tracing the muscle tendons. To identify individual finger muscles, repetitive active or passive finger movements may allow improved palpation of the target muscles. In target muscles, in which this is difficult, such as the iliopsoas or the piriformis muscles, or when selective injection of individual finger muscles is required, electromyographic guidance using a special injection needle may become necessary.

Simultaneous electric stimulation may be of additional help. General use of electromyography to place BT drugs is not necessary. Whether electromyography-guided BT placement would allow a dose reduction needs to be demonstrated. Tomographic imaging has also been suggested to improve drug placement. Since magnetic imaging may be disturbed by the presence of the injection needle, x-ray tomography has been applied. However, application of substantial radiation doses is problematic for chronic treatment. Ultrasound techniques seem to be more helpful, especially in children who may be uncooperative and particularly pain sensitive [53].

The effects of dilution and number of injection sites within a given target muscle have not been well studied so far. It is believed that increased dilution increases diffusion. The injection volumes at each injection site should be manageable. In our experience, injection volumes of 0.5–1.0 ml and 0.1–0.3 ml in facial muscles seem reasonable.

10.3.5 Monitoring of the Therapeutic Effect

Usually the therapeutic effect of BT therapy is evaluated punctually at 2–4 weeks after application. Apart from subjective scores reflecting overall outcome or dystonic pain, various objective dystonia rating scales, such as the Tsui Scale [54], the Toronto Western Spasmodic Torticollis Rating Scale (TWSTRS) [55] or the Burke-Fahn-Marsden Scale [56], have been introduced. The disadvantage of these scales is that they do not describe the dynamics of BT therapy, i.e. the time it takes for the therapeutic effect to build up and the time when it is waning. In order to construct the integral of the therapeutic effect, a new treatment calendar was recently introduced [57], (Fig. 10.3).

10.4 Specific Indications for Botulinum Toxin Therapy in Dystonia

10.4.1 Cranial Dystonias

Cranial dystonias can affect periocular muscles, mandibular muscles and perioral muscles. Rarely, scalp muscles or periauricular muscles are affected [58]. BT therapy can be used in all of these conditions successfully, either when they occur in isolation or when they occur in various combinations. Cranial target muscles with their recommended doses are shown in Tables 10.5 and 10.6. In periocular dystonia producing the classical clinical picture of blepharospasm, BT is injected into the orbicularis oculi muscle responsible for eyelid closure [3, 59]. Additional target muscles include the procerus and the corrugator supercilii muscles which form the horizontal and vertical nasal root folds and narrow the eyebrows. They may produce tension but do not have much influence on eyelid function. The nasalis muscle forms the longitudinal nasal dorsum fold and can be injected when the patient complains of irritation especially when wearing glasses. The frontalis muscle is an accessory eyelid-opening muscle and, therefore, should not be injected in blepharospasm contrary to occasional belief.

Patient Diary

Dear Patient,

Dystonia describes a group of very different and usually chronic conditions that include spasmodic torticollis (cervical dystonia), blepharospasm, and writer's cramp. In the past, the treatment of dystonia was usually frustrating for both the patient and the physician.

This situation has changed with the introduction of the botulinum toxin therapy. This therapy is now the most successful treatment for many types of dystonia.

To achieve optimal treatment results, the botulinum toxin therapy has to be individualised to match your particular symptoms. For this reason, it is necessary to exactly describe your dystonia, to meticulously monitor your response to the botulinum toxin therapy and to identify additional factors influencing your condition. Your personal experience with your dystonia is the most valuable source of information for all of this. This patient diary is designed to record your experience.

The patient diary can of course also be used to record other therapies.

Dirk Dressler, MD, PhD

Department of Neurology, Rostock University
Gehlsheimer Str. 20, D-18147 Rostock, Germany, dirk.dressler@med.uni-rostock.de

How to use this Patient Diary

1. Please fill in the Patient Diary every night before you go to bed.
2. Please circle the percentage which best describes the severity of all of your dystonia symptoms for that day. 0% describes the situation when you feel no dystonia symptoms at all, 100% when your dystonia is at its worst without any treatment.

   Please see examples A and B for assistance.
3. At first you will find it difficult to describe your dystonia symptoms to an accuracy of 5%. The fine gradation will however help you to accurately record changes in your dystonia symptoms from one day to the next.
4. Use the "Remarks" column to record all special observations of the day.

   Please see example C for assistance.
5. Please have your Patient Diary with you when you see your physician.
6. You can obtain additional copies of the patient diary from your physician.

Examples

A A woman is diagnosed as having cervical dystonia and has not yet been treated. Her dystonia symptoms are 100%. After initiation of a botulinum toxin therapy her dystonic symptoms are reduced to 20% on most days. After about three months the effect of the botulinum toxin therapy gradually fades and her dystonic symptoms are again 60%. About a week after repeating the botulinum toxin injections her dystonic symptoms are again reduced to 20% on most days.

B A man has dystonic symptoms of 50% from 7 am to 3 pm. By the time he goes to bed they have increased to 70%. Considering the entire day, his dystonic symptoms are probably best described by 60%.

C A woman plays tennis for 2 hours twice a week. She records that fact in her Patient Diary and realises that her dystonic symptoms on those days are twice as severe as on days without tennis. As a result of this, she reduces her playing time to 45 minutes and her dystonic symptoms are no longer worse on those days.

Fig. 10.3 Patient diary to monitor the effects of botulinum toxin therapy over time

DAY	DYSTONIA-SYMPTOMS (in percent of maximal untreated dystonia symptoms)																					REMARKS
1	0	5	10	15	20	25	30	35	40	45	50	55	60	65	70	75	80	85	90	95	100	
2	0	5	10	15	20	25	30	35	40	45	50	55	60	65	70	75	80	85	90	95	100	
3	0	5	10	15	20	25	30	35	40	45	50	55	60	65	70	75	80	85	90	95	100	
4	0	5	10	15	20	25	30	35	40	45	50	55	60	65	70	75	80	85	90	95	100	
5	0	5	10	15	20	25	30	35	40	45	50	55	60	65	70	75	80	85	90	95	100	
6	0	5	10	15	20	25	30	35	40	45	50	55	60	65	70	75	80	85	90	95	100	
7	0	5	10	15	20	25	30	35	40	45	50	55	60	65	70	75	80	85	90	95	100	
8	0	5	10	15	20	25	30	35	40	45	50	55	60	65	70	75	80	85	90	95	100	
9	0	5	10	15	20	25	30	35	40	45	50	55	60	65	70	75	80	85	90	95	100	
10	0	5	10	15	20	25	30	35	40	45	50	55	60	65	70	75	80	85	90	95	100	
11	0	5	10	15	20	25	30	35	40	45	50	55	60	65	70	75	80	85	90	95	100	
12	0	5	10	15	20	25	30	35	40	45	50	55	60	65	70	75	80	85	90	95	100	
13	0	5	10	15	20	25	30	35	40	45	50	55	60	65	70	75	80	85	90	95	100	
14	0	5	10	15	20	25	30	35	40	45	50	55	60	65	70	75	80	85	90	95	100	
15	0	5	10	15	20	25	30	35	40	45	50	55	60	65	70	75	80	85	90	95	100	
16	0	5	10	15	20	25	30	35	40	45	50	55	60	65	70	75	80	85	90	95	100	
17	0	5	10	15	20	25	30	35	40	45	50	55	60	65	70	75	80	85	90	95	100	
18	0	5	10	15	20	25	30	35	40	45	50	55	60	65	70	75	80	85	90	95	100	
19	0	5	10	15	20	25	30	35	40	45	50	55	60	65	70	75	80	85	90	95	100	
20	0	5	10	15	20	25	30	35	40	45	50	55	60	65	70	75	80	85	90	95	100	
21	0	5	10	15	20	25	30	35	40	45	50	55	60	65	70	75	80	85	90	95	100	
22	0	5	10	15	20	25	30	35	40	45	50	55	60	65	70	75	80	85	90	95	100	
23	0	5	10	15	20	25	30	35	40	45	50	55	60	65	70	75	80	85	90	95	100	
24	0	5	10	15	20	25	30	35	40	45	50	55	60	65	70	75	80	85	90	95	100	
25	0	5	10	15	20	25	30	35	40	45	50	55	60	65	70	75	80	85	90	95	100	
26	0	5	10	15	20	25	30	35	40	45	50	55	60	65	70	75	80	85	90	95	100	
27	0	5	10	15	20	25	30	35	40	45	50	55	60	65	70	75	80	85	90	95	100	
28	0	5	10	15	20	25	30	35	40	45	50	55	60	65	70	75	80	85	90	95	100	
29	0	5	10	15	20	25	30	35	40	45	50	55	60	65	70	75	80	85	90	95	100	
30	0	5	10	15	20	25	30	35	40	45	50	55	60	65	70	75	80	85	90	95	100	
31	0	5	10	15	20	25	30	35	40	45	50	55	60	65	70	75	80	85	90	95	100	

Fig. 10.3 (Continued)

Table 10.5 Recommended botulinum toxin doses in cranial muscles

Muscle	Function	Recommended dose [MU-A Botox®]
Orbicularis oculi	Eyelid closure	12–64
Procerus	Formation of horizontal nasal root fold	4–12
Corrugator supercilii	Eyebrow adduction	8–16
Nasalis (transversal part)	Formation of longitudinal nasal dorsum fold	8–12
Risorius	Corner of mouth abduction	4–8
Depressor anguli oris	Corner of mouth depression	4–12
Depressor labii inferioris	Stabilisation of lower lip	4–8
Mentalis	Formation of chin dimples	8–12

MU-A mouse unit of the mouse bioassay of Allergan Inc
Dilution: 100 MU-A in 2.5 ml 0.9 % Sodium chloride/H_2O

Table 10.6 Recommended botulinum toxin doses in mandibular muscles

Muscle	Function	Recommended dose [MU-A Botox®]
Masseter	Jaw closing	20–60
Temporalis	Jaw closing	40–80
Pterygoidei (per side)	Jaw protrusion Jaw lateralisation Jaw opening Jaw closing	20–40
Supra- and infrahyoid muscles (per side)	Jaw opening	20–60

MU-A mouse unit of the mouse bioassay of Allergan Inc
Dilution: 100 MU-A in 2.5 ml 0.9 % Sodium chloride/H_2O)

Doses, dilutions and injection points for the treatment of blepharospasm vary considerably, whereas results and adverse effects are surprisingly similar. Adverse effects include ptosis, double vision, lagophthalmus and hematoma. They are rare and transient. Ptosis can almost certainly be avoided by sparing the medial part of the upper eyelid. Some patients with blepharospasm, especially those with progressive supranuclear palsy, have a varying degree of additional apraxia of eyelid opening [60], i.e. a supranuclear impairment of the eyelid-opening mechanism. In those patients, additional BT injections close to the rim of the eyelid are helpful [61]. If this strategy does not produce satisfactory results, a bilateral suspension operation connecting the upper eyelid to the frontalis muscle by a subcutaneous non-resorbable thread is helpful [62]. In mild cases, a wire spring attached to a spectacle frame can produce similar effects.

In perioral dystonia, muscles above the oral orifice should generally be injected with special caution in order to avoid drooping of the mouth. BT injections into the upper lip may produce paraesthesias for unknown reasons. The risorius muscle can be injected safely 2 cm lateral to the corner of the mouth, whereas injections into the depressor labii inferioris bear the risk of instability of the lower lip.

In mandibular dystonia, jaw-opening and jaw-closing forms can be distinguished [63]. Combined activation of opening and closing muscles, however, is not infrequent. Additional jaw movements include jaw protrusion and lateral shifts of the jaw. Jaw closing is caused by activation of the masseter, the temporalis and – to a minor extent – the medial pterygoid muscles. Jaw opening is the result of activation of the lateral pterygoid muscle and the suprahyoid muscles forming the muscular floor of the cavity of the mouth. Protrusion and lateral shifts are caused by the pterygoid muscles, mainly the lateral ones. Whereas treatment of the jaw-closing type produces excellent results with only rare adverse effects, treatment of the jaw-opening type is less rewarding. Our experience indicates that BT injections into the pterygoid muscles through the incisura mandibulae together with injections of the suprahyoid muscles seem to work best in this situation. Attempts to inject the lateral and the medial pterygoid muscles individually would cause major technical problems and discomfort for the patient. Local spread and frequent co-activation of both muscles question the logic of this approach.

10.4.2 Pharyngolaryngeal Dystonia

Tonic or clonic dystonia of the pharynx can produce dysphagia and dyspnea [64]. They can occur spontaneously or in an action-induced fashion. BT injections into the posterior pharynx can easily be placed transorally and are effective. Doses range between 20 and 40 MU of Botox®. Laryngeal dystonia produces the clinical picture of spasmodic dysphonia, either in the adductor form with a strained-strangled voice or in the much less frequent abductor form with hypophonia [65]. In adductor forms, 2.5–10 MU of Botox® is administered into the thyroarytenoid (vocalis) muscle. Unilateral application appears to produce less adverse effects than when the same amount is distributed over both sides. In abductor forms, 2.5–10 MU of Botox® is administered into the posterior cricoarytenoid muscle unilaterally in order to avoid dyspnea. BT application can be performed perorally or transcutaneously using electromyographic guidance. The transoral approach allows detection of additional dystonic muscle activities in the pharynx or the larynx and, therefore, seems to be the superior method.

For the patient and for the physician, BT therapy of spasmodic dysphonia represents a highly satisfying indication [66]. Practically all patients benefit from BT therapy and the degree of improvement is astonishing. In many cases, almost normal speech patterns can be regained. In patients with abductor forms, the treatment results are less favourable. Adverse effects include difficulties with swallowing liquids or solid food. BT therapy can also induce weakness of coughing and some pain at the injection side. In treatment of adductor forms, hoarseness, breathiness of voice and hypophonia can occur, while in treatment of abductor forms, dyspnea may result.

Spasmodic laryngeal dyspnea describes spontaneously occurring or respiration-induced muscle hyperactivity of laryngeal muscles [67]. This condition is very rare and may influence both glottic and supraglottic muscles and can also be treated with BT therapy.

10.4.3 Cervical Dystonia

Cervical dystonia induces deviation of the neck and the head. Frequently shoulder elevation occurs. Whereas the neck can only be flexed and extended on a sagittal plane or frontal plane and rotated on a horizontal plane throughout its entire structure employing numerous interspinal joints, the head is flexed and extended on a sagittal plane, tilted on a frontal plane and rotated on a horizontal plane by using the singular atlanto-occipital joint with the dens as an additional stabilisator. Head and neck deviations can best be described by the scheme shown in Table 10.7. The

Table 10.7 Description of head and neck deviation in cervical dystonia

Head deviation	Neck deviation	Conventional description
Rotation		Torticollis
Flexion	Nil	Antecaput
	Flexion	Antecaput + retrocollis
	Extension	Antecaput + antecollis = posterior shift/retrusion
Extension	Nil	Retrocaput
	Flexion	Retrocaput + antecollis = anterior shift/protrusion
	Extension	Retrocaput + retrocollis
Lateral flexion	Nil	Laterocaput
	Flexion ipsilateral	Laterocaput + laterocollis
	Flexion contralateral	Lateral shift

I 0–30°
II 30–60°
III 60–90°

scheme allows an easy and semiquantitative, although comprehensive, description of the head and neck deviations elicited by cervical dystonia.

Isolated occurrence of head or neck deviations is rare. Most patients suffer from complex combinations of head deviations.

Head rotation occurs always together with *neck rotation*. It is caused by activation of the ipsilateral splenius capitis, the contralateral sternocleidomastoid muscle and the ipsilateral trapezius/semispinalis capitis muscle complex. Deep posterior neck muscles arising from the atlas and the axis including the obliquus capitis inferior, the rectus capitis posterior major and the rectus capitis posterior minor muscles are strong ipsilateral head rotators. The levator scapulae muscle is an additional but weaker ipsilateral head rotator. In head and neck rotation, the role of the sternocleidomastoid is often overestimated, whereas the role of the splenius capitis and the deep posterior neck muscles is often underestimated.

Head flexion is caused by activation of the supra- and infrahyoid muscles. *Neck flexion* is produced by the scaleni and deep anterior neck muscles including the longus colli, the longus capitis and the rectus capitis anterior muscles. *Head extension* originates from bilateral activation splenius capitis and the deep posterior neck muscles, and *neck extension* from bilateral activation of the trapezius/semispinalis capitis muscle complex. *Lateral head flexion* originates from ipsilateral activation of the sternocleidomastoid and the splenius capitis, and *lateral neck flexion* from ipsilateral activation of the scaleni, the levator scapulae and the trapezius/semispinalis capitis muscle complex. *Protrusion* is the consequence of neck flexion together with head extension, and *retrusion* of neck extension together with head flexion.

Planning of BT therapy for cervical dystonia includes careful examination of the spontaneous head position, the head position under motor and stress activation, the slow active head movement and passive head mobility. When dystonia occurs in waves or when it can be suppressed by gestes antagonistes, the time course of a

Table 10.8 Recommended botulinum toxin doses for cervical muscles

Muscle	Function	Recommended dose [MU-A Botox®]
Sternocleidomastoid	Contralateral horizontal head rotation Sagittal head flexion Frontal head flexion	20–60
Splenius capitis	Ipsilateral horizontal Head rotation	20–100
Scaleni	Frontal head flexion Sagittal head flexion	20–60
Levator scapulae	Shoulder elevation Ipsilateral head rotation Frontal head extension	20–60
Trapezius/semispinalis capitis complex	Sagittal head extension Ipsilateral head rotation	20–80
Trapezius, horizontal part	Shoulder elevation Frontal head flexion	40–80

MU-A mouse unit of the mouse bioassay of Allergan Inc
Dilution: 100 MU-A in 2.5 ml 0.9 % Sodium chloride/H_2O)

dystonic build-up helps to identify compensatory muscle activity. Slow active movements can identify trigger postures and therefore potential preventive postures. Testing for pre-existent dysphagia identifies patients in which BT application into anterior neck muscles should be performed with caution.

Target muscles and recommended BT doses for cervical muscles are shown in Table 10.8.

Dystonic pain as the leading complaint in most patients with cervical dystonia can almost always be markedly reduced by BT therapy. Residual pain may be caused by secondary degenerative processes or by radicular irritation. Head posture can also be improved substantially. Often, patients report the effects of the first BT applications enthusiastically, most likely due to the contrast to the sometimes prolonged period of insufficient treatment. Especially in the treatment of cervical dystonia, additional *physiotherapy* following the above-discussed guidelines is necessary [51]. Certain forms of cervical dystonia respond less favourably to BT therapy. Especially head flexion and neck flexion is difficult to treat when deep anterior neck muscles are involved, but also, alternating types of cervical dystonia may present a therapeutic challenge. In tremor types, sometimes reduced BT doses seem to be helpful.

The most common *adverse effect* of BT therapy of cervical dystonia is dysphagia. Depending on the definition of dysphagia and the effort to search for it, their incidence varies greatly. Applying current treatment standards, certainly less than 5 % of patients experience dysphagia constantly after each injection series. Mild dysphagia may be more frequent. Another adverse effect is head instability, especially due to impaired head and neck extension. When these adverse effects occur, their duration is usually limited to 1 or 2 weeks. Injection of the scaleni muscles can produce needle contacts with brachial plexus nerve fibres eliciting short-lasting electric sensations especially when injections are placed too close to Erb's point.

10.4.4 Arm Dystonia

Arm dystonia can be divided into *action-induced forms* and *spontaneous forms*. Action-induced forms occur only during certain activities which can sometimes be highly specific. Spontaneous forms are not associated with specific activities, although they may be increased by unspecific physical activity. Writer's cramp is the most common action-induced dystonia [68]. Other highly specific and sometimes peculiar activities can also trigger dystonia, such as playing musical instruments or performing sports. Especially, when these activities are performed under professional conditions, dystonia can result, thus explaining the term occupational cramp.

In *writer's cramp*, a wrist flexor and a wrist extensor type can be distinguished. Additionally, elbow and shoulder muscles may be involved. However, abnormal elbow and shoulder postures may be compensatory in order to change the writing position and to reduce dystonia. Planning of BT therapy for writer's cramp is based on careful examination of the clinical symptomatology. Electromyography is rarely contributory, since normal writing usually generates widespread muscle activation which can hardly be distinguished from dystonic muscle activity. Sometimes, asking the patient to write with the contralateral hand produces the dystonic pattern in the dominant hand without contamination of normal or compensatory muscle activity.

Target muscles and recommended BT doses for arm dystonia are shown in Table 10.9. Results of BT therapy of writer's cramp are limited, because of narrow therapeutic windows of the potential target muscles [69]. This is a problem especially in the finger extensors. Apart from this, writer's cramp frequently affects a large number of forearm muscles. BT therapy targeting all of these muscles would induce major paretic adverse effects. Additionally, distinction between physiological and dystonic muscle activity and identification of compensatory muscle activity may be difficult. Our experience indicates that even after several modifications of the injection scheme, only about half of the patients benefit from BT therapy and continue treatment. Results are better when the finger muscles are not involved, i.e. when the symptomatology is restricted to the wrist or elbow muscles. When finger muscles are involved, the outcome is better when individual finger muscles and when finger flexors rather than finger extensors are dystonic. If the symptomatology is restricted to individual finger muscles, electromyography possibly with additional electric stimulation may facilitate BT placement. If BT therapy is not successful in treatment of writer's cramp, the patient can shift writing to the contralateral hand. About half of the patients can permanently use the contralateral hand for writing, whereas the other half develops writer's cramp in this hand as well within 1 or 2 years. Increased use of keyboards is also one option to circumvent writer's cramp. Retraining exercises may also become a therapeutic option in the future [70].

Table 10.9 Recommended botulinum toxin doses for arm muscles

Muscle	Function	Recommended dose [MU-A Botox®]
Deltoideus	Shoulder abduction	40–120
Biceps brachii	Elbow flexion	40–100
Triceps brachii	Elbow extension	40–100
Brachialis	Elbow flexion	40–80
Brachioradialis	Elbow flexion	40–80
Flexor digitorum profundus et superficialis	Finger flexion	40–120
Flexor carpi radialis	Wrist flexion	40–80
Extensor digitorum profundus et superficialis	Finger extension	20–60
Extensor carpi radialis	Wrist extension	20–40
Interosseous	Metacarpophalangeal adduction Metacarpophalangeal abduction	20
Abductor digiti quinti	Little finger metacarpophalangeal abduction	20–40
Flexor pollicis longus	Thumb flexion	20–60

MU-A mouse unit of the mouse bioassay of Allergan Inc
Dilution: 100 MU-A in 2.5 ml 0.9 % Sodium chloride/H_2O)

Treatment of other action-induced arm dystonias, especially when they are occupational, is even more problematic, since the motor performance expected by the patient is usually so high that it cannot be met, either due to dystonic residues or due to therapy-induced paresis [71].

Spontaneous induced arm dystonia usually occurs as part of a spasticity-dystonia syndrome or as idiopathic dystonia. Typical postures include finger flexion, thumb flexion, wrist flexion, elbow flexion and shoulder adduction or shoulder abduction. In spasticity-dystonia syndrome, treatment is focused on pain, prevention of contractures and eased care. Functional improvement may result but is often restricted by the underlying paresis.

10.4.5 Leg Dystonia

Leg dystonia can occur in idiopathic as well as in symptomatic dystonia, mostly as part of a spasticity-dystonia syndrome due to stroke or cerebral palsy. Action-induced forms would be extremely rare. Typical postures include hip adduction; knee flexion; equinovarus posture, i.e. the combination of ankle plantar flexion, foot supination and toe flexion; ankle plantar flexion; and toe flexion. Hip adduction is caused by activation of the adductor muscle group (adductor magnus, minimus, longus, brevis and gracilis muscles), knee flexion by activation of the hamstrings (semimembranosus, semitendinosus, biceps femoris muscles) and

Table 10.10 Recommended botulinum toxin doses for leg muscles

Muscle	Function	Recommended dose [MU-A Botox®]
Adductor muscle group	Hip adduction	100–300
Quadriceps femoris	Knee extension	100–300
Hamstring muscles	Knee flexion	100–300
Triceps surae	Ankle plantar flexion	100–200
Tibialis posterior	Foot supination	60–100
Flexor hallucis longus	Great toe flexion Ankle plantar flexion	60–100
Tibialis anterior	Ankle dorsiflexion	40–100
Extensor digitorum longus	Toe extension Ankle dorsiflexion	40–100
Extensor hallucis longus	Great toe extension Ankle dorsiflexion	40–60

MU-A mouse unit of the mouse bioassay of Allergan Inc
Dilution: 100 MU-A in 2.5 ml 0.9 % Sodium chloride/H_2O)

equinovarus posture by activation of the tibialis posterior, triceps surae, flexor hallucis and digitorum longus muscles. Ankle plantar flexion is the result of activation of triceps surae, peroneus longus and brevis and flexor digitorum longus muscles, and toe flexion of the activation of the flexor digitorum longus and brevis muscles.

Target muscles and recommended BT doses for leg dystonia are shown in Table 10.10.

Hip adduction, ankle plantar flexion and equinovarus postures respond well to BT therapy. BT doses, however, may be high, especially when bilateral injections are necessary. BT therapy for knee extension bears the risk of knee weakness, especially in patients with additional paresis as in spasticity-dystonia syndrome after stroke. Toe and great toe flexion often requires combined treatment of short and long toe and great toe flexors.

10.4.6 Segmental Dystonia and Generalised Dystonia

BT therapy of extended dystonic symptomatologies usually requires selection of those target muscles which play a major role in functional impairment, pain and prevention of complications. Less relevant target muscles may need to be left untreated in order not to exceed total BT doses which are safe with respect to toxicological and immunological adverse effects. Paravertebral muscles require 40–60 MU Botox® per segment per side, the rectus abdominis 40–80 MU Botox® per segment per side and the abdominal wall muscle complex (obliquus internus abdominis, obliquus externus abdominis and transversus abdominis muscles) 80–200 MU Botox® per side.

Toxicological and immunological safety margins have been discussed above. When safety margins are exploited to their full extent, BT therapy can improve even extended symptomatologies substantially. In patients requiring excess BT doses, deep brain stimulation may offer a treatment alternative. Combinations of BT therapy and deep brain therapy or intrathecal baclofen are possible.

10.5 Outlook

BT therapy presents a novel therapeutic concept. It has revolutionised many medical fields. In dystonia, for the first time, it offers help to a large numbers of patients with focal dystonia [72]. In segmental and generalised dystonia, high BT doses are necessary. Emerging experience demonstrates that high BT doses are immunologically and toxicologically safe [73–75]. The recent introduction of low-antigenicity BT drugs may allow booster injections for rapid dose adaptation, reduced interinjection intervals for an improved dynamic adjustment and increased BT dose for treatment of extended BT symptomatologies [76, 77]. High-affinity BT drugs may improve antigenicity even further. They may reduce the threshold for systemic toxicology so that higher total BT doses may be applied.

References

1. Scott AB. Botulinum toxin injection into extraocular muscles as an alternative to strabismus surgery. J Pediatr Ophthalmol Strabismus. 1980;17:21–5.
2. Scott AB, Rosenbaum A, Collins CC. Pharmacologic weakening of extraocular muscles. Invest Ophthalmol. 1973;12:924–7.
3. Frueh BR, Felt DP, Wojno TH, Musch DC. Treatment of blepharospasm with botulinum toxin. A preliminary report. Arch Ophthalmol. 1984;102:1464–8.
4. Tsui JK, Eisen A, Mak E, Carruthers J, Scott A, Calne DB. A pilot study on the use of botulinum toxin in spasmodic torticollis. Can J Neurol Sci. 1985;12:314–6.
5. Aurora SK, Winner P, Freeman MC, Spierings EL, Heiring JO, DeGryse RE, VanDenburgh AM, Nolan ME, Turkel CC. OnabotulinumtoxinA for treatment of chronic migraine: pooled analyses of the 56-week PREEMPT clinical program. Headache. 2011;51:1358–73.
6. Truong D, Dressler D, Hallett M, Zachary C. Manual of botulinum toxin therapy. 2nd ed. Cambridge: Cambridge University Press; 2013.
7. Frevert J, Dressler D. Complexing proteins in botulinum toxin type A drugs – a help or hindrance? Biologics. 2010;4:325–32.
8. Dressler D. Five-year experience with incobotulinumtoxinA (Xeomin®): the first botulinum toxin drug free of complexing proteins. Eur J Neurol. 2012;19:385–9.
9. Dong M, Richards DA, Goodnough MC, Tepp WH, Johnson EA, Chapman ER. Synaptotagmins I and II mediate entry of botulinum neurotoxin B into cells. J Cell Biol. 2003;162:1293–303.
10. Dong M, Yeh F, Tepp WH, Dean C, Johnson EA, Janz R, Chapman ER. SV2 is the protein receptor for botulinum neurotoxin A. Science. 2006;312:592–6.
11. Mahrhold S, Rummel A, Bigalke H, Davletov B, Binz T. The synaptic vesicle protein 2C mediates the uptake of botulinum neurotoxin A into phrenic nerves. FEBS Lett. 2006;580:2011–4.
12. Rummel A, Karnath T, Henke T, Bigalke H, Binz T. Synaptotagmins I and II act as nerve cell receptors for botulinum neurotoxin G. J Biol Chem. 2004;279:30865–70.

13. Pellizzari R, Rossetto O, Schiavo G, Montecucco C. Tetanus and botulinum neurotoxins: mechanism of action and therapeutic uses. Philos Trans R Soc Lond B Biol Sci. 1999;354:259–68.
14. Brin MF, Dressler D, Aoki R. Pharmacology of botulinum toxin therapy. In: Jankovic J, Comella C, Brin MF, editors. Dystonia: etiology, clinical features, and treatment. Philadelphia: Lippincott Williams & Wilkins; 2004. p. 93–112.
15. Dressler D, Eckert J, Kukowski B, Meyer BU. Somatosensorisch Evozierte Potentiale bei Schreibkrampf: Normalisierung pathologischer Befunde unter Botulinum Toxin Therapie. Z EEG EMG. 1993;24:191.
16. Filippi GM, Errico P, Santarelli R, Bagolini B, Manni E. Botulinum A toxin effects on rat jaw muscle spindles. Acta Otolaryngol. 1993;113:400–4.
17. Rosales RL, Arimura K, Takenaga S, Osame M. Extrafusal and intrafusal muscle effects in experimental botulinum toxin-A injection. Muscle Nerve. 1996;19:488–96.
18. Kaji R, Kohara N, Katayama M, Kubori T, Mezaki T, Shibasaki H, Kimura J. Muscle afferent block by intramuscular injection of lidocaine for the treatment of writer's cramp. Muscle Nerve. 1995;18:234–5.
19. Kaji R, Rothwell JC, Katayama M, Ikeda T, Kubori T, Kohara N, Mezaki T, Shibasaki H, Kimura J. Tonic vibration reflex and muscle afferent block in writer's cramp. Ann Neurol. 1995;38:155–62.
20. Caleo M, Schiavo G. Central effects of tetanus and botulinum neurotoxins. Toxicon. 2009;54:593–9.
21. Wiegand H, Erdmann G, Wellhoner HH. 125I-labelled botulinum A neurotoxin: pharmacokinetics in cats after intramuscular injection. Naunyn Schmiedebergs Arch Pharmacol. 1976;292:161–5.
22. Cui M, Li Z, You S, Khanijou S, Aoki R. Mechanisms of the antinociceptive effect of subcutaneous Botox: inhibition of peripheral and central nociceptive processing. Arch Pharmacol. 2002;365:R17.
23. Ishikawa H, Mitsui Y, Yoshitomi T, Mashimo K, Aoki S, Mukuno K, Shimizu K. Presynaptic effects of botulinum toxin type A on the neuronally evoked response of albino and pigmented rabbit iris sphincter and dilator muscles. Jpn J Ophthalmol. 2000;44:106–9.
24. McMahon H, Foran P, Dolly J. Tetanus toxin and botulinum toxins type A and B inhibit glutamate, gamma-aminobutyric acid, aspartate, and met-enkephalin release from synaptosomes: clues to the locus of action. J Biol Chem. 1992;267:21338–43.
25. Morris J, Jobling P, Gibbins I. Differential inhibition by botulinum neurotoxin A of cotransmitters released from autonomic vasodilator neurons. Am J Physiol Heart Circ Physiol. 2001;281:2124–32.
26. Purkiss J, Welch M, Doward S, Foster K. Capsaicin-stimulated release of substance P from cultured dorsal root ganglion neurons: involvement of two distinct mechanisms. Biochem Pharmacol. 2000;59:1403–6.
27. Shone CC, Melling J. Inhibition of calcium-dependent release of noradrenaline from PC12 cells by botulinum type-A neurotoxin. Long-term effects of the neurotoxin on intact cells. Eur J Biochem. 1992;207:1009–16.
28. Welch MJ, Purkiss JR, Foster KA. Sensitivity of embryonic rat dorsal root ganglia neurons to Clostridium botulinum neurotoxins. Toxicon. 2000;38:245–58.
29. Dressler D, Rothwell JC. Electromyographic quantification of the paralysing effect of botulinum toxin. Eur Neurol. 2000;43:13–6.
30. Dressler D, Tacik P, Adib Saberi F. Botulinum toxin therapy of cervical dystonia: duration of therapeutic effects. J Neural Transm. 2013;120:1142.
31. Dressler D, Tacik P, Adib Saberi F. Botulinum toxin therapy of cervical dystonia: comparing Botox® and Xeomin®. J Neural Transm. 2014;121:29–31.
32. Benecke R, Jost WH, Kanovsky P, Ruzicka E, Comes G, Grafe S. A new botulinum toxin type A free of complexing proteins for treatment of cervical dystonia. Neurology. 2005;64:1949–51.

33. Dressler D, Adib Saberi F. Safety aspects of high dose Xeomin® therapy. J Neurol. 2006;253(2):II/141.
34. Dressler D, Mander X, Fink K. Measuring the potency labelling of onabotulinumtoxinA (Botox®) and Xeomin® in an LD50 assay. J Neural Transm. 2012;119:13–5.
35. Dressler D, Muenchau A, Bhatia KP, Quinn NP, Bigalke H. Antibody induced botulinum toxin therapy failure: can it be overcome by increased botulinum toxin doses? Eur Neurol. 2002;47:118–21.
36. Dressler D, Dirnberger G. Botulinum toxin therapy: risk factors for therapy failure. Mov Disord. 2000;15(2):51.
37. Dressler D, Benecke R. Xeomin® eine neue therapeutische botulinum toxin typ A-Präparation. Aktuelle Neurol. 2006;33:138–41.
38. Jankovic J, Vuong KD, Ahsan J. Comparison of efficacy and immunogenicity of original versus current botulinum toxin in cervical dystonia. Neurology. 2003;60:1186–8.
39. Pickett A, Panjwani N, O'Keeffe RS (2003) Potency of Type A botulinum toxin preparations in clinical use. 40th annual meeting of the Interagency Botulism Research Coordinating Committee (IBRCC), Nov. 2003; Atlanta.
40. Setler P. The biochemistry of botulinum toxin type B. Neurology. 2000;55(5):S22–8.
41. Dressler D. Complete secondary botulinum toxin therapy failure in blepharospasm. J Neurol. 2000;247:809–10.
42. Dressler D. New formulation of BOTOX®: complete antibody-induced therapy failure in hemifacial spasm. J Neurol. 2004;251:360.
43. Dressler D. Dysport produces intrinsically more swallowing problems than Botox: unexpected results from a conversion factor study in cervical dystonia. J Neurol Neurosurg Psychiatry. 2002;73:604.
44. Probst TE, Heise H, Heise P, Benecke R, Dressler D. Rare immunologic side effects of botulinum toxin therapy: brachial plexus neuropathy and dermatomyositis. Mov Disord. 2002;17(5):S49.
45. Dressler D. Subclinical myasthenia gravis causing increased sensitivity to botulinum toxin therapy. J Neural Transm. 2010;117:1293–4.
46. Erbguth F, Claus D, Engelhardt A, Dressler D. Systemic effect of local botulinum toxin injections unmasks subclinical Lambert-Eaton myasthenic syndrome. J Neurol Neurosurg Psychiatry. 1993;56:1235–6.
47. Naumann M, Albanese A, Heinen F, Molenaers G, Relja M. Safety and efficacy of botulinum toxin type A following long-term use. Eur J Neurol. 2006;13 Suppl 4:35–40.
48. Dressler D, Benecke R. Autonomic side effects of botulinum toxin type B treatment of cervical dystonia and hyperhidrosis. Eur Neurol. 2003;49:34–8.
49. Dressler D, Eleopra R. Clinical use of non-A botulinum toxins: botulinum toxin type B. Neurotox Res. 2006;9:121–5.
50. Adib Saberi F, Dressler D. Interdisziplinärer Arbeitskreis Bewegungsstörungen (IAB): a new approach for promoting interdisciplinary therapy of movement disorders. J Neural Transm. 2013;120:705–10.
51. Barth K, Dressler D. Die krankengymnastische Zusammenarbeit mit Schiefhalspatienten nach der Behandlung mit Botulinum Toxin. Krankengymnastik. 1993;45:134–42.
52. Dressler D. Electromyographic evaluation of cervical dystonia for planning of botulinum toxin therapy. Eur J Neurol. 2000;7:713–8.
53. Berweck S, Feldkamp A, Francke A, Nehles J, Schwerin A, Heinen F. Sonography-guided injection of botulinum toxin A in children with cerebral palsy. Neuropediatrics. 2002;33:221–3.
54. Tsui JK, Eisen A, Stoessl AJ, Calne S, Calne DB. Double-blind study of botulinum toxin in spasmodic torticollis. Lancet. 1986;2(8501):245–7.
55. Consky ES, Basinki A, Belle L, Ranawaya R, Lang AE. The Toronto Western Spasmodic Torticollis Rating Scale (TWSTRS): assessment of validity and inter-rater reliability. Neurology. 1990;40(1):445.

56. Burke RE, Fahn S, Marsden CD, Bressman SB, Moskowitz C, Friedman J. Validity and reliability of a rating scale for the primary torsion dystonias. Neurology. 1985;35:73–7.
57. Dressler D, Kupsch A, Paus S, Seitzinger A, Minnasch P. The Dystonia Discomfort Scale (DDS): a novel instrument to monitor the temporal profile of botulinum toxin therapy in cervical dystonia. Eur J Neurol. 2014;21:459–62.
58. Alonso-Navarro H, Puertas I, Cabrera-Valdivia F, de Toledo-Heras M, García-Albea E, Jiménez-Jiménez FJ. Posterior auricular muscle 'dystonia'. Eur J Neurol. 2007;14:e14–5.
59. Costa J, Espírito-Santo C, Borges A, Ferreira JJ, Coelho M, Moore P, Sampaio C (2005) Botulinum toxin type A therapy for blepharospasm. Cochrane Database Syst Rev. (1):CD004900.
60. Aramideh M, Ongerboer de Visser BW, Koelman JH, Bour LJ, Devriese PP, Speelman JD. Clinical and electromyographic features of levator palpebrae superioris muscle dysfunction in involuntary eyelid closure. Mov Disord. 1994;9:395–402.
61. Jankovic J. Pretarsal injection of botulinum toxin for blepharospasm and apraxia of eyelid opening. J Neurol Neurosurg Psychiatry. 1996;60:704.
62. Roggenkämper P, Nüssgens Z. Frontalis suspension for essential blepharospasm unresponsive to botulinum toxin therapy. First results. Ger J Ophthalmol. 1993;2:426–8.
63. Marsden CD. The problem of adult-onset idiopathic torsion dystonia and other isolated dyskinesias in adult life (including blepharospasm, oromandibular dystonia, dystonic writer's cramp, and torticollis, or axial dystonia). Adv Neurol. 1976;14:259–76.
64. Zwirner P, Dressler D. Dystonie als Ursache pharyngo-laryngealer Motilitätsstörungen. HNO. 1995;43:498–501.
65. Marsden CD, Sheehy MP. Spastic dysphonia, Meige disease, and torsion dystonia. Neurology. 1982;32:1202–3.
66. Whurr R, Nye C, Lorch M. Meta-analysis of botulinum toxin treatment of spasmodic dysphonia: a review of 22 studies. Int J Lang Commun Disord. 1998;33(Suppl):327–9.
67. Zwirner P, Dressler D, Kruse E. Spasmodic laryngeal dyspnea: a rare manifestation of laryngeal dystonia. Eur Arch Otorhinolaryngol. 1997;254:242–5.
68. Sheehy MP, Rothwell JC, Marsden CD. Writer's cramp. Adv Neurol. 1988;50:457–72.
69. Das CP, Dressler D, Hallett M. Botulinum toxin therapy of writer's cramp. Eur J Neurol. 2006;13(1):55–9.
70. Zeuner KE, Bara-Jimenez W, Noguchi PS, Goldstein SR, Dambrosia JM, Hallett M. Sensory training for patients with focal hand dystonia. Ann Neurol. 2002;51:593–8.
71. Jabusch HC, Zschucke D, Schmidt A, Schuele S, Altenmüller E. Focal dystonia in musicians: treatment strategies and long-term outcome in 144 patients. Mov Disord. 2005;20:1623–6.
72. Dressler D. Botulinum toxin therapy. Stuttgart: Thieme Verlag; 2000.
73. Dressler D. Clinical features of secondary failure of botulinum toxin therapy. Eur Neurol. 2002;48:26–9.
74. Dressler D, Benecke R. Pharmacology of therapeutic botulinum toxin preparations. Disabil Rehabil. 2007;29:1761–8.
75. Dressler D, Benecke R, Bigalke H. Botulinum toxin type B (NeuroBloc®) in patients with botulinum toxin type A antibody-induced therapy failure. J Neurol. 2003;250:967–9.
76. Dressler D, Bigalke H. Antibody-induced failure of botulinum toxin type B therapy in de novo patients. Eur Neurol. 2004;52:132–5.
77. Moore AP, Naumann M, editors. Handbook of botulinum toxin treatment. 2nd ed. Malden: Blackwell Science; 2003.

11 Long-Term Treatment of Focal Dystonias (Cervical and Writer's Cramp): Clinical and Quality of Life Impact

Martin Bareš, Marek Baláž, and Pavel Filip

11.1 Introduction

Dystonias are a group of disorders with many different clinical manifestations but with certain common features. Affected patients share involuntary sustained or intermittent muscle contractions that cause abnormal postures and/or repetitive movements. The most common forms of adult-onset focal dystonias are cervical dystonia, blepharospasm, oromandibular dystonia, laryngeal dystonia, and limb dystonias such as writer's cramp [1]. In this chapter, we were interested in certain aspects of the long-term treatment of two focal dystonias, cervical dystonia and limb dystonia (writer's cramp), with botulinum toxin A (BoNT-A).

11.2 Cervical Dystonia

Cervical dystonia is a chronic neurological condition with a prevalence of approximately 5.7–8.9 per 100,000 [2]. Cervical dystonia is characterised by involuntary sustained contractions of the cervical muscles, resulting in spasmodic head

M. Bareš, MD, PhD (✉) • P. Filip
1st Department of Neurology, St. Anne's University Hospital, Faculty of Medicine, Masaryk University, Brno, Czech Republic

Behavioural and Social Neuroscience Research Group, CEITEC – Central European Institute of Technology, Masaryk University, Brno, Czech Republic
e-mail: bares@muni.cz

M. Baláž
1st Department of Neurology, Faculty of Medicine, St. Anne's University Hospital, Masaryk University Medical School, Brno, Czech Republic

Multimodal and Functional Imaging Research Group, CEITEC – Central European Institute of Technology, Masaryk University, Brno, Czech Republic
e-mail: marek.balaz@fnusa.cz

P. Kanovsky et al. (eds.), *Dystonia and Dystonic Syndromes*,
DOI 10.1007/978-3-7091-1516-9_11

movements and abnormal postures [3]. When isolated, cervical dystonia is classified as a focal dystonia; it can also be a part of more complex dystonias. The majority of cervical dystonias are focal and primary; they have no other associated clinical or neuroimaging features, and they remain restricted to the neck and shoulder girdle muscles [2]. Pathophysiologically, there is an increase in the motor cortex activity and a decrease in the activity of the cortical inhibition system [4, 5].

11.3 Diagnosis of Cervical Dystonia

The diagnosis of cervical dystonia is based on the recognition of the above-stated clinical features. After establishing the clinical diagnosis, further examinations may be performed, including imaging studies and eventually neuropsychological tests. Clinical assessment should be anchored in a validated rating scale, with about 10 rating scales for cervical dystonia generally available [6]. The most frequently used scales are the Tsui scale, the Toronto Western Spasmodic Torticollis Rating Scale, and the Cervical Dystonia Severity Scale [7, 8]. Nevertheless, dystonia is a complex disorder, and despite the undoubted clinical value of rating scales, only a careful clinical examination can reveal all the important domains relevant to the patient.

Writer's Cramp Writer's cramp is a task-specific focal hand dystonia [9] that typically affects the distal muscles of the forearm and hand, generally of one's dominant writing hand, and is usually related to overuse [10]. It appears to be the most common focal hand dystonia affecting the general population [11]. Occupational dystonia was described in the early nineteenth century, but the initial reports of writer's cramp came from British Civil Service communications [12], originally published by Sir Charles Bell as early as 1833 [13].

Two basic categories are recognised: simple writer's cramp, in which dystonic posturing of the hand and arm occurs only during writing, and complex or dystonic writer's cramp, in which the condition also manifests during other manual tasks and sometimes with spontaneous abnormal posturing [6, 14, 15].

The prevalence of writer's cramp is relatively low, varying between 3 and 14 per 100,000 [16–18], but the condition may be responsible for considerable morbidity in terms of working impairment, pain, embarrassment, low self-esteem, and poor social interaction [19].

11.4 Diagnosis of Writer's Cramp

Writer's cramp often initially presents with muscle tightness, accompanied by fatigue and/or pain during continued writing. Writing becomes slow. Paper destruction and difficulty in holding a pen or an exceptionally tight grip may occur. The development of pain is usually not correlated with the severity of writing [15].

Writer's cramp typically progresses over months and years into involuntary muscular contractions and motor performance impairment [11]. Abnormal posturing of the fingers is common and may further involve the wrist and elbow and occasionally the shoulder [20]. The most severe cases present with dystonic posturing at rest [10].

In approximately 50 % of those with writer's cramp, the dystonia only affects the task-specific activity of writing, leaving other tasks, even those performed with the same muscles, unaffected [15, 21]. The pattern of dystonia in writer's cramp typically affects distal muscles more than proximal muscles and flexors more than extensors. The abnormality of sensorimotor integration may be due to impaired afferent activity; a sensorimotor training programme can modify this disorder [22].

To evaluate the extent of writer's cramp, patients have to be examined at rest and then with a sustained posture during writing that provokes the dystonia and also while performing other purposeful activities. Routine electromyogram (EMG) may be useful in ruling out neuropathy or radiculopathy and other peripheral nerve conditions.

11.5 Treatment of Focal Dystonias with Botulinum Toxin A

Botulinum toxin A (BoNT-A) was introduced into the treatment of focal dystonias, including cervical dystonia and writer's cramp, 20 years ago. It has achieved widespread use for this chronic condition, which affects the patient's quality of life and sociological and employment status [23]. Although nowadays there is no question about the use of BoNT-A as a first-line treatment in terms of its efficacy and safety, we lack data about the long-term course of the illness under BoNT-A treatment and about related issues of quality of life and financial costs, as well as data addressing the dropout rate of treated patients.

11.5.1 Treatment of Cervical Dystonia with Botulinum Toxin A

Botulinum toxin (BTX) injections have been shown to be an effective and safe treatment for most patients with cervical dystonia. Based on the recommendations of the Therapeutics and Technology Assessment Subcommittee of the American Academy of Neurology, botulinum neurotoxin (BoNT) should be offered as a treatment option for cervical dystonia (Level A) [24].

With its chronic use and expanding list of therapeutic indications, the need to carefully examine the long-term efficacy of BoNT treatment has become more pressing. However, reports on the long-term treatment of cervical dystonia remain sparse, with most of the earlier studies focusing on short-term follow-up ranging from several months to 2 years. Relatively few studies have provided information about patients receiving regular BoNT injections over a period of more than 5 years (see Table 11.1).

Generally, long-term BoNT treatment is associated with good patient satisfaction, a favourable safety profile with mild adverse reactions, and low immunogenicity, with the most dramatic improvements in subjective pain assessment (up to 93 %) and head position (up to 90 %).

Table 11.1 Long-term studies of cervical dystonia patients treated with BoNT for 5 years and more

Study	No. of patients	Mean follow-up (years)	BoNT type
Haussermann et al. (2004) [25]	100	5.1 ± 4.5	BoNT-A
Skogseid and Kerty (2005) [26]	78	5.5 (range 1.5–10)	BoNT-A
Mejia et al. (2005) [27]	45	15.8 ± 1.5	BoNT-A + BoNT-B
Mohammadi et al. (2009) [28]	207	6.7 ± 3.5	BoNT-A
Vivancos-Matellano et al. (2012) [29]	37	7.0 (range 1–17)	BoNT-A
Ramirez-Castaneda and Jankovic (2013) [30]	51	18.5 ± 3.1	BoNT-A + BoNT-B

Cervical dystonia patients usually show a high adherence to the treatment over time. Care discontinuation is mostly caused by a suboptimal response to BoNT therapy – excessive cost, travel burden, and side effects are cited by less than half of the patients as the primary factor. The main reasons for a lack of response to BoNT treatment include inappropriate muscle selection, inadequate dose, and development of neutralising antibodies, with immunogenicity being the least probable factor [31].

11.5.2 Treatment of Writer's Cramp by Botulinum Toxin A

The primary goal in treatment is to improve function. BoNT as a treatment for focal hand dystonia was first reported by Cohen and colleagues [32]. BTX acts peripherally by inhibiting acetylcholine release from the presynaptic neuromuscular terminals [33, 34]. Studies in patients with dystonia show that BoNT-A acts at the intrafusal fibre level, thereby reducing muscle spindle afferent discharges [35].

BoNT-A is recognised as a first-line therapy and an effective treatment in patients with focal hand dystonias [36–38], reducing neuromuscular junction activity and relieving clinical symptoms. A good response to BoNT-A is reported in 50–80 % of patients [19, 39, 40].

The general contraindications to BoNT-A therapy, such as myasthenia gravis and concomitant aminoglycoside therapy, also apply to its use in writer's cramp. Weakness of the target muscle is the most common side effect. There have been no reports of long-term adverse effects, other than muscle atrophy, with repeated BoNT-A injections [11]. According to some authors [10], most patients experience stable responses to the treatment after three to four injections and maintain benefits for many years (in the absence of antibody formation). However, from our clinical experience, it appears that quite a sizeable share of patients choose to stop BoNT-A treatment after a short time. The dropout rate of patients with focal hand dystonia has not recently been thoroughly studied.

According to some studies, 50 % of patients with writer's cramp return for injections over longer periods of time [41]. Some other long-term results showed that only approximately 33 % of patients continued treatment with BoNT-A after 2 years [42]. According to a long-term study by Lung et al. [43], only 20 out of 214 patients with focal hand dystonia who were treated with BoNT-A at least once continued with the treatment for 10 years or more. This number suggests quite a high dropout rate: more than 90 % within 10 years. Out of the 20 patients who were followed in this study for 10 or more years, nine subjects stopped the treatment for various reasons. The authors reported that two patients discontinued with the treatment due to insufficient response after 5 and 26 visits, two moved out of the area, and five were lost to follow-up.

Mari and co-workers studied why patients who stopped treatment with BoNT-A had done so [44]. Almost 50 % of these patients experienced a loss of satisfaction with BoNT-A injection therapy. The authors noted that the limited availability of trained local providers constituted an important factor in the lack of long-term treatment adherence.

Our Study We assessed the long-term treatment of (1) idiopathic cervical dystonia in patients regularly treated with BoNT-A, including changes in the BoNT-A dose, the possible impact on patient quality of life, and financial costs. Additionally, (2) in patients with writer's cramp, we were interested in their reasons for discontinuing long-term treatment with BoNT-A. We selected the patients with cervical dystonia and writer's cramp who were regularly followed up with relevant and precise documentation of their illness and treatment in their medical records from the Movement Disorders Centre Brno database.

11.5.3 Cervical Dystonia Study Group

Selection criteria:

(i) Diagnosis of primary focal cervical dystonia based on recent recommendations [37]
(ii) Regular treatment with BoNT-A for at least 10 years
(iii) Continuous good clinical response to the BoNT-A injections

The studied group consisted of 103 patients [63 females, 40 males; mean age 55 years (SD = ±12.1), mean disease duration 18.8 years (SD = ±9.40), and mean age at disease onset 38.4 years (SD = ±12.99)].

All the patients gave their informed consent before the initiation of the BoNT-A treatment. Before the first treatment with BoNT-A, all the patients underwent polymyographic recordings with needle electrodes to detect the neurophysiologically dystonic patterns of their cervical dystonia and to find the leading dystonic muscle. The injections of BoNT-A were performed under visual guidance with careful palpation of the targeted muscles; all the clinical cases of cervical dystonia were focal, not complex, forms. Before treatment with BoNT-A was initiated, the patients were instructed to carefully monitor the stages of improvement (none, minimal, mild, moderate, marked, striking, or complete relief of the cervical dystonia symptoms) and the onset of clinical response after the BoNT-A therapeutic session, including its duration and any possible side effects and their duration. BoNT-A (the brand Dysport® or Botox®) was diluted in 0.9 % NaCl/H_2O: 500 units of Dysport in 2.5 ml and 100 units of Botox in 2.0 ml [45]. The initial dose was adapted to the individual case, taking various parameters into account: muscle size, spasm intensity, degree of the weakness to obtain, and type of the torticollis. In all of the cases, the dose administered was always the minimum effective dose in order to avoid risks of antibody formation, high costs, and side effects. The efficacy of the treatment was assessed with the Clinical Global Impression (CGI) scale, the Tsui scale [7], and patients' self-evaluation [46]. Our decision to use the Tsui scale instead of the more complex Toronto scale was done historically: the Tsui scale had been established at the time when we started treating the patients. The Toronto scale is definitely better, but it takes too long to use in daily practice and is more suited to clinical studies. After 10 years of treatment, the questions regarding some aspects of quality of life were introduced to the patients [47]. The cost of the BoNT-A treatment per year was compared to previous peroral treatments (baclofen and clonazepam in most of the cases). The peroral medication given for the indication of cervical dystonia was stopped at the time of the BoNT-A initiation. The patients continued with their regular physiotherapy.

11.5.4 Writer's Cramp Study Group

Selection criteria:

(i) All the patients with focal hand dystonia or writer's cramp, who were examined at the Movement Disorders Centre Brno after January 2002
(ii) Availability of all necessary data for retrospective assessment

The patients were referred to the Movement Disorders Centre by their primary care physicians and neurologists with requests to initiate treatment with BoNT-A. There were 34 women and 25 men, with a median age of 48.4 (SD ± 10.6) years, in the studied group. We reviewed the medical records and tried to establish contact with patients who had discontinued the BoNT-A treatment.

During a 10-year period, 71 patients with focal dystonia were examined at our centre. Fifty-nine of them were started on BoNT-A treatment. Patients were treated for an average of 62 (SD ± 29) months. Each patient had at least two application sessions. We used EMG for diagnostic purposes only and did not use it during the BTX application itself. The treatment benefit was assessed according to the Clinical Global Impression of Change (CGIC) scale and the Patient Global Impression of Change (PGIC) scale.

11.5.5 Results: Cervical Dystonia Study Group

The mean follow-up period of the patients was 12.6 years (range 10–17 years). The total number of therapeutic sessions was 2,852; the cumulative dose of Dysport/

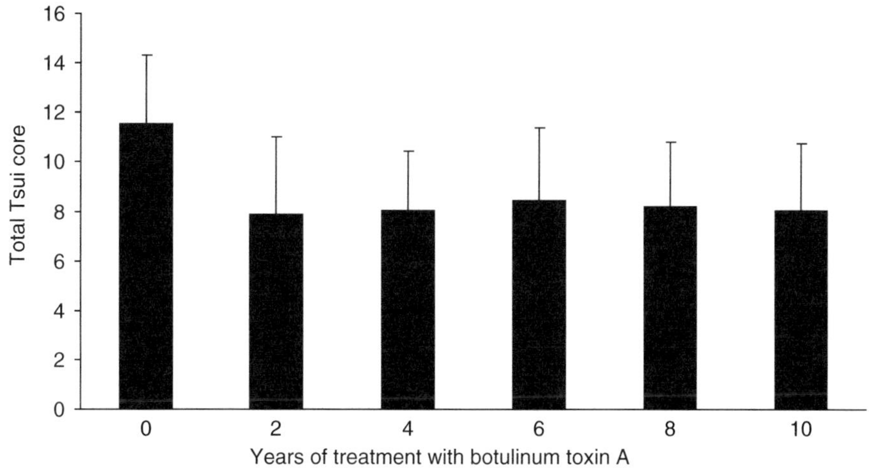

Fig. 11.1 Course of Tsui score

Botox was 755,589/193,741 units. The mean number of BoNT-A injections per patient was 38.5 (range 22–69); the mean dose per treatment session was 550 units of Dysport® (SD=±145) and 141 units of Botox® (SD=±29). The ratio between Dysport® and Botox® was 3.9. The mean period between two consecutive treatment sessions was 105 days (SD=±60.8). The number of the therapeutic sessions decreased from 9.1 (SD=±2.75) measured at year 2 to 6.5 (= ±1.2) measured at year 10. The mean Tsui score before the treatment was 11.5 (SD=±2.76); after 10 years, the Tsui score declined to 8.0 (SD=±2.75) (see Fig. 11.1, $p=0.00002$). It is a very important point that the Tsui scores were collected at the time of BoNT-A sessions, when the patient was experiencing the significant decline of the previous BoNT-A injection and the clinician had decided to perform another injection. We did not systematically collect Tsui scores at the time of the maximal clinically and subjectively observed improvements, so we do not report it here. After 10 years of treatment, the patients' mean CGI score was 1.9 (SD=±0.55) which corresponds to the significant improvement. The statistically significant need to change the BoNT-A dose is very interesting: Dysport/Botox mean dose increased from 500/130 units at year 0 to 620/163 units at year 10 (the increase of 24 % for Dysport® and 25 % for Botox®, respectively). Adverse events (dysphagia, local pain, weakness) were temporal and rare, occurring in less than 1 % of treatment sessions. Dysphagia was most commonly associated with the injection to the sternocleidomastoid muscle with doses exceeding 300/75 units of Dysport®/Botox® per therapeutic session. Based on the patient questionnaires, the most positive impact on their quality of life was in social communication (see Fig. 11.2). The cost of the BoNT-A treatment per one patient and year was approximately 32,000 Czech crowns (1,280 euros); the cost of the commonly used peroral treatment was 6,000 Czech crowns (240 euros).

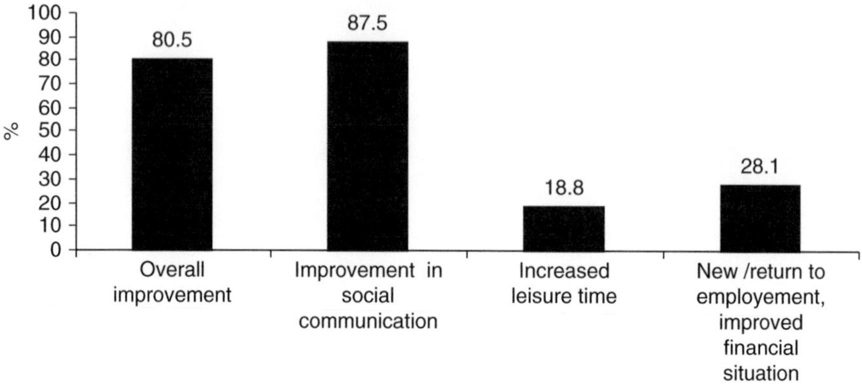

Fig. 11.2 Questionnaire outcomes

11.5.6 Results: Writer's Cramp Study Group

Of the treated patients, 83 % reported at least a partial improvement of their condition. There were no serious adverse events. The most frequent side effects were the weakness of muscles adjacent to the injection site and pain at the injection site.

Altogether, 30 (55 %) patients chose to discontinue the treatment: 20 patients (34 %) rejected further therapy after two sessions and ten patients (17 %) discontinued treatment after three to five treatment sessions. There were no further discontinuations. In all, 16 patients continued with the treatment past five and more sessions, and ten patients continued for 8 or more years.

There were several types of focal hand dystonia present among the followed patients. Thirty-three patients presented with writing-type focal hand dystonia (writer's cramp), 15 patients had mixed-type dystonia (not only present during writing but also during other purposeful hand movements), 5 patients had typing dystonia, and 6 suffered from focal hand dystonia during other tasks (such as eating with cutlery).

We asked the patient to state the most relevant factor that led them to decide to stop the BTX therapy.

The reasons for dropping out of the therapy were as follows:

Side effects (32 %)
Low efficacy (30 %)
Relocation (7 %)
Change of writing habits (7 %)
New symptoms and change of treatment (7 %)
Other and unknown reasons (17 %) (for details, see Fig. 11.3)

The most common side effects were muscle weakness and pain at the injection site, as already mentioned. All of the patients who stopped the treatment declared the muscle weakness limiting the performance of other tasks with the (usually

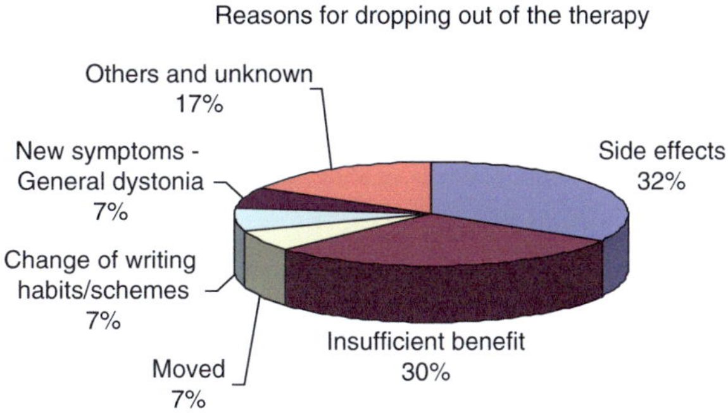

Fig. 11.3 Reasons for stopping botulinum toxin A (BoNT-A) therapy for writer's cramp ($N=59$)

dominant) hand as the primary reason for the discontinuation. Insufficient benefits occurred in nine (30 %) patients. Two patients relocated to other regions and were lost to follow-up. Two patients changed their writing habits (from handwriting to typing) with good results and no burdensome dystonia occurred with this task. In two patients, the dystonia became more generalised, and eventually the patients were switched to peroral treatment. One patient reported a good benefit after he switched BoNT-A for the oral treatment (clonazepam). We were not able to contact three of the patients and included them in the category of unknown reasons.

11.6 Discussion

11.6.1 Cervical Dystonia Study Group

We are aware of the weakness of the presented study: it is not a double-blind, controlled study [48–51]. However, there is a lack of controlled studies for many conditions, and more clinical trials in many different areas are warranted. Meanwhile, long-term open-label studies dealing with real clinical scenarios in routine clinical care, outside the rigid clinical studies' protocols, are needed, and many physicians are asking for them. The observed improvement of the clinical outcome evaluated and verified by Tsui score after 2 years of BoNT-A treatment with approximately the same course over subsequent years can address the issue of anticipated clinical outcome, which is very important for patients. The clinical measures obtained after 2 years of regular BoNT-A treatment are realistic for the next 8–10 years in this patient population. We can conclude that after several years of regular treatment, major fluctuations are limited and the condition becomes relatively stable. The literature presents biased information about the waning effects of BoNT-A injections. Some authors claim that it is rare to have to increase doses significantly as sessions progress [52], but others do not agree with this conclusion [53]. Based on our experience, there is a need to increase the total BoNT-A dose by approximately 25 % (see results section) which becomes apparent after 4 years of the treatment ($p=0.001$).

The presented study offers new insight into patient feelings about the treatment. We were somewhat surprised by their answers, which showed that the efficacy of the BoNT-A treatment for the severity of the condition (cervical dystonia) is not the single end point for their therapy outcome but that other topics (which are not often mentioned in other studies or the physicians do not think of them) are also important for patients: briefly, quality of life issues and improved financial situation. Spontaneous responses reflecting this were presented by the patients in the questionnaires, such as "I think that due to the botulinum toxin treatment I can work and live my normal life", "significant improvement in my body outlook", "improvement during driving, writing, reading", "significantly decreased pain, I can perform my home related activities much faster, I can walk longer", and "I had hoped for a complete recovery from my condition, but I am very satisfied with the current situation".

Further, more detailed evaluations of the long-term BoNT-A responders compared to the BoNT-A nonresponders are necessary to define a good response to BoNT-A treatment. BoNT treatment failure is a limiting factor for disabled patients with cervical dystonia because BoNT-A is the first-line and most effective treatment for patients. Then, surgery such as peripheral denervation or, recently, deep brain stimulation (DBS) of the globus pallidus can be considered in limited cases.

11.6.2 Writer's Cramp Study Group

The dropout rate for BTX treatment appears to be quite high. Lung et al. [43] reported insufficient therapy benefit as the most common reason for discontinuation (in 62.5 % of patients) [43]. Mari and co-workers reported a loss of satisfaction with BTX in about 50 % of patients. In our group of patients, we observed that side effects were the most common reason (33 %) for discontinuation, followed by insufficient benefits (30 %).

One of the weaknesses of our study may be that we did not use EMG guidance for application. But when comparing our study with another study where EMG was used, there appears to be no difference in the discontinuation rates [43].

It may be that even with the accurate targeting, the prevalence of side effects and other factors limit the usage of BoNT-A.

It may be useful to refine the dosage and application schemes for our patients with writer's cramp in order to improve the dropout rate and to offer them the optimal benefit in the future.

It may be that after living through the side effects, the patients re-evaluate their priorities and elect not to receive the therapy, coping with the writer's cramp by themselves by using other writing techniques, avoiding activities that cause dystonic posturing, etc.

Conclusions

Long-term treatment of cervical dystonia over a period of 10 years with BoNT-A is safe [54–57]. Chemical denervation with BoNT-A improves the quality of life and improves social communication [23, 58, 59]. Although the treatment with BoNT-A is more expensive than the peroral treatment, the indirect costs have to be strongly considered when initiating therapy for cervical dystonia. We would like to emphasise quality of life issues, a very important aspect which is vastly overlooked and should be strongly considered when initialising BoNT-A treatment together with precise goal setting in order to accommodate the patients' needs and expectations [60].

A relatively high proportion (55 %) of the patients with writer's cramp may choose to discontinue the treatment with BTX. The main reasons for discontinuation in our patient group were side effects (32 %), low efficacy (30 %), relocation (7 %), changed writing habits (7 %), new symptoms and change of treatment (7 %), and other and unknown reasons (17 %).

While BoNT-A remains an effective treatment, the dropout rate may be higher than in other focal dystonias, especially due to the incidence of side effects and the subjectively assessed low efficacy in some patients.

Acknowledgements This work was supported by the Ministry of Health of the Czech Republic/the Ministry of Health Departmental Research and Development Programme III (2010–2015) NT/13437 and by the "CEITEC – Central European Institute of Technology" project (CZ.1.05/1.1.00/02.0068) from the European Regional Development Fund.

We wish to express our thanks to Ing. Zdenek Novotny for the statistical analysis and to Anne Johnson for grammatical assistance.

We would like to thank our colleagues who treated and followed some of the patients, Petr Kaňovský, Hana Klajblová, Dagmar Kubová, Eduard Minks, Igor Nestrašil, Jiří Pulkrábek, Irena Rektorová, Silvie Sedláčková, Martina Bočková, Hana Streitová, and Jiří Zicha; our nursing staff, Vlasta Dírerová, Anita Hlučková, Jana Bártová, and Petra Krainerová; and Ludmila Vápeníková, who helped with part of the patient database.

References

1. Fahn S. Concept and classification of dystonia. Adv Neurol. 1988;50:1–8.
2. Singer C, Velickovic M. Cervical dystonia: etiology and pathophysiology. Neurol Clin. 2008;26 Suppl 1:9–22.
3. Molho E, Jankovic J, Lew M. Role of botulinum toxin in the treatment of cervical dystonia. Neurol Clin. 2008;26 Suppl 1:43–53.
4. Kaňovský P, Streitová H, Bareš M, Hortová H. Treatment of facial and orolinguomandibular tardive dystonia by botulinum toxin A: evidence of a long lasting effect. Mov Disord. 1999;14(5):886–8.
5. Kaňovský P, Bareš M, Streitová H, Klajblová H, Daniel P, Rektor I. Abnormalities of cortical excitability and cortical inhibition in cervical dystonia. Evidence from somatosensory evoked potentials and paired transcranial magnetic stimulation. J Neurol. 2003;250:42–50.
6. Cano SJ, Hobart JC, Fitzpatrick R, Bhatia K, Thompson AJ, Warner TT. Patient-based outcomes of cervical dystonia: a review of rating scales. Mov Disord. 2004;19:1054–9.
7. Tsui JK, Eisen A, Stoessl AJ, Calne S, Calne DB. Double-blind study of botulinum toxin in spasmodic torticollis. Lancet. 1986;2:245–7.
8. O'Brien C, Brashear A, Cullis P, Truong D, Molho E, Jenkins S, Wojcieszek J, O'Neil T, Factor S, Seeberger L. Cervical dystonia severity scale reliability study. Mov Disord. 2001;16:1086–90.
9. Sheehy MP, Rothwell JC, Marsden CD. Writer's cramp. Adv Neurol. 1988;50:457–72.
10. Dashtipour K, Pender RA. Evidence for the effectiveness of botulinum toxin for writer's cramp. J Neural Transm. 2008;115:653–6.

11. Karp BI. Limb dystonia. In: Stacy MA, editor. Handbook of dystonia. New York: Informa Healthcare; 2007. p. 155–77.
12. Pearce JMS. A note on scrivener's palsy. J Neurol Neurosurg Psychiatry. 2005;76:513.
13. Bell C. Partial paralysis of the muscles of the extremities. The nervous system of the human body. London: Taylor and Francis; 1833. p. 57–8.
14. Marsden CD, Sheehy MP. Writer's cramp. Trends Neurosci. 1990;13(4):148–53.
15. Jedynak PC, Tranchant C, de Beyl DZ. Prospective clinical study of writer's cramp. Mov Disord. 2001;16:494–9.
16. The Epidemiological Study of Dystonia in Europe (ESDE) Collaborative Group. A prevalence study of primary dystonia in eight European countries. J Neurol. 2000;247:787–92.
17. Butler AG, Duffey PO, Hawthorne MR, Barnes MP. An epidemiologic survey of dystonia within the entire population of northeast England over the past nine years. Adv Neurol. 2004;94:95–9.
18. Nutt JG, Muenter MD, Aronson A, Kurland LT, Melton 3rd LJ. Epidemiology of focal and generalized dystonia in Rochester, Minnesota. Mov Disord. 1988;3:188–1894.
19. Kruisdijk JJ, Koelman JH, Ongerboer de Visser BW, et al. Botulinum toxin for writer's cramp: a randomised, placebo-controlled trial and 1-year follow-up. J Neurol Neurosurg Psychiatry. 2007;78:264–70.
20. Das CP, Prabhakar S, Truong D. Clinical profile of various sub-types of writer's cramp. Parkinsonism Relat Disord. 2007;10:1016.
21. Sheehy MP, Marsden CD. Writers' cramp–a focal dystonia. Brain. 1982;105(Part 3):461–80.
22. Zeuner KE, Shill HA, Sohn YH, Molloy FM, Thornton BC, Dambrosia JM, Hallett M. Motor training as treatment in focal hand dystonia. Mov Disord. 2005;20(3):335–41.
23. Streitova H, Bares M. Long-term therapy of benign essential blepharospasm and facial hemispasm with botulinum toxin A: retrospective assessment of the clinical and quality of life impact in patients treated for more than 15 years. Acta Neurol Belg. 2014;114(4):285–91.
24. Simpson DM, Blitzer A, Brashear A, Comella C, Dubinsky R, Hallett M, Jankovic J, Karp B, Ludlow CL, Miyasaki JM, Naumann M, So Y; Therapeutics and Technology Assessment Subcommittee of the American Academy of Neurology. Assessment: Botulinum neurotoxin for the treatment of movement disorders (an evidence-based review): report of the Therapeutics and Technology Assessment Subcommittee of the American Academy of Neurology. Neurology. 2008;70(19):1699–706.
25. Haussermann P, Marczoch P, Klinger C, Landgrebe M, Conrad B, Ceballos-Baumann A. Long-term follow-up of cervical dystonia patients treated with botulinum toxin A. Mov Disord. 2004;19:303–8.
26. Skogseid IM, Kerty E. The course of cervical dystonia and patient satisfaction with long-term botulinum toxin A treatment. Eur J Neurol. 2005;12(3):163–70.
27. Mejia NI, Vuong KD, Jankovic J. Long-term botulinum toxin efficacy, safety, and immunogenicity. Mov Disord. 2005;20:592–7.
28. Mohammadi B, Buhr N, Bigalke H, Krampfl K, Dengler R, Kollewe K. A long-term follow-up of botulinum toxin A in cervical dystonia. Neurol Res. 2009;31(5):463–6.
29. Vivancos-Matellano F, Ybot-Gorrin I, Diez-Tejedor E. A 17-year experience of abobotulinumtoxin A in cervical dystonia. Int J Neurosci. 2012;122:354–7.
30. Ramirez-Castaneda J, Jankovic J. Long-term efficacy and safety of botulinum toxin injections in dystonia. Toxins (Basel). 2013;5(2):249–66.
31. Brin MF, Comella CL, Jankovic J, Lai F, Naumann M. Long-term treatment with botulinum toxin type A in cervical dystonia has low immunogenicity by mouse protection assay. Mov Disord. 2008;23:1353–60.
32. Cohen LG, Hallett M, Geller BD, Hochberg F. Treatment of focal dystonias of the hand with botulinum toxin injections. J Neurol Neurosurg Psychiatry. 1989;52:355–63.
33. Abbruzzese G, Berardelli A. Neurophysiological effects of botulinum toxin type A. Neurotox Res. 2006;9:109–14.
34. Valls-Sole J, Tolosa ES, Ribera G. Neurophysiological observations on the effects of botulinum toxin treatment in patients with dystonic blepharospasm. J Neurol Neurosurg Psychiatry. 1991;54:310–3.

35. Trompetto C, Curra A, Buccolieri A, et al. Botulinum toxin changes intrafusal feedback in dystonia: a study with the tonic vibration reflex. Mov Disord. 2006;21:777–82.
36. Karp BI. Botulinum toxin treatment of occupational and focal hand dystonia. Mov Disord. 2004;(Suppl 8):S116–9.
37. Albanese A, Barnes AP, Bhatia KP, Fernandez-Alvarez E, Filippini G, Gasser T, Krauss JK, Newton A, Rektor I, Savoiardo M, Valls-Sole J. A systematic review on the diagnosis and treatment of primary (idiopathic) dystonia and dystonia plus syndromes: report of an EFNS/MDS-ES Task Force. Eur J Neurol. 2006;13:433–44.
38. Blackie JD, Lees AJ. Botulinum toxin treatment in spasmodic torticollis. J Neurol Neurosurg Psychiatry. 1990;53:640–3.
39. Djebbari R, du Montcel ST, Sangla S, Vidal JS, Gallouedec G, Vidailhet M. Factors predicting improvement in motor disability in writer's cramp treated with botulinum toxin. J Neurol Neurosurg Psychiatry. 2004;75(12):1688–91.
40. Hsiung GY, Das SK, Ranawaya R. Long-term efficacy of botulinum toxin A in treatment of various movement disorders over a 10 years period. Mov Disord. 2002;17:1288–93.
41. Turjanski N, Pirtosek Z, Quirk J. Botulinum toxin in the treatment of writer's cramp. Clin Neuropharmacol. 1996;19:314–20.
42. Karp BI, Cole RA, Cohen LG, Grill S, Lou JS, Hallett M. Long-term botulinum toxin treatment of focal hand dystonia. Neurology. 1994;44:70–6.
43. Lungu C, Karp BI, Alter K, Zolbrod R, Hallett M. Long-term follow-up of botulinum toxin therapy for focal hand dystonia: outcome at 10 years or more. Mov Disord. 2011;26(4):750–3.
44. Mari Z, Bruno M, Lee B, Karp BI, Hallett M. Long-term botulinum toxin treatment for focal hand dystonia. Neurology. 2004;62(7, Suppl 5):A512.
45. Ranoux D, Gury C. Practical handbook on botulinum toxin. Marseille: Solal Editeurs; 2007.
46. Dressler D. Botulinum toxin therapy. Stuttgart/New York: Thieme-Verlag; 2000.
47. Hilker R, Schischniaschvili M, Ghaemi M, Jacobs A, Rudolf J. Health related quality of life is improved by botulinum neurotoxin type A in long term treated patients with focal dystonia. J Neurol Neurosurg Psychiatry. 2001;71(2):193–9.
48. Jankovic J, Orman J. Botulinum A toxin for cranial-cervical dystonia A double-blind, placebo-controlled study. Neurology. 1987;37:616–23.
49. Greene P, Kang U, Fahn S, Brin M, Moskowitz C, Flaster E. Double-blind, placebo-controlled trial of botulinum toxin injections for the treatment of spasmodic torticollis. Neurology. 1990;40:1213–8.
50. Wissel J, Kaňovský P, Růžička E, Bareš M, Hortová H, Streitová H, Jech R, Roth J, Brenneis C, Müller J, Schnider P, Auff E, Richardson A, Poewe W. Efficacy and safety of a standardised 500 unit of Dysport® (clostridium botulinum type A haemaglutinin complex) in a heterogeneous cervical dystonia population: results of a prospective, multicentre, randomised, double-blind, placebo-controlled, parallel group study. J Neurol. 2001;248(12):1073–8.
51. Charles D, Brashear A, Hauser RA, Li HI, Boo LM, Brin MF. Efficacy, tolerability, and immunogenicity of onabotulinumtoxin A in a randomized, double-blind, placebo-controlled trial for cervical dystonia. Clin Neuropharmacol. 2012;35:208–14.
52. Jankovic J. Botulinum toxin therapy for cervical dystonia. Neurotox Res. 2006;9(2–3):145–8.
53. Vogt T, Lüssi F, Paul A, Urban P. Long-term therapy of focal dystonia and facial hemispasm with botulinum toxin A. Nervenarzt. 2008;79(8):912–7.
54. Kessler KR, Skuta M, Benecke R. for German Dystonia Study Group. Long-term treatment of cervical dystonia with botulinum toxin A: efficacy, safety, and antibody frequency. J Neurol. 1999;246:265–74.
55. Mohammadi B, Kollewe K, Wegener M, Bigalke H, Dengler R. Experience with long-term treatment with albumin-supplemented botulinum toxin type A. J Neural Transm. 2009;116(4):437–41.
56. Truong D, Brodsky M, Lew M, Brashear A, Jankovic J, Molho E, Orlova O, Timerbaeva S. Long-term efficacy and safety of botulinum toxin type A (Dysport) in cervical dystonia. Parkinsonism Relat Disord. 2010;16:316–23.

57. Gill CE, Manus ND, Pelster MW, Cook JA, Molinari AL, Charles D. Continuation of long-term care for cervical dystonia at an Academic Movement Disorders Clinic. Toxins (Basel). 2014;5:776–83.
58. Kaňovský P, Bareš M, Streitová H, Pospíšilová D, Dufek M. Socioeconomic aspects of the treatment of facial dyskinesias with botulinum toxin A. Czech Slov Neurol Neurochir. 1996;59/92(3):148–51.
59. Skogseid IM, Røislien J, Claussen B, Kerty E. Long-term botulinum toxin treatment increases employment rate in patients with cervical dystonia. Mov Disord. 2005;20(12):1604–9.
60. Jinnah HA, Berardelli A, Comella C, DeFazio G, DeLong MR, Factor S, Galpern WR, Hallett M, Ludlow CL, Perlmutter JS, Rosen AM. for the Dystonia Coalition Investigators. The focal dystonias: current views and challenges for future research. Mov Disord. 2013;28(7):926–43.

12 Surgical Treatment of Dystonia

Jürgen Voges

12.1 Surgical Techniques in the Treatment of Dystonia: Historical Overview

Prior to the introduction of botulinum toxin (BTX), medical treatment options for patients with focal or segmental dystonia were markedly limited. With this historical background in mind, it is reproducible that for a long time, surgical ablative techniques either used extracerebrally (e.g. myectomy, myotomy, rhizotomy or denervation) or intracerebrally (e.g. thalamotomy, pallidotomy) were the only therapeutic alternatives to improve disabling and deforming dystonic symptoms.

12.1.1 Extracerebral Treatment of Dystonia

Already in the nineteenth century, surgeons attempted to treat cervical dystonia (CD), which represents the most frequent phenotype of dystonia [1]. Superior goal of surgery was reduction of the pathologically strengthened muscle tone. Because myotomy of the sternocleidomastoid muscle (SCM) alone turned out to be ineffective for complete improvement, myotomy was extended and various denervation techniques were developed [2]. In 1891, Keen proposed rhizotomy using a unilateral intraspinal approach with section of the dorsal proportion of C1–C3 [3]. Finney and Hughson recognised the benefit of expanded bilateral procedures and favoured bilateral extraspinal rhizotomy together with resection of the spinal accessory nerve, removal of the greater occipital nerve and of the second and third cervical nerves at their emergence from the vertebral foramina [4]. In 1930, Dandy introduced a more radical, bilateral intraspinal approach with division of the anterior and posterior roots

J. Voges
Department of Stereotactic Neurosurgery, Otto von Guericke University,
University Hospital and Leibniz Institute for Neurobiology, Magdeburg, Germany
e-mail: juergen.voges@med.ovgu.de

P. Kanovsky et al. (eds.), *Dystonia and Dystonic Syndromes*,
DOI 10.1007/978-3-7091-1516-9_12

from C1 to C3, section of the spinal accessory nerve at the level of the foramen magnum and division of the spinal accessory nerves in the neck [5]. McKenzie modified this procedure in this respect that in the first place the spinal accessory branch to the SCM involved in dystonia was sectioned (preserving the branch to the trapezius) and that the intraspinal approach including division of the anterior roots C1–C4 at the more affected side and of C1–C3 on the less involved side was added if necessary [6]. Bertrand favoured extradural denervation only and coined the term "selective peripheral denervation" (SPD) to characterise a procedure disconnecting only rami and nerves to the dystonic muscles at the same time sparing the muscles, which are not involved [7–9]. Whilst these techniques were used rather inflexible even if the surgeons mixed the different approaches, Grossmann and Hamilton applied them in a more flexible, individualised way performing selective extradural denervation and/or rhizotomies according to the patient's individual symptomatology [10].

12.1.2 Intracerebral Treatment of Dystonia

With the evolution of stereotactic neurosurgery starting in the 1950s, circumscribed tissue lesions in intracerebral relay nuclei or fibre connections were applied also in dystonia patients. Although the pallidum and its connections to the thalamus were important anatomical substrates in pathophysiological concepts of dystonia, most patients were treated with thalamic lesions and not with ablative surgery of the pallidum, from 1950 to 1980. Other non-thalamic targets used at this time were the caudate nucleus, dentate nucleus, longitudinal medial fascicle, globus pallidus (internal segment), interstitial nucleus of Cajal (INC), nucleus ruber and olivary nucleus [11–13].

Hassler and Dieckmann preferred as targets for the treatment of dystonia the ventrolateral thalamus (n. ventrooralis anterior (v.o.a.), n. ventrooralis posterior (v.o.p.) and n. ventrointermedius (v.im.) according to Hassler's classification) and the white matter below the thalamus (Forel's field H1 and H2) aiming at the selective reduction of thalamic input from the interstitiothalamic and pallidofugal pathways as well on the elimination of vestibular influence on head posture and neck muscle [14]. Hassler and Dieckmann were probably the first who choose in CD patients the target according to the individual leading symptom [15, 16]. Cooper proposed for patients with CD ablation of the somatosensory thalamus (ventral posterolateral (VPL) nucleus or ventral posteromedial (VPM) nucleus according to the Jones classification) and the centrum medianum (CM) in order to destroy all pallidal and cerebellar afferents to the thalamus [17]. In the case of refractory symptoms, he extended surgery by ablation of the pulvinar thalami [18]. Sano introduced lesioning of the INC for the treatment of CD because in his experience stimulation at distinct points in this part of the thalamus could elicit anteflexion and retroflexion of the patient's head [19].

It can be assumed that Cooper performed the largest number of thalamotomies for the treatment of dystonia. Between 1957 and 1977, he treated 160 patients for CD. Copper classified the outcome as "satisfactory" in 60 % and "not improved" in

40 % of the cases. Most frequent side effects were "dysphonic symptoms" (20 %) [20]. In patients with generalised dystonia (226 patients, operated between 1955 and 1974), Cooper extended the thalamotomy caudally to destroy all pallidal and cerebellar afferents to the thalamus. Patients with still persisting and disabling symptoms underwent repeated surgery with up to 13 lesions per patient including also pulvinotomy. Most patients had follow-up times of approximately 8 years. The response to surgery was "good" in 25 % and "moderate" in 45 % of the cases. The operative mortality of 0.7 %, however, increased to 2 % when surgery was repeated or performed bilaterally. Bilateral ablations caused also speech disturbances in 15 % of the patients [18].

Loher et al. referred to an unpublished series of Mundinger with 162 patients treated for CD between 1972 and 1986 [12]. A doctoral thesis summarised the long-term outcome of 111 patients, who underwent unilateral (92 patients) or bilateral surgery (19 patients) [21]. In most cases, lesions were made in the zona incerta (ZI) alone (27 %) or in combination with destruction of the v.o.a. thalamic nucleus (56 %). Infrequently used targets were the pulvinar thalami or the INC. Approximately half of the patients had a long-lasting benefit from surgery, which was in 30 % of the cases classified as "excellent" or "good". Most frequent side effects were limb paresis (16 %) and dysarthria (15 %). The rate of adverse events was higher after bilateral surgery (68 %) compared to unilateral ablation (40 %) [21]. The latter point addresses an important problem of bilateral lesioning, in particular of bilateral thalamotomy, because its higher efficacy, namely, on axial symptoms, had to be balanced against the higher risk. In most series side effects after bilateral surgery developed in 20–70 % of the patients [22].

In the late 1970s, the number of lesional stereotactic procedures, particularly of those performed for CD, dropped for several reasons. Besides a general decline of movement disorder surgery at that time, the introduction of selective peripheral denervation with a lower rate of severe side effects and the efficacy of BTX-therapy, which was proven thereafter, contributed mainly to this development.

The internal segment of the pallidum (GPi) as target for the treatment of dystonia gained renewed interest when results after the treatment of parkinsonian patients showed that pallidotomy not only improved L-DOPA-induced dyskinesia but also off-dystonia [23]. However, even in the late 1990s, only a few publications addressed pallidotomy for the treatment of dystonia. Symptoms improved in four patients after unilateral pallidotomy in average by 74 % (preoperative BFMDRS movement scores, 26–41). Another four patients who were more severely disabled (preoperative movement scores, 48–75) underwent bilateral pallidotomy. The average percentage improvement was also 74 % [24–26].

Notably prior to the 1990s, a period regarded as the onset of systematic application of deep brain stimulation (DBS), there were already attempts to improve dystonic symptoms by direct electrical stimulation of subcortical structures supposed to be involved in the pathophysiology of this disease. In 1977, Mundinger from Freiburg, for instance, summarised in a German journal his experiences in stimulating the v.o.a. and v.o.i. thalamic nuclei as well as the ZI. Unilateral (six patients) or bilateral (one patient) intermittent stimulation with frequencies of 5 Hz or 25–150 Hz improved CD in all seven patients for a period of 4–6 h [27]. Electrical stimulation of the internal capsule as performed by Cooper in one CD patient improved also the position of head and neck [28]. At the same time, Andy used successfully unilateral thalamic stimulation in patients with CD [29]. However, mainly influenced by the positive results achieved with pallidal lesioning in patients with Parkinson's disease, the ventral posterolateral (VPL) segment of the GPi became finally the primary target for chronic intracerebral electrical stimulation of dystonia.

12.2 Myotomy, Rhizotomy and Peripheral Denervation

Myotomies and/or myectomies were almost exclusively performed in patients with CD. This surgical technique has also been used to treat focal dystonia such as blepharospasm (single frontal suspension or frontal myectomy) or spasmodic dysphonia (myectomy of the thyroarytenoid muscle) (overview in [30]). These rare indications, however, are not the focus of this chapter.

In CD patients, myotomy or myomectomy is aimed at disconnecting the sternocleidomastoid muscle (SCM) from its distal attachments (sternal and clavicular) or at the reduction of muscle stretch through multiple incisions including also other muscle groups responsible for the dystonic symptoms such as the trapezius muscle or several posterior cervical muscle groups. Already at the beginning of standardised peripheral resective surgery for CD, myotomy was in some cases accomplished by excision of spinal accessory nerves. According to descriptive data from early small case series, the symptoms improved markedly in all patients [2].

In 1981, Chen published a larger series of CD patients ($N=60$), who were predominantly treated with selective resection of cervical muscles but also with sparsely applied rhizotomies depending on individual symptoms. In cases with rotational components of CD, unilateral removal of the splenius capitis muscle (contralateral to chin turning) was combined with sectioning of the spinal accessory nerve root to reduce also the muscle tone in the ipsilateral SCM and trapezius. Anterocollis was treated by bilateral section of the spinal accessory nerves only and laterocollis by bilateral resection of parts of the suboccipital muscles. Patients presenting with lateroflexion of the head underwent resection of the splenius capitis and levator scapulae muscles on the flexed side plus section of the spinal accessory nerve. Following surgery, symptoms had completely disappeared in 33 cases (55 %) and had markedly improved in 17 patients (28.3 %). Surgery gained partial improvement in nine cases (15 %) and was ineffective in only one patient (1.7 %). Complications did not occur [31].

Since the 1990s, SPD became the procedure of choice for the peripheral functional treatment of CD. Several publications reported results from monocentric studies addressing the efficacy of SPD (Table 12.1). For the following reasons, however, a conclusive appraisal of these data is difficult: (1) most studies were retrospective, (2) the outcome was assessed using different measures, (3) most series contained patients operated in the pre-BTX era, and (4) factors determining essentially the outcome of CD patients such as selection of patients, severity of dystonia, impact of disability and pain were only addressed in the minority of the studies.

Table 12.1 Selective peripheral denervation for the treatment of cervical dystonia

Citation/study type	FU	No. pts.	Symptoms/previous therapy (no. pts.)	Surgery (no. pts.)	Outcome		Adverse events
Bertrand (1993)/ retrospective	n. g.	260	n. g.	SPD	Excellent	106 pts. (40 %)	Sensory loss: N. occ. major Atrophy: post. muscle group
					Very good	124 pts. (48 %)	
					Fair	27 pts. (10 %)	
					Poor	3 pts. (1 %)	
Braun et al. (1994)/ retrospective	25 m (mean)	49	Symptoms not specified BTX: 13/49 pts.	SPD 2nd surgery: 6	Sign. improvement (pain + dystonia)	37 pts. (76 %)	Perm. sensory loss: N. occ. major (variable degree) Transient sensory loss/ neuralgia (other nerves): 18 %
					Minor effect/no improvement	12 pts. (24 %)	
Kraus et al. (1997)/ retrospective	6.5 yr (mean)	46	1 symptom/1 vector (10)[a] 2 symptoms/2 vectors (27)[a] 3 symptoms/3 vectors (9)[a] Medication therapy: 29 pts. BTX: 24 pts.	70 individually staged procedures in 46 pts.: Intradural (33) Extradural (21) Muscle sections (22)	Excellent	9 pts. (21 %)	Transient/mild: 10 % Persistent (disturbed balance): 2.1 %
					Marked	12 pts. (27 %)	
					Moderate	9 pts. (21 %)	
					Mild	9 pts. (21 %)	
					No response	5 pts. (11 %)	
					TWSTRS improvement (34 pts. evaluated)		
					Severity	51.8%**	
					Function	57.3%**	
					Pain	44.8%**	

(continued)

Table 12.1 (continued)

<table>
<tr><th>Citation/study type</th><th>FU</th><th>No. pts.</th><th>Symptoms/previous therapy (no. pts.)</th><th>Surgery (no. pts.)</th><th colspan="2">Outcome</th><th>Adverse events</th></tr>
<tr><td>Ford et al. (1998)/ retrospective</td><td>5.2 yr (mean)</td><td>16</td><td>Idiopathic dyst. (15)
Tardive dyst. (1)
Complex dystonic movements (11)
Mainly pain (14)
BTX-A: all pts</td><td>SPD
2nd surgery: 6
3rd surgery: 1</td><td colspan="2">Moderate/complete return of normal head function: 6/16 pts.
Minor improvement: 10 pts.
TWSTRS (mean): 15.4 % improvement (n.s.)</td><td>n. g.</td></tr>
<tr><td rowspan="6">Münchau et al. (2001)/ prospective</td><td rowspan="6">16.7 m (mean)</td><td rowspan="6">37</td><td rowspan="6">CD with primary or secondary BTX failure</td><td rowspan="6">SPD + myectomy-upper 1/3 of SCM
2nd surgery: 2</td><td colspan="2">TWSRTS (at 12 months) – functionally relevant improvement in 68% of the patients</td><td>Wound infection: 1/37
Progressive spread of dystonia: 1/37
Perm. sensory loss: all pts.
Transient deficits: 8/37</td></tr>
<tr><td>Total score</td><td>30 %**</td><td rowspan="4"></td></tr>
<tr><td>Severity</td><td>20 %*</td></tr>
<tr><td>Disability</td><td>40 %*</td></tr>
<tr><td>Pain</td><td>30 %*</td></tr>
<tr><td colspan="2">Global outcome score: not significantly improved</td><td></td></tr>
</table>

Cohen-Gadol et al. (2003)/ retrospective	3.4 yr	168	Rotational torticollis: 84 % Laterocollis: 35 % Retrocollis: 12 % Anterocollis: 7 % 1 vector: 12% (20)[a] 2 vectors: 42% (70)[a] BTX_ 144/168 Additional medication: almost all pts.	SPD 2nd surgery: 17	At last FU (130/168 pts.)		Wound infection: 1 Myocard infarction: 1 Pulmonary embolism: 1 Respiratory arrest/ tracheostoma: 1 Death after respiratory arrest: 1 Shoulder weakness Transient: 1 Permanent: 2
					Head position		
					Moderate/marked improved	75 pts. (58%)	
					Slight improved	28 pts. (22%)	
					No benefit	21 pts. (16%)	
					Continued to worsen	6 pts. (5%)	
					Pain		
					Moderate/marked improved	76 pts. (58%)	
					Slight improvement	20 pts. (15%)	
					No benefit	25 pts. (19%)	
					Continued to worsen	9 pts. (7%)	
Huh et al. (2010)/ retrospective[b]	29.5 m (mean)	16	Dystonic symptoms not specified separately for SPD patients/previous therapy not mentioned	SPD	TWSRTS score: improvement Subjective outcome Excellent: 4 pts Good: 10 pts Fair: 2 pts	59%	Wound infection: 1 Occipital neuralgia: 3 Delirium: 1 Venous air embolism: 1

FU follow-up, *No.* number, *pts.* patients, *BTX* botulinum toxin, *m* months, *n.g.* not given, *n.s.* not significant, *pts.* patients, *SPD* selective peripheral denervation, *SCM* sternocleido muscle, *TWSTRS* Toronto Western Spasmodic Torticollis Rating Scale, *yr.* years

*($P<0.01$); **($P<0.0001$)

[a]Symptoms/axis/vectors: indicate patterns of abnormal head posture (rotation, tilt, antero-or retrocollis). Presence of only one position is marked "1", combinations are marked "2" or "3"

[b]Study comparing patients treated with SPD with a second group treated for CD with deep brain stimulation

In 1993, Bertrand reported the data of 260 patients operated with SPD for CD over a period of approximately 15 years. Patients were considered for surgery, when conservative therapy was insufficient including also BTX in a subset of patients. Preoperatively, most patients presented with rotational torticollis (involvement of the SCM on one side and of posterior cervical muscle group (PCG) on the other side). Essential steps of the surgical technique were denervation of the PCG saving the deep paraspinous muscles (extradural section of the roots of C1 and C2 together with section of the posterior primary divisions (rami) of C3, C4, C5 and C6 and denervation of the SCM). The side of SCM denervation and extent of posterior sectioning depended on the individual symptoms. According to a self-developed classification system, improvement after surgery was in 88 % of the patients either excellent (no detectable abnormal movements) or very good (slight deviation or slight residual movements) [7].

Fifty patients of a series published by Braun and Richter were operated over a time period of 5 years. Different from the technique of Bertrand, the authors modified the surgical approach to expose the posterior nerve branches [9, 32]. Candidates considered for surgery should have symptoms lasting for a minimum of 12 months and being refractory to conservative treatment. In most patients, the ipsilateral or contralateral splenius capitis and SCM contributed to the torticollis. Postoperatively, patients were asked to self-estimate the overall effect of surgery. Substantial improvement of dystonia and pain in the order of 76 % was roughly in the same range as reported by Bertrand [7, 32]. In two cases with severe iatrogenic injury of the trapezius branch of the spinal accessory nerve, end-to-end anastomosis of this structure leads to full recovery of muscle function [32].

Krauss et al. assessed retrospectively the outcome of 46 out of 52 CD patients (classified idiopathic in 35 and secondary in 11 patients) [33] treated with an individually adopted surgical technique combining intradural rhizotomy, peripheral denervation and myotomy [10]. Inclusion criteria for surgery were existence of disabling and/or socially embarrassing symptoms, no response to or intolerance of medication therapy, no adequate benefit to BTS or loss of efficacy with repeated injections. Postoperatively, the global outcome determined for all 46 patients was excellent or marked in 48 % of the cases. The Toronto Western Spasmodic Torticollis Rating Scale (TWSTRS) subscores for severity, functional disability or pain had improved significantly (determined in 36/46 patients). After surgery, 18/46 patients required BTX injections and 20/46 patients additional medical treatment [33].

Ford et al. analysed retrospectively the long-term course of 16 CD patients [34]. Surgery included ramisectomy ipsilateral to SCM denervation in the case of laterocollis and bilateral section of cervical roots and posterior rami (C1 to C5) in patients with retrocollis [7]. Surgery was repeated in seven cases due to declining benefit. The authors obtained clinical information from medical records, records of BTX injections and operative reports, from videotape examinations (14/16 patients with baseline and follow-up records) and from standardised interviews. In the long-term course, dystonia had improved moderately (enables the patient to function with only minor limitation or discomfort) or completely (resolution of torticollis and return to normal function) in 37.5 % of the patients. The video-based improvement of the TWSTRS score, however, was not significant. In contrast, patients in a control group (BTX treatment only) improved statistically significant by 26.5 %. Despite improvements of head position, the functional improvement was in most cases not sufficient to return to full-time work after surgery [34].

Münchau et al. published the only study investigating prospectively the efficacy of SPD for botulinum resistant dystonia to date [35]. Forty of 62 preoperatively assessed patients were included and 37/40 cases underwent surgery. The surgical procedure was basically standardised with selective denervation of SCM and posterior ramisectomy (C2–C5, occasionally C6 or C7 and C1) in combination with myectomy of the upper third of SCM and division of the tendons of the levator scapulae. Pre- and postoperative assessments included detailed clinical examination together with video documentation and completion of a booklet consisting of questionnaires quantifying different psychosocial functions. Postoperative blinded video-based rating of CD severity was complicated because of inconsistent documentation, but pre- and postoperative ratings by the examiner and video reviewer were highly correlated. In 31 patients with a follow-up of 12 months, surgery had improved the TWSTRS global score and all subscores significantly in comparison to baseline values. Referred to different clinical aspects, surgery had reduced the degree of abnormal dystonic head rotation and tilt but did not change the direction of abnormal head posture and did not improve head tremor.

The primary failure to BTX therapy (six patients) seemed to be a factor associated with no significant improvement of TWSTRS scores if compared to patients treated with SPD for CD after secondary failure (26 patients). Data obtained from the completed booklet questionnaires (12/37 patients) showed significant postoperative improvement in most tests measuring psychosocial function. At an average of 16 weeks postoperatively, surgery had not substantially affected anxiety, acceptance of illness and self-esteem. Scores obtained on an average of 24 months after surgery were, in general, not different if compared to the 16-week results. Only self-reported depression had deteriorated significantly over time. Eleven of 22 patients (50 %) without preoperative dysphagia reported mild or moderate swallowing problems at 3 months postoperatively, and 4/37 patients (10.8 %) presented with prominent side shift of the head caused by ipsilateral denervation of the SCM [35].

The primary aim of the retrospective long-term analysis (follow-up period range, 3 months–11.4 years) published by Cohen-Gadol et al. was the identification of preoperative factors, which may be predictive for the outcome of CD patients after peripheral functional surgery. Within 8 years, 168 consecutive patients were operated with SPD for torticollis by one neurosurgeon. Denervation was repeated in 17 patients (10 %) for recurrence of their symptoms. Cervical dystonia was in most patients characterised as idiopathic. Additional symptoms observed in some cases were phasic head movements (29 patients) and head tremor (43 patients). Prior to surgery, most patients were treated with BTX-A and varying medications. At the most recent follow-up, postoperative improvement was rated as "marked" (almost normal head position with significant reduction in pain and/or spasms) or "moderate" (some residual dystonic head deviation and pain/or spasms) in more than 50 % of the cases. None of the investigated variables (e.g. preoperative head position, patient gender, severity of abnormal head posture, etc.) were a predictor for the postoperative outcome. Surgery caused a complication rate in a total of 4.8 %. In particular, the mortality rate was 0.6 %, and the frequency of permanent (moderate) morbidity was 1.2 % [36].

Huh et al. compared retrospectively the outcome after SPD (16 patients) with postoperative results of 8 patients treated in the same institution for CD with bilateral high-frequency GPi-DBS [37]. One DBS case underwent SPD prior to stimulation therapy. Monosymptomatic CD was present in eight cases, a combination of rotational movement or lateral-, retro- or anterocollis in 16 cases and associated neck pain in 18 patients. Electrical stimulation was the preferred method in patients with additional tremor or tremulous CD, whilst SPD was favoured in cases which had primarily responded to BTX. Peripheral surgery improved the TWSTRS score by 59 % and GPI-DBS by 64 % ($P=0.326$). GPI-DBS, however, seemed to have a stronger effect on pain reduction than SPD [37].

12.3 Deep Brain Stimulation (DBS)

The first small case series reported at the end of the 1990s covered different types of dystonia. Three patients with CD treated by Krauss et al. responded gradually within the first 3 months after initiation of DBS according to the modified TWSTRS total score and subscores (improvement of severity, 41–65 %; pain, 25–71 %; disability, 41–71 %) and presented with some more improvement at later follow-up (6–15 months) [38]. In 1999, Coubes et al. published a case of a young girl with severe primary generalised dystonia (PGD) who underwent pallidal surgery in 1996 and responded dramatically already after 6 weeks of electrical stimulation [39]. One year later, a second report of the same group confirmed the efficacy of bilateral GPI-DBS in seven cases (six children, one adult) with DYT1-positive PGD with a mean improvement of 90.4 % according to the Burke-Fahn-Marsden Dystonia Rating Scale (BFMDRS) [40]. At the same time, Tronnier et al. presented three dystonia patients. Two cases with primary dystonia responded significantly to DBS (BFMDRS score improvements, 58.9 and 34.4 %, respectively); the patient with secondary dystonia had only minor benefit from the therapy (14.3 % improvement) [41].

To get more insight into the pathophysiology of dystonia and changes of cerebral activity under GPi-DBS. Kumar et al. investigated one patient (primary generalised dystonia) with positron emission tomography using the tracer [^{15}O]H_2O.

Bilateral GPi-DBS improved dystonia by 67 %. Whilst the patient performed joystick movements, electrical pallidal stimulation reduced PET activation bilaterally in the primary and association motor cortex [42]. Excessive activation of these brain areas is documented for idiopathic dystonia as well as for acquired hemidystonia [43, 44].

12.3.1 DBS for Primary Dystonia

12.3.1.1 Primary Generalised Dystonia

Two controlled studies confirmed on the highest evidence level for patients suffering from primary segmental or generalised dystonia improvements of motor symptoms, motor disability and quality of life (QoL) in response to bilateral GPi-DBS [45, 46]. A prospective multicentric study conducted in France (the French SPIDY study group) included 22 patients with generalised dystonia (mean age at surgery, 30 years; mean age at the onset of dystonia, 8 years; mean disease duration, 18 years, seven cases with DYT1 mutation) [46]. Basing upon videotapes, an independent rater assessed the motor scores prior to surgery and 3, 6 and 12 months postoperatively. At 3 months after surgery, evaluation was performed with and without DBS adjusted in random order on separate days. In all patients, the positive effect of neurostimulation on the movement subscale was significant (34.6 ± 22.3 (stim-OFF) vs. 24.6 ± 17.7 (stim-ON)). Also 12 months postoperatively, both movement and disability scores of the BFMDRS (primary outcome measures) improved significantly by 50 %. Post hoc analysis gave a clear hint that patients with predominantly phasic dystonic problems responded to a much greater extent than those with severe tonic posturing. Preoperatively, 20 patients were under medical treatment for dystonia, and at the last follow-up, 18 patients still received specific medication [46].

The 3-year follow-up study of these patients (video-based scoring by an independent expert, intention to treat analysis) showed that the overall motor improvement once achieved after 1-year DBS (mean improvement, 51 %) remained stable (mean improvement, 58 %). Total movement scores and subscores for lower and upper limbs were even significantly better at 3 years compared to the 1-year scoring. Also other variables or skills such as QoL, cognition, concept formation and reasoning or memory did not change between the 1-year and 3-year follow-up study. Additional improvement during this time period was observed for some subitems of the QoL SF-36 questionnaire addressing general health, physical functioning and pain as well as for executive functions. Five adverse events occurred including three serious adverse events (lead fracture, removal of DBS system because of infection) [47].

The second controlled multicentric study with participating academic centres from Germany ($n = 10$), Austria ($n = 1$) and Norway ($n = 1$) recruited 40 patients with segmental (16/40 patients) or generalised (24/40 patients) dystonia and used bilateral GPi-DBS according to a randomised, double-blind, sham-controlled clinical protocol [45]. Three months after surgery (end of the blinded study phase, 20 patients randomly assigned for effective DBS), two independent investigators assessed the severity of dystonia (BFMDRS movement score) on the basis of videotapes. At this time point, stimulated patients had a significant greater mean improvement than non-stimulated patients (movement score, 39.3 % vs. 4.3 %; disability score, 37.5 % vs. 8.3 %). In the following 6 months (open study phase), also those patients not assigned for DBS in the blinded study phase were stimulated and responded with an improvement comparable to the outcome of the cases already stimulated during the blinded phase. Compared to baseline values, the final improvement of all patients was 16.7 ± 13.0 (45.9 %, $p < 0{,}001$) for the BFMDRS movement score and 4.1 ± 3.6 (41.0 %, $p < 0{,}001$) for the disability score. Twenty patients received specific medication for dystonia at study inclusion. Under GPi-DBS, medication therapy was entirely discontinued in five cases and the dosage reduced (average reduction, 37 %) in the remaining cases [45].

Volkmann et al. compared the outcome of this patient cohort at 3 years (31 patients) and 5 years (32 patients) with baseline values (intention to treat analysis). Unmasked investigators in the participating centres performed the BFMDRS ratings. The mean improvement after 3-year (movement score, +64.1 % (−27.8 points); disability score, +41.0 % (−4.1 points) and 5-year DBS (movement score, +60.5 % (−20.0 points); disability score, +47.0 % (−4.7 points) was significant compared to baseline. Between the third postoperative year and the fifth postoperative year, the scores remained stable. Beneficial response defined as ≥25 % improvement of BFMDRS movement score under GPi-DBS was observed in 29/31 patients (94 %) at 3 years and in 26/32 patients (81 %) at 5 years. Depression and anxiety ratings were significantly improved at 6 months and remained stable until 5 years. Stimulation parameters at 5 years (3.2 (SD, 0.8) V, 129.0 (SD, 35.7) μs, 162.9 (SD, 34.6) Hz) were not statistically different from the parameters at 6 months (3.5 (SD, 1.8) V, 125.0 (SD, 39.0) μs, 146.9 (SD, 28.9) Hz). Forty-nine new adverse events (21 of them classified as serious) were registered. Dysarthria was the most frequent non-serious event related to DBS. In contrast to the SPIDY study, neither hardware-related nor therapy-related adverse events led to permanent discontinuation of neurostimulation [48]. In both controlled studies, neurostimulation had no negative effects on mood or cognition [45, 46]. Other investigations addressing this issue in detail confirmed the safety of GPi-DBS regarding cognitive and neuropsychiatric functions [49, 50].

Prognostic Factors

Factors predicting the therapeutic response to neurostimulation are not explicitly defined to date. Isaias et al. confirmed in a multicentre, long-term analysis of 44 patients treated for primary dystonia with bilateral GPi-DBS previous findings indicating that younger age and shorter symptom duration at the time of surgery predict better clinical outcomes independent from skeletal deformities. Moreover, these authors elucidated that older age at surgery does not negatively influence the possible maximum benefit but does prolong the time after surgery necessary to achieve an optimal result [51]. DYT1 gene mutation was previously thought to be associated with favourable outcome of pallidal surgery [52]. Other groups, however, did not confirm the relevance of this factor [47, 48, 51]. In addition, the controlled studies cited here could not identify any single clinical factor predicting the therapeutic response to neurostimulation [47, 48]. Volkmann et al. hypothesised that a treatment failure of pallidal neurostimulation might be the result of an individual combination of many possible factors including suboptimal lead placement, programming issues or disease heterogeneity [48].

12.3.1.2 Cervical Dystonia

Data from studies addressing GPi-DBS of CD are summarised in Table 12.2. In retrospective studies (24 treated patients), the authors observed improvements of the TWSTRS total scores ranging from 54 to 59 %. Improvements of the subscores ranged from 55 to 63 % for severity, from 62 to 69 % for disability and from 50 to 58 % for pain [53–56]. Kiss et al. published in 2007 results of a prospective multicentre pilot study (ten patients), which basically confirmed the data from the retrospectively analysed case series [57]. Comparable to generalised dystonia, also in CD patients, the initial outcome remained stable in the long-term run. In a prospective analysis of eight patients, the response to DBS (blinded video rating) showed a significant additional improvement of the total, severity and disability TWSTRS scores from the 6-month follow-up to the assessment 12–18 months postoperatively with final score percentage improvements exceeding those reported from other studies (total TWSTRS improvement of 83 %). The pain score did not change during this observation period [58].

12.3.1.3 Meige Syndrome

Five single case reports quantified the outcome after GPi-DBS for Meige syndrome [59–63]. Overall improvement according to the BFMDRS ranged from 66 to 80 % [59, 60, 63]. One group specified the results for BFMDRS movement subscores (eyes, 92 %; mouth, 75 %; speech and swallowing, 33 %) [61]. Also other global scores such as the Unified Dystonia Rating Scale (UDRS) (69 or 85 % improvement) or the Global Dystonia Scale (80 % improvement) confirmed the efficacy of DBS [62, 63].

Ostrem et al. treated in an open-label study seven patients with medically refractory Meige syndrome. Six months after the onset of bilateral GPi-DBS, the preoperative BFMDRS movement score (22 (SD, 8.3)) improved significantly by 72 % to 6.1 (SD, 4.2) points. The difference in the disability score was not significant. Whilst individual movement scores for the eye, mouth and neck responded significantly, those for speech and swallowing showed a trend only. The improvement of the mean total TWSTRS score from 39 (SD, 20) at baseline to 17 (SD, 16) at 6 months (52 % improvement) was significant. Stimulation programming parameters were in average 3.7 V (voltage), 198 μs (pulse-width) and 141 Hz (frequency) [64].

Two publications addressed retrospectively the outcome after longer follow-up times [65, 66]. Ghang et al. documented a steadily increasing amelioration of the BFMDRS movement score from 59.8 % after 3 months GPi-DBS (11 patients) to 85.5 % after 24 months GPi-DBS (five patients). Even the subscores for speech and swallowing responded by 68.4 % after this time period [65]. Rees et al. reported data from 11 patients with a follow-up range of 12–78 months (mean, 38.8 (SD, 21.7) months). Compared to baseline improvements of the BFMDRS movement score (53 %), the Blepharospasm Movement Scale (BMS) (38 %) and the Abnormal Involuntary Movement Scale (AIMS) (49 %) were significant [66].

12.3.1.4 Myoclonus Dystonia

In single case series, patients were treated with thalamic (v.im.) DBS or GPi-DBS [67–73]. Thalamic DBS seemed to be most effective for jerky myoclonic components, whilst dystonic problems in particular cervical dystonia did respond to a minor

Table 12.2 Bilateral GPi-DBS for the treatment of cervical dystonia

Citation/study type	No. pts.	FU (month)	Neurological outcome			Stimulation parameters Ampl (V)	PW (μs)	Frequ. (Hz)	Adverse events/ complications
Krauss et al. (2002)/ retrospective	5	20	TWSTRS improvement at last FU			3.8 (3–4.5)	210.0	135 (130–145)	Lead fracture: 2/5 pts.
			Severity	63 % (diff.: 13.0)					
			Disability	69 % (diff.: 27.8)					
			Pain	50 % (diff.: 3.0)					
Bereznai (2002)/ retrospective	2	3	Case 1: retro-/torticollis completely improved, residual laterocollis			Not mentioned for pts. with CD			n.g.
			Case 2: complete improvement						
Bittar et al. (2005) retrospective	6	33	TWSTRS improvement – total score			5.8 (0.6)	168.8 (66.4)	143.8 (66.4)	Not specified for pts. with CD
			At 1 year	62.3 % (diff.: 36.0)					
			At 2 years	59.0 % (diff.: 34.1)					
			TWSTRS improvement at 2 years – subscores						
			Severity	58%					
			Disability	62%					
			Pain	58%					
Hung et al. (2007)/ retrospective	10	32	TWSTRS improvement – total score			3.1 (0.7)	71.4 (17.7)	135.2 (21.3)	Infection, system removal: 1/10 pts.
			6 m	56.8 % (diff.: 31.5)					Dysarthria: 2/10 pts.
			Last FU	54.3 % (diff.: 29.2)					
			TWSTRS improvement						
				6 m	last FU				
			Total	56.8 %	54.3 %				
			Severity	48.9 %	54.8 %				
			Disability	64.4 %	58.9 %				
			Pain	52.1 %	50.4 %				

Kiss et al. (2007) [57] / prospective	10	12	TWSTRS improvement at last FU		3.3 (0.7)	204.0 (49.0)	170.0 (20.0)	Side effects due to stimulation: 4 pts. (transient)
			Severity	42.9 %				Chronic SDH + transient hemiparesis: 1 pt.
			Disability	63.8 %				
			Pain	65.4 %				
Skogseid et al. (2012) [58]/prospective	8	30	TWSTRS improvement at last FU		3.5 (2.5–3.8)	120.0	130.0 (130–180)	*Surgical complications*
			Total	83% (diff.: 51)				Brain oedema + mild perm neurologic impairment (1 pt.)
			Severity	73 % (diff.: 21)				Bow-stringing of extension cable (1 pt.)
			Disability	91 % (diff.: 20)				
			Pain	92 % (diff.: 13)				

FU follow-up, *No* number, *pts.* patients, *n.g.* not given, *TWSTRS* Toronto Western Spasmodic Torticollis Rating Scale, *Ampl* amplitude, *PW* pulse width, *Frequ.* frequency *SDH* subdural hematoma degree only [67, 68, 72]. Gruber et al. analysed retrospectively data of ten myoclonic patients (nine ε-sarcoglycan mutation positive) with bilateral active stimulation electrodes in the v.im. (one patient), GPi (two patients) or in both targets (seven patients, four electrodes in each case). In the latter group, assessment of all four possible stimulation conditions was performed 3–6 months postoperatively. In general, both thalamic and pallidal DBS improved the symptoms equally effective. The overall improvement at long-term evaluation (Unified Myoclonus Rating Scale (UMRS), 65.5 %; Tsui score, 35.1 %; BFMDRS, 48.2 %; mean follow-up, 62.3 months) was stable compared to the immediate postoperative period with no negative effects on affective status and cognition. The use of all four stimulation electrodes seemed not to be superior compared to single target DBS. At the last follow-up, only four of seven patients with four electrodes were stimulated in both target areas. Stimulation-induced adverse events were more frequently registered for thalamic DBS [74].

12.3.2 Secondary Dystonia

Reports of patients treated with DBS for secondary dystonia subtypes are rare. According to a meta-regression analysis of individual patient outcome published in 2010, DBS was applied in 41 patients for tardive dystonia and in 22 patients for cerebral palsy. For all other secondary forms, <10 cases were reported for each condition only [52].

12.3.2.1 Dystonia Secondary to Heterodegenerative Diseases

In total, six case reports of patients treated for Huntington's disease (follow-up, 8 months–4 years) documented an averaged improvement of choreatiform movements after bilateral GPi-DBS of 60 % (range, 39–77 %) on the Unified Huntington's Disease Rating Scale (UHDRS). Four patients gained the best outcome with high frequencies only (130–180 Hz), one patient required low-frequency DBS (40 Hz), and in one patient 160 or 40 Hz was equally effective (overview in [75]).

Comparable to Huntington's disease, also in patients treated for dystonia secondary to chorea-acanthocytosis (eight published cases), the stimulation frequency required for the best response varied interindividually and was 130–160 Hz in four cases and 40 Hz in three of them. The average reduction of involuntary movements was 51 % (range, 37–73 %) on the UHDRS [75–77]. There was one case report of a patient who did not respond despite correct placement of the brain electrodes and use of different stimulation parameters [78].

The overall response to bilateral GPi-DBS for dystonia secondary to pantothenate kinase-associated neurodegeneration (PKAN) was variable. In six single case series, the BFMDRS movement score improved within a range of 23.3–80.0 % [79–84] (ZITATE). One of these patients deteriorated in the long-term run after initial remarkable response to DBS [82]. Castelnau et al. with the largest single centre series (6 patients; mean FU, 21 months) observed a mean improvement of the BFMDRS movement score of 74.6 % (range, 46.0–91.5 %) and of the disability score of 53 % (range, 21–82 %) [84].

Timmermann et al. performed a worldwide systematic survey addressing DBS applied in patients with brain iron accumulation (NBIA). Twenty-three patients and data from 16 centres were included in the analysis. Electrical stimulation improved the mean severity of dystonia (28.5 % after 2–6 months, 25.7 % after 9–15 months) significantly. Referred to the 9–15-month FU period, the percentage of patients with a clinically relevant motor response (=BFMDRS movement score improvement of ≥20 %) was 66.7 %. The difference of the BFMDRS disability scores compared to baseline, however, was not significant (mean improvement, 15.8 %). The median global QoL ratings improved significantly (83.3 % at 9–15 months), and 67 % of the caregivers rated the positive outcome of the patients as clinically relevant (≥20 % improvement) [85].

12.3.2.2 Tardive Dystonia

Among all types of secondary dystonia, GPi-DBS seems to be most effective in medication-induced tardive dystonia. The cited meta-regression reported for 39 patients treated with GPi-DBS a 67.4 % mean improvement of motor/severity scores [52]. According to a more recently published systematic review of the literature (50 patients included; follow-up, 3–76 months), GPi-DBS improved significantly the mean BFMDRS total score and movement and disability subscores by 77.5 % (95 % CI, 71.4–83.3 %), 80.9 % (95 % CI, 74.7–86.6 %) and 74.0 % (95 % CI, 65.0–83.3 %), respectively [86]. Also scores other than the BFMDRS documented a significant response with an average improvement on the AIMS of 71.5 % and on the Extrapyramidal Symptom Rating Scale of 67.2 %. Even though 44/50 patients had a psychiatric diagnosis prior to DBS, only in two cases (5 %) previous symptoms exacerbated. Correction of the analysis for those five individuals without psychiatric medication did not significantly change the motor outcome after DBS [86].

12.3.2.3 Cerebral Palsy (CP)

Andrews et al. identified in their meta-regression analysis 22 patients treated with DBS for this condition. The mean percentage improvement of motor/severity scores was 24 % (SD, 21.3 %; IQR, 7.6–31.8 %) [52]. Marks et al. treated in a noncontrolled, non-blinded study 13 patients with bilateral GPi-DBS and one patient with unilateral GPi-DBS for CP. After a 6-month DBS, the improvement of the BFMDRS movement score (25.5 % (SD, 24.0 %)), the disability score (8.9 % (SD, 13.4 %)) and the Barry-Albright scores (11.7 % (SD, 19.3 %)) was significant ($p<0.05$) compared to baseline. Differentiated for the age at surgery, only younger patients (mean age, 10.7 years (SD, 2.9)) without skeletal deformities responded significantly to DBS, whilst older patients (mean age, 20.7 years (SD, 3.6)), who had reached skeletal maturity, did not [87].

12.3.2.4 Other Secondary Forms of Dystonia

Comparable to CP, also other secondary forms of dystonia responded less well compared to primary dystonia. For instance, the mean percentage improvement was 45 % (SD, 31.5 %) for postanoxic dystonia (nine patients), 17 % (SD, 37.7 %) for postencephalic/infectious subtypes (six patients) and 32.6 % (SD, 37.2 %) in the case of posttraumatic dystonia (five patients) [52].

References

1. Jankovic J, Chrispim FS. Dystonic disorders. In: Jankovic JTE, editor. Parkinson's disease and movement disorders. 2nd ed. Baltimore: Williams and Wilkons; 1993. p. 337–74.
2. Vogel TD, Pendleton C, Quinones-Hinojosa A, Cohen-Gadol AA. Surgery for cervical dystonia: the emergence of denervation and myotomy techniques and the contributions of early surgeons at The Johns Hopkins Hospital. J Neurosurg Spine. 2010;12(3):280–5.
3. Keen III WW. A new operation for spasmodic wry-neck. Ann Surg. 1891;13(1):44–7.
4. Finney JM, Hughson W. Spasmodic torticollis. Ann Surg. 1925;81(1):255–69.
5. Dandy W. An operation for the treatment of spasmodic torticollis. Arch Surg. 1930;20: 1021–31.
6. McKenzie KG. The surgical treatment of spasmodic torticollis. Clin Neurosurg. 1954;2: 37–43.
7. Bertrand CM. Selective peripheral denervation for spasmodic torticollis: surgical technique, results, and observations in 260 cases. Surg Neurol. 1993;40(2):96–103.
8. Bertrand CM, Lenz FA. Surgical treatment of dystonias. In: Tsui JKC, Calne DB, editors. Handbook of dystonia. New York: Marcel Dekker; 1995. p. 329–45.
9. Bertrand C, Molina Negro P, Martinez SN. Technical aspects of selective peripheral denervation for spasmodic torticollis. Appl Neurophysiol. 1982;45(3):326–30.
10. Grossman R, Hamilton WJ. Surgery for movement disorders. In: Jankovic JTE, editor. Parkinson's disease and movement disorders. 2nd ed. Baltimore: Willliams and Wilkons; 1993. p. 531–48.
11. Krack P, Vercueil L. Review of the functional surgical treatment of dystonia. Eur J Neurol. 2001;8(5):389–99.
12. Loher TJ, Pohle T, Krauss JK. Functional stereotactic surgery for treatment of cervical dystonia: review of the experience from the lesional era. Stereotact Funct Neurosurg. 2004;82(1): 1–13.
13. Ondo WG, Krauss JK. Surgical therapies for dystonia. In: Brin MF, Cornelia CC, Jancovic J, editors. Dystonia monograph. Philadelphia: Lippincott WIlliams & Williams; 2004. p. 125–47.
14. Hassler R, Dieckmann G. Stereotaxic treatment for spasmodic torticollis. In: Schaltenbrand G, Walker AE, editors. Stereotaxy of the human brain. Stuttgart: Thieme; 1982. p. 522–31.
15. Hassler R, Dieckmann G. Stereotactic treatment of different kinds of spasmodic torticollis. Confin Neurol. 1970;32(2):135–43.
16. Hassler R, Dieckmann G. Stereotaxic treatment of torticollis according to animal experiment experiences about direction determined movement. Nervenarzt. 1970;41(10):437–87.
17. Cooper IS. Effect of thalamic lesions upon torticollis. N Engl J Med. 1964;270:567–72.
18. Cooper I. 20 year follow-up study on the neurosurgical treatment of dystonia musculorum deformans. In: Eldrige RFS, editor. Advances in neurology. New York: Raven Press; 1976. p. 423–52.
19. Sano K, Yoshioka M, Mayanagi Y, Sekino H, Yoshimasu N. Stimulation and destruction of and around the interstitial nucleus of Cajal in man. Confin Neurol. 1970;32(2):118–25.
20. Cooper IS. Neurosurgical treatment of the dyskinesias. Clin Neurosurg. 1977;24:367–90.
21. Hiltl A. Spasmodic torticollis: long-term results of stereotactic treatment Freiburg: the Albert Ludwig University Freiburg, Germany; 1990.
22. Meares R. Natural history of spasmodic torticollis, and effect of surgery. Lancet. 1971;2(7716): 149–50.
23. Vitek JL, Bakay RA. The role of pallidotomy in Parkinson's disease and dystonia. Curr Opin Neurol. 1997;10(4):332–9.
24. Vitek JL, Zhang J, Evatt M, Mewes K, DeLong MR, Hashimoto T, et al. GPi pallidotomy for dystonia: clinical outcome and neuronal activity. Adv Neurol. 1998;78:211–9.
25. Ondo WG, Desaloms JM, Jankovic J, Grossman RG. Pallidotomy for generalized dystonia. Mov Disord. 1998;13(4):693–8.
26. Lozano AM, Kumar R, Gross RE, Giladi N, Hutchison WD, Dostrovsky JO, et al. Globus pallidus internus pallidotomy for generalized dystonia. Mov Disord. 1997;12(6):865–70.

27. Mundinger F. New stereotactic treatment of spasmodic torticollis with a brain stimulation system (author's transl). Med Klin. 1977;72(46):1982–6.
28. Cooper IS, Upton AR, Amin I. Chronic cerebellar stimulation (CCS) and deep brain stimulation (DBS) in involuntary movement disorders. Appl Neurophysiol. 1982;45(3):209–17.
29. Andy OJ. Thalamic stimulation for control of movement disorders. Appl Neurophysiol. 1983;46(1–4):107–11.
30. Delnooz CC, van de Warrenburg BP. Current and future medical treatment in primary dystonia. Ther Adv Neurol Disord. 2012;5(4):221–40.
31. Chen X, Ma A, Liang J, Ji S, Pei S. Selective denervation and resection of cervical muscles in the treatment of spasmodic torticollis: long-term follow-up results in 207 cases. Stereotact Funct Neurosurg. 2000;75(2–3):96–102.
32. Braun V, Richter HP. Selective peripheral denervation for the treatment of spasmodic torticollis. Neurosurgery. 1994;35(1):58–62; discussion -3.
33. Krauss JK, Toups EG, Jankovic J, Grossman RG. Symptomatic and functional outcome of surgical treatment of cervical dystonia. J Neurol Neurosurg Psychiatry. 1997;63(5):642–8.
34. Ford B, Louis ED, Greene P, Fahn S. Outcome of selective ramisectomy for botulinum toxin resistant torticollis. J Neurol Neurosurg Psychiatry. 1998;65(4):472–8.
35. Munchau A, Palmer JD, Dressler D, O'Sullivan JD, Tsang KL, Jahanshahi M, et al. Prospective study of selective peripheral denervation for botulinum-toxin resistant patients with cervical dystonia. Brain. 2001;124(Pt 4):769–83.
36. Cohen-Gadol AA, Ahlskog JE, Matsumoto JY, Swenson MA, McClelland RL, Davis DH. Selective peripheral denervation for the treatment of intractable spasmodic torticollis: experience with 168 patients at the Mayo Clinic. J Neurosurg. 2003;98(6):1247–54.
37. Huh R, Han IB, Chung M, Chung S. Comparison of treatment results between selective peripheral denervation and deep brain stimulation in patients with cervical dystonia. Stereotact Funct Neurosurg. 2010;88(4):234–8.
38. Krauss JK, Pohle T, Weber S, Ozdoba C, Burgunder JM. Bilateral stimulation of globus pallidus internus for treatment of cervical dystonia. Lancet. 1999;354(9181):837–8.
39. Coubes P, Echenne B, Roubertie A, Vayssiere N, Tuffery S, Humbertclaude V, et al. Treatment of early-onset generalized dystonia by chronic bilateral stimulation of the internal globus pallidus. Apropos of a case. Neurochirurgie. 1999;45(2):139–44.
40. Coubes P, Roubertie A, Vayssiere N, Hemm S, Echenne B. Treatment of DYT1-generalised dystonia by stimulation of the internal globus pallidus. Lancet. 2000;355(9222):2220–1.
41. Tronnier VM, Fogel W. Pallidal stimulation for generalized dystonia. Report of three cases. J Neurosurg. 2000;92(3):453–6.
42. Kumar R, Dagher A, Hutchison WD, Lang AE, Lozano AM. Globus pallidus deep brain stimulation for generalized dystonia: clinical and PET investigation. Neurology. 1999;53(4):871–4.
43. Ceballos-Baumann AO, Passingham RE, Marsden CD, Brooks DJ. Motor reorganization in acquired hemidystonia. Ann Neurol. 1995;37(6):746–57.
44. Ceballos-Baumann AO, Passingham RE, Warner T, Playford ED, Marsden CD, Brooks DJ. Overactive prefrontal and underactive motor cortical areas in idiopathic dystonia. Ann Neurol. 1995;37(3):363–72.
45. Kupsch A, Benecke R, Muller J, Trottenberg T, Schneider GH, Poewe W, et al. Pallidal deep-brain stimulation in primary generalized or segmental dystonia. N Engl J Med. 2006;355(19): 1978–90.
46. Vidailhet M, Vercueil L, Houeto JL, Krystkowiak P, Benabid AL, Cornu P, et al. Bilateral deep-brain stimulation of the globus pallidus in primary generalized dystonia. N Engl J Med. 2005;352(5):459–67.
47. Vidailhet M, Vercueil L, Houeto JL, Krystkowiak P, Lagrange C, Yelnik J, et al. Bilateral, pallidal, deep-brain stimulation in primary generalised dystonia: a prospective 3 year follow-up study. Lancet Neurol. 2007;6(3):223–9.
48. Volkmann J, Wolters A, Kupsch A, Muller J, Kuhn AA, Schneider GH, et al. Pallidal deep brain stimulation in patients with primary generalised or segmental dystonia: 5-year follow-up of a randomised trial. Lancet Neurol. 2012;11(12):1029–38.

49. Halbig TD, Gruber D, Kopp UA, Schneider GH, Trottenberg T, Kupsch A. Pallidal stimulation in dystonia: effects on cognition, mood, and quality of life. J Neurol Neurosurg Psychiatry. 2005;76(12):1713–6.
50. Pillon B, Ardouin C, Dujardin K, Vittini P, Pelissolo A, Cottencin O, et al. Preservation of cognitive function in dystonia treated by pallidal stimulation. Neurology. 2006;66(10): 1556–8.
51. Isaias IU, Volkmann J, Kupsch A, Burgunder JM, Ostrem JL, Alterman RL, et al. Factors predicting protracted improvement after pallidal DBS for primary dystonia: the role of age and disease duration. J Neurol. 2011;258(8):1469–76.
52. Andrews C, Aviles-Olmos I, Hariz M, Foltynie T. Which patients with dystonia benefit from deep brain stimulation? A metaregression of individual patient outcomes. J Neurol Neurosurg Psychiatry. 2010;81(12):1383–9.
53. Krauss JK, Loher TJ, Pohle T, Weber S, Taub E, Barlocher CB, et al. Pallidal deep brain stimulation in patients with cervical dystonia and severe cervical dyskinesias with cervical myelopathy. J Neurol Neurosurg Psychiatry. 2002;72(2):249–56.
54. Bereznai B, Steude U, Seelos K, Botzel K. Chronic high-frequency globus pallidus internus stimulation in different types of dystonia: a clinical, video, and MRI report of six patients presenting with segmental, cervical, and generalized dystonia. Mov Disord. 2002;17(1): 138–44.
55. Bittar RG, Yianni J, Wang S, Liu X, Nandi D, Joint C, et al. Deep brain stimulation for generalised dystonia and spasmodic torticollis. J Clin Neurosci. 2005;12(1):12–6.
56. Hung SW, Hamani C, Lozano AM, Poon YY, Piboolnurak P, Miyasaki JM, et al. Long-term outcome of bilateral pallidal deep brain stimulation for primary cervical dystonia. Neurology. 2007;68(6):457–9.
57. Kiss ZH, Doig-Beyaert K, Eliasziw M, Tsui J, Haffenden A, Suchowersky O. The Canadian multicentre study of deep brain stimulation for cervical dystonia. Brain. 2007;130(Pt 11): 2879–86.
58. Skogseid IM, Ramm-Pettersen J, Volkmann J, Kerty E, Dietrichs E, Roste GK. Good long-term efficacy of pallidal stimulation in cervical dystonia: a prospective, observer-blinded study. Eur J Neurol. 2012;19(4):610–5.
59. Vercueil L, Pollak P, Fraix V, Caputo E, Moro E, Benazzouz A, et al. Deep brain stimulation in the treatment of severe dystonia. J Neurol. 2001;248(8):695–700.
60. Muta D, Goto S, Nishikawa S, Hamasaki T, Ushio Y, Inoue N, et al. Bilateral pallidal stimulation for idiopathic segmental axial dystonia advanced from Meige syndrome refractory to bilateral thalamotomy. Mov Disord. 2001;16(4):774–7.
61. Capelle HH, Weigel R, Krauss JK. Bilateral pallidal stimulation for blepharospasm-oromandibular dystonia (Meige syndrome). Neurology. 2003;60(12):2017–8.
62. Foote KD, Sanchez JC, Okun MS. Staged deep brain stimulation for refractory craniofacial dystonia with blepharospasm: case report and physiology. Neurosurgery. 2005;56(2):E415; discussion E.
63. Houser M, Waltz T. Meige syndrome and pallidal deep brain stimulation. Mov Disord. 2005;20(9):1203–5.
64. Ostrem JL, Marks Jr WJ, Volz MM, Heath SL, Starr PA. Pallidal deep brain stimulation in patients with cranial-cervical dystonia (Meige syndrome). Mov Disord. 2007;22(13): 1885–91.
65. Ghang JY, Lee MK, Jun SM, Ghang CG. Outcome of pallidal deep brain stimulation in meige syndrome. J Kor Neurosurg Soc. 2010;48(2):134–8.
66. Reese R, Gruber D, Schoenecker T, Bazner H, Blahak C, Capelle HH, et al. Long-term clinical outcome in meige syndrome treated with internal pallidum deep brain stimulation. Mov Disord. 2011;26(4):691–8.
67. Trottenberg T, Meissner W, Kabus C, Arnold G, Funk T, Einhaupl KM, et al. Neurostimulation of the ventral intermediate thalamic nucleus in inherited myoclonus-dystonia syndrome. Mov Disord. 2001;16(4):769–71.

68. Kupsch A, Trottenberg T, Meissner W, Funk T. Neurostimulation of the ventral intermediate thalamic nucleus alleviates hereditary essential myoclonus. J Neurol Neurosurg Psychiatry. 1999;67(3):415–6.
69. Cif L, Valente EM, Hemm S, Coubes C, Vayssiere N, Serrat S, et al. Deep brain stimulation in myoclonus-dystonia syndrome. Mov Disord. 2004;19(6):724–7.
70. Magarinos-Ascone CM, Regidor I, Martinez-Castrillo JC, Gomez-Galan M, Figueiras-Mendez R. Pallidal stimulation relieves myoclonus-dystonia syndrome. J Neurol Neurosurg Psychiatry. 2005;76(7):989–91.
71. Jog M, Kumar H. Bilateral pallidal deep brain stimulation in a case of myoclonus dystonia syndrome. Mov Disord. 2009;24(10):1547–9.
72. Kuncel AM, Turner DA, Ozelius LJ, Greene PE, Grill WM, Stacy MA. Myoclonus and tremor response to thalamic deep brain stimulation parameters in a patient with inherited myoclonus-dystonia syndrome. Clin Neurol Neurosurg. 2009;111(3):303–6.
73. Oropilla JQ, Diesta CC, Itthimathin P, Suchowersky O, Kiss ZH. Both thalamic and pallidal deep brain stimulation for myoclonic dystonia. J Neurosurg. 2010;112(6):1267–70.
74. Gruber D, Kuhn AA, Schoenecker T, Kivi A, Trottenberg T, Hoffmann KT, et al. Pallidal and thalamic deep brain stimulation in myoclonus-dystonia. Mov Disord. 2010;25(11):1733–43.
75. Edwards TC, Zrinzo L, Limousin P, Foltynie T. Deep brain stimulation in the treatment of chorea. Mov Disord. 2012;27(3):357–63.
76. Li P, Huang R, Song W, Ji J, Burgunder JM, Wang X, et al. Deep brain stimulation of the globus pallidus internal improves symptoms of chorea-acanthocytosis. Neurol Sci. 2012; 33(2):269–74.
77. Shin H, Ki CS, Cho AR, Lee JI, Ahn JY, Lee JH, et al. Globus pallidus interna deep brain stimulation improves chorea and functional status in a patient with chorea-acanthocytosis. Stereotact Funct Neurosurg. 2012;90(4):273–7.
78. Wihl G, Volkmann J, Allert N, Lehrke R, Sturm V, Freund HJ. Deep brain stimulation of the internal pallidum did not improve chorea in a patient with neuro-acanthocytosis. Mov Disord. 2001;16(3):572–5.
79. Umemura A, Jaggi JL, Dolinskas CA, Stern MB, Baltuch GH. Pallidal deep brain stimulation for longstanding severe generalized dystonia in Hallervorden-Spatz syndrome. Case report. J Neurosurg. 2004;100(4):706–9.
80. Isaac C, Wright I, Bhattacharyya D, Baxter P, Rowe J. Pallidal stimulation for pantothenate kinase-associated neurodegeneration dystonia. Arch Dis Child. 2008;93(3):239–40.
81. Shields DC, Sharma N, Gale JT, Eskandar EN. Pallidal stimulation for dystonia in pantothenate kinase-associated neurodegeneration. Pediatr Neurol. 2007;37(6):442–5.
82. Krause M, Fogel W, Tronnier V, Pohle S, Hortnagel K, Thyen U, et al. Long-term benefit to pallidal deep brain stimulation in a case of dystonia secondary to pantothenate kinase-associated neurodegeneration. Mov Disord. 2006;21(12):2255–7.
83. Mikati MA, Yehya A, Darwish H, Karam P, Comair Y. Deep brain stimulation as a mode of treatment of early onset pantothenate kinase-associated neurodegeneration. Eur J Paediatr Neurol. 2009;13(1):61–4.
84. Castelnau P, Cif L, Valente EM, Vayssiere N, Hemm S, Gannau A, et al. Pallidal stimulation improves pantothenate kinase-associated neurodegeneration. Ann Neurol. 2005;57(5): 738–41.
85. Timmermann L, Pauls KA, Wieland K, Jech R, Kurlemann G, Sharma N, et al. Dystonia in neurodegeneration with brain iron accumulation: outcome of bilateral pallidal stimulation. Brain. 2010;133(Pt 3):701–12.
86. Mentzel CL, Tenback DE, Tijssen MA, Visser-Vandewalle VE, van Harten PN. Efficacy and safety of deep brain stimulation in patients with medication-induced tardive dyskinesia and/or dystonia: a systematic review. J Clin Psychiatry. 2012;73(11):1434–8.
87. Marks WA, Honeycutt J, Acosta Jr F, Reed M, Bailey L, Pomykal A, et al. Dystonia due to cerebral palsy responds to deep brain stimulation of the globus pallidus internus. Mov Disord. 2011;26(9):1748–51.

Pathophysiology of Dystonia: Models and Mechanisms

13

Claudia M. Hendrix and Jerrold L. Vitek

13.1 Introduction

Dystonia is a movement disorder characterized by sustained muscle contractions leading to twisting, repetitive movements, and abnormal postures [1]. The diagnosis and treatment of dystonia is complicated by the intricacy of etiology, therapeutic responsiveness, and clinical manifestation of the disease. This raises the question of whether the underlying pathophysiology of dystonia can be linked to the clinical manifestation and therapeutic responsiveness of the disease. It also brings into question whether there is a unifying model that can underlie all forms of dystonia or whether dystonia is a symptom that may manifest from one of a number of physiological changes in multiple circuits or locations across the peripheral and central nervous system. Most hypotheses regarding the pathophysiology of dystonia emphasize alterations in the basal ganglia thalamocortical (BGTC) motor circuit. Indeed this pathway is largely implicated in the pathophysiology of both hypokinetic (e.g., Parkinson's disease, PD) and hyperkinetic (e.g., dystonia, drug-induced dyskinesia, and hemiballismus) movement disorders [2, 3]. Current and emerging hypotheses characterizing the role of the basal ganglia (BG) in motor control are central to our understanding of BG involvement in movement disorders such as dystonia.

This chapter will review BG function and its role within the BGTC motor circuit followed by a discussion of the potential mechanisms underlying the pathophysiology of dystonia and related conditions. These mechanisms are based on observed changes in physiological properties of neurons within key nodal structures of the BGTC circuit recorded from patients with and animal models of dystonia. Pathophysiological properties of dystonia include changes in neuronal discharge

C.M. Hendrix • J.L. Vitek (✉)
Department of Neurology, University of Minnesota,
2001 6th Street SE, Minneapolis, MN 55455, USA
e-mail: vitek004@umn.edu

P. Kanovsky et al. (eds.), *Dystonia and Dystonic Syndromes*,
DOI 10.1007/978-3-7091-1516-9_13

rates and patterns as well as changes in receptive field properties of neurons and plasticity within the network. Mechanisms underlying the pathophysiology of dystonia are discussed within the context of normal function but also within the context of other seemingly different but related hypokinetic disorders such as PD. Network-level properties, including parallel cerebellar (CB) motor circuits and linking pathways, will also be addressed. When pertinent, we also discuss potential contributions of the knowledge obtained from observed pathophysiology in dystonia to the development of rational-based DBS therapies. Finally, we propose a model of dystonia that accounts for the changes in physiological properties discussed, including change in mean discharge rates, distorted receptive fields, and alterations in patterns of neuronal discharge observed in human and animal models of dystonia. Pathways that may contribute to certain forms of dystonia including the cerebellum and brainstem structures are also incorporated into this model.

13.2 The Basal Ganglia Thalamocortical Motor Circuit

We have previously described models of BGTC function and dysfunction as it relates to movement disorders [4–6]. A brief review of the key points is provided below followed by a more detailed review of striatal function and the potential mechanisms by which striatal dysfunction may contribute to the pathophysiology of dystonia. The BG (Fig. 13.1a) are viewed as components of multiple segregated circuits including, oculomotor, associative, limbic, and motor circuits. Input to the BG motor circuit (originating from pre- and postcentral sensorimotor fields in the cortex) engages specific portions of the putamen, external and internal segments of the globus pallidus (GPe, GPi), substantia nigra pars reticulata (SNr), and the subthalamic nucleus (STN). BG output involves portions of the motor thalamus (ventralis lateralis pars oralis, VLo; ventralis anterior, VA; and ventralis anterior magnocellularis, VAmc [7]), before returning to the same precentral motor fields from which it originated [8] (Fig. 13.1b). All intrinsic connections within the BG are inhibitory (red arrows), except for the excitatory outputs (green arrows) from the STN. Input to and output from the BG are largely associated with the striatum and GPi, respectively. A second major input to the BG, known as the hyperdirect pathway, directs sensorimotor-based glutaminergic drive to the STN [9, 10]. The striatum influences the output from the GPi through pathways arising from two separate subpopulations of inhibitory neurons. There is a "direct" pathway to the GPi (red arrow) and an "indirect" pathway involving the GPe (red arrow) and STN (green arrow). The dopaminergic nigrostriatal pathway from the substantia nigra pars compacta (SNc) modulates activity of the two striato-pallidal pathways differentially by activation of different dopamine (DA) receptors. Specifically, DA appears to facilitate transmission in the "direct" pathway (Fig. 13.1, SNc, green trace) via D_1 receptors containing GABA, substance P, and dynorphin while inhibiting transmission in the "indirect" pathway (Fig. 13.1, SNc, red trace) via D_2 receptors containing GABA and encephalin [3]. BG output from the GPi/SNr in turn exerts tonic inhibition on thalamocortical neurons [11].

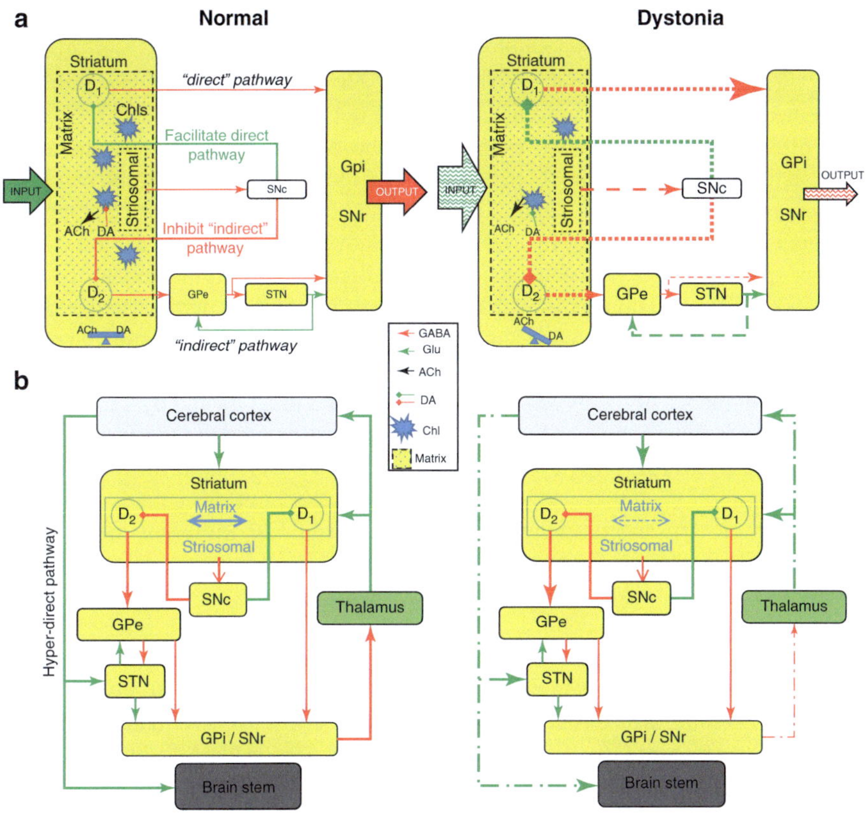

Fig. 13.1 Motor circuits within BG (**a**) and BGTC (**b**) structures. (**a**) BG structures (*yellow*) receive major input via the striatum and output via the GPi/SNr. Intrinsic connections within the BG are inhibitory (*red*) with the exception of excitatory projections (*green*) from the STN. Within the BG, direct (*top*) and indirect (*bottom*) pathways are regulated via SNc input that both facilitate (*green*) and inhibit (*red*) "direct" and "indirect" pathways, respectively. (**b**) Models of the BGTC in the normal state (*left panel*) and in dystonia (*right panel*). These models are based on changes in discharge rate (line width, thickness increases/decreases with rate) and temporal patterns (*line pattern*). *Lines* represent inhibitory (*red*) or excitatory (*green*) projections. The basal ganglia (*yellow*) include the striatum (dopamine receptor subtypes represented as D1 and D2). Early preferential loss in density or abnormal function of striatal projection neurons (matrix and striosomal, change in line width) and of ChIs and/or the disruption in DA-ACh coupling (*blue* balance) are related to the development of dystonia. *GPe* and *GPi* globus pallidus pars externa and interna, respectively, *SNc* and *SNr* substantia nigra pars compacta and reticulata, respectively, *STN* subthalamic nucleus, *BG* basal ganglia, *ChI* cholinergic interneurons, *ACh* acetylcholine, and *DA* dopamine

Several theories have emerged concerning the functional role of the "direct" and "indirect" pathways in the BGCT motor circuit. The most popular of which is the "scaling and focusing" hypothesis suggesting that these motor pathways scale and focuses motor activity. *Scaling* of movement could occur by inhibition of GPi/SNr output in the "direct" pathway to facilitate movement, while activation of the same

GPi/SNr neurons via the "indirect" pathway would cause inhibition of movement [2, 3, 12]. *Focusing* of motor activity, on the other hand, may occur via activation of the "direct" pathway which reduces GPi/SNr output, thereby disinhibiting movement-facilitating thalamic neurons. Thalamic disinhibition would then activate prime mover muscles and execution of the desired movement. At the same time, activation of the "indirect" pathway would increase GPi/SNr output, resulting in increased inhibition on a population of thalamocortical neurons which suppress antagonistic movements [10, 13]. Thus, phasic reductions of GPi/SNr output via the "direct" pathway appear to gate or facilitate cortically initiated movements by disinhibition of the thalamus, while phasic increases in GPi/SNr output via the "indirect" pathway inhibit antagonistic or unwanted movements. A major function of the BG according to this hypothesis is to inhibit competing motor commands. In recent years, the influence of the hyperdirect pathway in BG function and dysfunction has received more attention [4, 14, 15]. The hyperdirect pathway is the fastest, most direct route by which cortical signaling can influence BG output nuclei [16, 17]. This direct and fast, monosynaptic pathway to the BG bypasses striatal inhibitory nodes (Fig. 13.1b) and is thought to inhibit unwanted or irrelevant motor commands, while the "direct" pathway originating in the BG adjusts the selected motor programs based on movement context [15].

Models of BG motor control function (and dysfunction) have predominantly focused on DA and the regulation of the "direct" and "indirect" pathways via the striatum. More recent models involving striatal influence over motor control have focused on two intrinsic circuits originating within striosomal and matrix compartments of the striatum [18, 19]. Striosomal compartment medium spiny projection neurons (MSNs) project directly (Fig. 13.1a, striosomal, black arrow) to SNc dopaminergic neurons, thereby directly influencing SNc modulation of the "direct" and "indirect" BG pathways. These monosynaptic striosome-SNc connections are thought to originate uniquely from the striosomal compartment and have a substantial influence over signaling within the midbrain [20, 21]. While both striosomal and matrix MSNs project directly to the output nuclei of the BG, MSN projections originate predominantly from the larger matrix compartment and project to nuclei (GPe, GPi, SNr) in both the "indirect" and "direct" BG pathways (Fig. 13.1a, matrix, D_1 and D_2 red arrows, respectively). In addition to the differential expression of MSNs originating from striosomal and matrix compartments, the compartments also receive different input. While there is considerable evidence that matrix and striosomal compartments are part of largely segregated input-output striatal circuits, the contribution of each compartment to striatal function remains unclear. It is generally understood that matrix MSNs, locally part of both the "direct" and "indirect" BG pathways, are globally associated with the sensorimotor and associative circuits, while striosomal MSNs, regulating the "direct" and "indirect" pathway via direct projections to SNc dopamine-containing neurons, are functionally related to the limbic circuit [20, 21].

Striatal influence over motor control has also been strongly associated with cholinergic interneurons (ChIs) (Fig. 13.1a, blue) [22–25]. The central role ChIs in BG function is well known as anticholinergic drugs have long been a treatment of choice for movement disorders [26–28]. While ChIs only represent a small percentage (<2 %) of striatal interneurons, they provide one of the greatest concentrations of acetylcholine (ACh) in the brain [29, 30]. Striatal ChIs activity is highly regulated via complex interactions between glutamatergic drive from cortical and thalamic inputs as well as strong dopaminergic innervation from the SNc [31–33]. Modulation of ACh is predominantly attributed to DA [34]; while glutamatergic inputs drive ACh release from striatal ChIs, DA activation of D_2-like receptors regulates the release through inhibition (Fig. 13.1a, striatum, ACh black arrow, DA red trace, respectively). In turn, ACh release regulates MSN activity via cholinergic receptors which modulate glutamate, DA, and ACh neurotransmission. Postsynaptic effects of ACh on MSNs, via muscarinic ACh (mACh) receptors, inhibit MSN neurotransmitter release by providing a presynaptic inhibitory influence over the excitatory glutamatergic drive onto MSNs [35]. Presynaptic effects of ACh, via nicotinic ACh (nACh) receptors, regulate both glutamate and GABA release from striatal afferents, thus driving MSN neurotransmitter release [36]. ACh release on nACh receptors located on dopaminergic terminals also modulates DA neurotransmission [36]. Models of BGTC control of movement generally emphasize DA involvement via SNc in the regulation of "direct" and "indirect" pathways. However, there is strong support for the idea that the coupling of dopaminergic and cholinergic signaling within the striatum is the key to striatal regulation of "direct" and "indirect" BG pathways [23].

13.3 Physiological Properties of Dystonia

13.3.1 Changes in Discharge Rates

Early models of movement disorders were based on observed changes in mean discharge rates occurring at key nodal points in the BGTC. These models provided a potential pathophysiological basis underlying the development of PD in particular and of other movement disorders in general [2, 3]. The "rate" model of PD postulates that the depletion of dopamine in the SNc leads to subsequent changes in the BGTC via an excessive increase in activity in the GPi, causing excessive inhibition in the pallidal-receiving areas of the motor thalamus. The excessive inhibition of this area of the thalamus was postulated to drive a corresponding reduction in thalamocortical activity resulting in slower and/or fewer movements. By extension, the "rate" model would therefore predict that hyperkinetic disorders would result from opposite changes in the BGTC; decreased GPi activity disinhibits thalamocortical activity resulting in faster and more frequent non-volitional movements. However, recorded neuronal activity from the GPi and motor thalamus in humans and animal models of dystonia [37], were not always consistent with proposed "rate" model predictions. Furthermore, the rate model failed to explain why drug-induced dyskinesia, dystonia, and other hyperkinetic disorders were improved following pallidotomy [45–47]. GPi lesioning would remove inhibitory output from the pallidum which, according to the "rate" model, should induce hyperkinetic movements not improve them. The observation that pallidotomy did improve dystonia lent further support to the hypothesis that the BG plays an implicit role in mediating the development of dystonia.

Studies generally agree that mean discharge rates of the GPi and GPe in patients with dystonia are either similar to or lower than those found in normal nonhuman primates and lower than that found in PD patients [38, 40, 43, 48, 49] with similar reductions in mean discharge rates reported in GPe and GPi [38, 41, 44, 50]. Our previous observations, however, have shown rates in GPe that are significantly lower than those found for GPi [5, 40]. In contrast, preliminary observations of patients with prior thalamotomies showed similar firing rates between GPi and GPe, suggesting strong thalamic influence over the "indirect" pathway in dystonia [5]. Lowered mean discharge rates in both GPe and GPi may be an indication that the "direct" pathway plays a predominant role in lowering GPi rates, while the "indirect" pathway disrupts somatosensory processing [4, 40]. Changes in both the "indirect" and "direct" pathways would disrupt function in the BGTC motor circuit and ultimately the ability to properly focus movements.

Studies relating changes in neuronal activity to development, severity, and phenotypic distribution of dystonia are few and inconsistent. An inverse correlation has been reported between the severity of dystonic symptoms (severity was based on the total score of the Burke-Fahn-Marsden Dystonia Rating Scale) and the mean discharge rate of GPi neurons [38]. Importantly, the patient population was comprised of a combination of primary (including focal, segmental, and generalized) as well as secondary dystonia (tardive, trauma, post-stroke, and metabolic). This association of change in mean discharge rate to disease severity also supported the observation made earlier in a single case report by Suarez et al. [51] finding mean discharge rates recorded from neurons later in the procedure as the patient's dystonic symptoms worsened were lower than those collected in the earlier stages of the procedure when the patient was less dystonic. Tang and colleagues [43], similar to Starr et al. [38], examined the relationship between mean discharge rate of neurons in GPi in patients with CD and severity based on the Toronto Western Spasmodic Torticollis Rating Scale but did not find a correlation between severity and mean discharge rate [43]. The primary difference between these reports is that Starr et al. [38] studied patients with PGD while Tang et al. [43] studied patients with CD. We have previously hypothesized that the number and distribution of "affected" GPi neurons determines the severity and distribution of dystonia. As such, a relationship between mean discharge rate and severity of CD could have been washed out in the Tang study by averaging the mean discharge rates of all cells in GPi, including those that were in a physiologically normal region of GPi. Consistent with the above hypothesis, Tang et al. argued that, because CD is a focal disorder, it was possible that only a small portion of the GPi in the CD patients was affected, and therefore many of the recordings could have been made in the relatively unaffected portions of GPi [43].

Overall, most studies support the finding that GP rates are generally reduced in dystonia and, in some studies, correlate with disease onset and severity. The exact contribution of rate change to the pathophysiology of dystonia, however, remains unclear. Models of dystonia based on rate changes alone are limited and not supported by the observation that lesions in the GPi improve dystonia. Therefore, other pathophysiological properties must be explored to further elucidate the neural mechanisms involved in dystonia and are discussed below. Although we tend to disregard the rate changes observed in dystonia and other movement disorders with the acceptance of new models incorporating changes in pattern and receptive field properties, we may still gain some understanding of the pathophysiology of dystonia by exploring how these changes occur. We may, in addition, develop refinements in DBS therapeutic stimulation parameters. Although early studies suggested that wide pulse widths (450 us) and high frequencies (185 Hz) [52] were required for improvement in dystonic symptoms, more recent studies have found improvement in dystonia using low frequencies and shorter pulse widths [53–56]. Alterman et al. have argued that frequencies in the range of 60 Hz are equally effective as high frequencies in the treatment of PGD [54]; the rational for this approach being that one needs to use a frequency range that is higher than the mean discharge rate in GPi to overdrive the majority of neurons into a more regular pattern. Advantages to this approach include a comparable therapeutic effect that is achieved with less current drain and thus prolonged battery life and fewer side effects. This approach although reported effective for some has not been effective for all patients with dystonia. Cases which did not respond to lower-frequency stimulation one might wonder whether the mean discharge rates in GPi were faster in those cases and thus not susceptible to override by 60 Hz stimulation.

13.3.2 Receptive Fields and Plasticity

Along with mean discharge rates, early studies began characterizing the pathophysiology of dystonia in terms of the somatosensory response (SSR) of neurons. Evidence of altered SSRs at the level of the BG in dystonia led to the idea that dystonia may be, at least in part, a "sensory" disorder [57–61]. The association of an altered sensory system to the pathophysiology of dystonia is robust in both animals models [60–62] and patients [59, 63, 64]. However, the relationship between the incidence and types of SSRs and the development, severity, and phenotypic distribution of dystonic movements remains to be determined. The SSRs of neurons in the GPi and the thalamus have been reported as widened and less specific in patients with dystonia [40, 42, 65], and the extent of the distortion may be associated with disease severity. Widened receptive fields may not be uniquely related to dystonia, as similar findings have been reported in PD, but could instead reflect altered activity in the "indirect" pathway, which does occur in both disorders [66, 67]. An example of SSRs of GPi neurons is shown (Fig. 13.2) along comparable parasagittal planes in patients with PD (left), PGD (center), and focal, orofacial dystonia (right). The PGD patient's SSRs cover a wide area of GPi while little distortion is noted to the arm and leg regions of the patient with a more focalized, orofacial dystonia. Additionally, the patient with the focal, orofacial dystonia revealed a disproportionate number of GPi cells that responded to orofacial structures (36 %, noted by the filed squares) compared to the PGD patient where <4 % of cells responded to neck and orofacial manipulation. Altered somatotopy is not limited to the BG; expanded digit representation was also

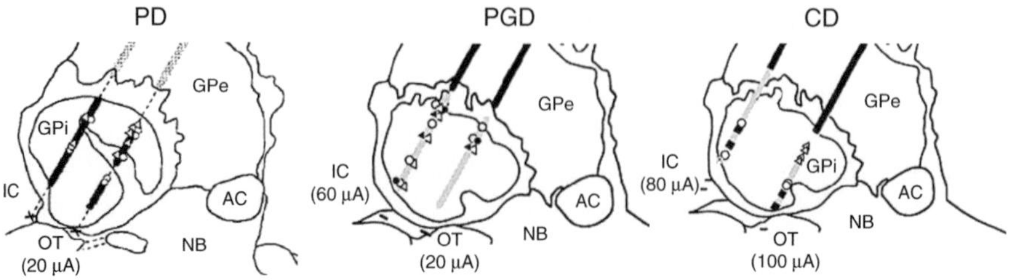

Fig. 13.2 Topographic map representing the receptive fields of neurons in the GPi in PD (**a**) and dystonia (**b**). A patient with PGD (*middle*) and focal, orofacial dystonia (*right*). These patients were mapped using multiple microelectrode tracks through several planes. The tracks represented here best fit the lateral 21.5 parasagittal plane of the Schaltenbrand and Bailey Atlas. Each track is numbered at the *top*. Different portions of the basal ganglia including the putamen (PUT), GPe, and GPi are denoted by different shades along each electrode track. Laminae are characterized by the spaces between the *shaded lines*. Receptive fields of each isolated neuron in the GPi are represented by symbols as follows: Δ proximal leg, ▲ distal leg, ○ proximal arm, ● distal arm, and ■ orofacial. Symbols side by side represent a cell with responses to multiple body parts. For the sake of clarity, only cells with receptive fields to active or passive movement are plotted in the figure. Cells examined that did not respond to active or passive manipulation are not plotted. The numbers in parentheses at the *bottom* of each track indicate the threshold for microstimulation-evoked movement, identifying the internal capsule (*IC*), or phosphenes, identifying the optic tract (*OT*) (Figure adapted from previous publications by Vitek [4])

found in the somatosensory cortex in focalized hand dystonia in the non-human primate (NHP) [62, 68] as well as in humans [69]. Furthermore, in the primate the extent of the cortical area affected was found to be proportionate to abnormalities found in thalamic representations [42]. Extended somatosensory representation alone, however, appears to be less predictive of dystonic behavior than the presence of combined receptive fields and abnormal columnar overlap function [42]. The limited association of widened SSRs to dystonic behavior along with the finding of widened SSRs in other movement disorders may suggest that altered somatotopy may be secondary to the development of the pathophysiology of dystonia.

The presence of altered SSR receptive fields, however, does not necessarily result in explicit manifestations of dystonia. Previous studies have demonstrated decreased striatal dopamine receptor (D_2) binding, impaired sequence learning, and changes in regional metabolism, as well as reduced integrity of cerebellothalamocortical (CBTC) fiber tracts in asymptomatic carriers of the DYT1 gene mutation [70–73]. Furthermore, the finding of abnormalities in subcortical-cortical motor circuits is not necessarily unexpected in "clinically normal" patients or in portions of the body that are considered "clinically normal" in patients with focal dystonia [70]. Thus, even a "clinically normal" individual or limb may have physiological changes in cortical and subcortical structures that fit the characteristic pathophysiological profile. These findings support the concept that manifestation of overt dystonic movement may not occur until a sufficient number of cells, or a "critical mass," express the pathophysiological profile. This hypothesis could explain the gradual evolution of symptoms that initially only appear under certain movements or stressful conditions, evolving to a point where even the simplest movement evokes dystonic posturing or where dystonic movement occurs at rest.

The overt manifestation of dystonic movements resulting from altered somatotopy in the BGTC circuit is not fully understood. The disruption of sensory processing of proprioceptive input has been hypothesized to contribute to the lack of focusing and overflow of EMG activity to non-prime mover muscles, leading to the erratic movements and loss of motor control associated with dystonia. Traditional kinesiological studies [74–76] have demonstrated that patients with dystonia exhibit varying levels of difficulty with force control. Goal-directed movements are typically characterized by a bi- or triphasic EMG patterns of the agonist and antagonist muscles, with the amplitude of this activity highly correlated to the velocity of the movement [76, 77]. Unlike patients with PD who require multiple agonist bursts to reach a target force or position, patients with dystonia typically have relatively normal force time patterns and force variability in patients with focal hand dystonia has not been found to differ from age-matched controls [78, 79]. However, as a result of the co-activation of agonist and antagonist muscles, these patients typically overshoot the target during functional activities and produce excessive grip forces during the simple unimanual lifting of objects [80, 81], thus diminishing the performance of fine motor activities.

More recent models of BG function relate dystonia to abnormal coordination or imbalance in striatal function. There is some evidence that dystonia and other hypo- and hyperkinetic movement disorders may result from abnormal sensorimotor coordination at the level of the striatum [2, 3, 82, 83]. Goto et al. suggest that dystonia results from early preferential loss of striosomal projection neurons and to a lesser extent from matrix projection neurons [18]. More specifically, they hypothesize that dystonia may result from an imbalance in activity between the matrix-based pathways, regulating the "direct" and "indirect" pathways, and the striosomal pathway which projects predominately to dopaminergic SNc neurons (Fig. 13.1a, right panel). ChIs control striosomal to matrix activity ratios [84] and changes in ChI signaling may contribute to the imbalance in the striatal-matrix pathways. Albin et al. [83] reported widespread reduction in [(123)I] iodobenzovesamicol (IBVM) binding in the putamen and total striatum. These data suggested that in primary dystonia there is a decrease in the density of striatal ChIs (Fig. 13.1a, right panel). Consistent with the imaging studies showing little evidence of structural changes, it is unlikely that abnormalities in striatal structure would be detected in PGD as ChIs represent less than 2 % of striatal neurons. Abnormal cholinergic signaling within the striatum has been linked to altered DA signaling in a mouse model of DYT1 dystonia [85]. Pisani et al. found that D_2 receptor activation in these mice induces an excitatory effect on ChIs, increasing interneuronal autonomous spiking (Fig. 13.1a, right panel), rather than an inhibitory effect as observed in normal striatal function (Fig. 13.1a, left panel). Increased ChI tone would likely result in increased ACh release in dystonia and would be consistent with the well-established therapeutic benefit of anticholinergic drugs in the treatment of dystonia. Pisani et al. [22, 24] suggest that it is an imbalance in striatal DA-ACh coupling that underlies the pathophysiology of dystonia (as illustrated in Fig. 13.1a, right panel, imbalance of ACh and DA).

Along with altered SSRs, there is strong evidence of changes in synaptic plasticity in dystonia. Increased paired associative plasticity responses, an electrophysiological measure of long-term potentiation-like synaptic plasticity in cortical circuits, have been reported in patients with dystonia compared to normal subjects [86]. Ruge et al. hypothesize that dystonic movements become "set" within the sensorimotor system because of increased synaptic plasticity [87]. There is considerable evidence linking abnormal corticostriatal synaptic plasticity to striatal DA and ACh neuromodulatory systems [24, 88, 89]. Cholinergic tone determines the direction of corticostriatal plasticity with higher levels of endogenous ACh enhancing LTP while lower levels encourage LTD [90]. Furthermore, the pathophysiology in both PD and dystonia has been characterized by abnormal cholinergic function and by association abnormal plasticity [24, 91] as well as altered receptive fields. However, differences in the physical manifestation and treatment response may be an indication that abnormal synaptic plasticity in PD is different or less involved than in dystonia. For example, unlike DBS in PD where immediate clinical improvements are noted, improvement and maximal clinical benefits are not immediately evident in patients with dystonia [56, 92]. Furthermore, short- and long-term maintenance of DBS therapeutic benefit after stimulation is turned off is often noted, as is an immediate return of benefit upon reactivation of DBS therapy [93, 94]. Changes in plasticity may explain both the delay in therapeutic onset and offset of therapeutic DBS in dystonia. Patients with higher levels of measured plasticity while on long-term DBS therapy are more likely to retain clinical benefit when DBS therapy is subsequently discontinued or reestablished [87]. Ruge et al. suggest that the clinical benefit obtained during long-term DBS therapy may be more efficiently stored in patients with more plasticity.

Plasticity may also play a role in the reorganization of receptive fields that seems to propagate throughout the BGTC. In focal dystonia, the dystonic body part may be determined by the somatotopic region of the BGTC structure that is affected while PGD may be determined by extensive distortions involving many somatotopic regions. The involvement of key nodes throughout the BGTC in the manifestation of altered receptive fields and abnormal plasticity leads to the question of what mechanism and structure underlies the development of this pathophysiology. Evidence points towards a disruption in coupling in striatal DA and ACh neuromodulatory systems. In dystonia, the development of future patient-specific stimulation patterns based on measured levels of plasticity may be possible, leading to more effective DBS therapies. DBS may induce long-lasting beneficial changes in the organization of the striatum and could maintain clinical benefit with intermittent, rather than continuous, periods of stimulation.

13.3.3 Discharge Patterns

In addition to reduced mean discharge rates, altered somatotopy, and abnormal plasticity, the pattern of neuronal activity in the pallidum is significantly altered. The importance of altered discharge patterns of BG output neurons in the expression of dystonia has been demonstrated in an animal model of idiopathic dystonia [95]. Spontaneous discharge patterns of GABAergic entopeduncular neurons

(GPi equivalent in primates) during age-related periods of paroxysmal dystonia expression were extremely irregular and showed altered burst characteristics to non-dystonic controls. Most importantly, the subsequent spontaneous remission of dystonic behaviors coincided with a normalization of discharge patterns and more tonic activity in entopeduncular neurons. Interestingly, discharge rates and patterns of GABAergic nigral neurons (SNr) did not change with the expression and subsequent remission of dystonia and were the same as in non-dystonic controls. In dystonia, most GPi neurons do not discharge tonically as reported in normal NHP but in irregularly grouped discharges with intermittent bursts [37–41, 43, 48, 49, 96]. Oscillatory activity has also been found in dystonia patients and is characterized by increased power in both the delta (1–3 Hz) and theta (3–8 Hz) range [43, 96, 97]. Silberstein et al. recorded local field potentials (LFP) in the pallidum from implanted DBS electrodes and reported an increase in power between 4 and 10 Hz in patients with PGD, CD, and multi-segmental dystonia [97]. Using the same method, Tang et al. reported a peak in the power spectrum in CD patients at still lower frequencies ranging from 1 to 3 Hz [43]. Consistent with these observations, Liu et al. found that in both PGD and CD patients active or passive movements were associated with an increase in synchronized oscillatory LFP activity in the 1–3 Hz range coincident with a reduction in oscillatory activity in the 8–16 Hz range [96]. Furthermore, patients with little dystonic activity at rest also had little oscillatory LFP activity. During movement, however, the dystonia worsened and the power of oscillatory activity in the 1–3 Hz range increased in these patients. This change in power was most closely correlated to electromyography (EMG) activity in the dystonic body parts, whereas EMG activity from non-affected regions of the body showed no correlation to the low-frequency oscillations [96]. We have previously described a model for primary dystonia relating BGTC asynchronous and synchronous activity during rest and movement, respectively (Fig. 13.3) [4]. Briefly, the pathophysiology of dystonic movements is characterized by changes in mean discharge rates (line width) and altered discharge patterns (broken lines) with asynchronous or synchronous neuronal activity (multiple parallel broken lines). During movement pallidal activity is further reduced, leading to increased thalamic activity and the development of uncontrolled synchrony throughout the subcortical network (parallel lines of equal length). A disruption in cortical and brainstem output ensues, provoking disordered dystonic movements. These findings support the view that enhanced low-frequency oscillatory activity is correlated to the development of dystonia. Also, the similar degree and time course of the therapeutic response of focal and generalized dystonic movements to DBS suggests that phasic dystonic movements, whether generalized or focal, arise from the same physiological changes [96, 98].

Neural recordings from GPi neurons in patients with cervical dystonia (CD) and PGD are consistent with the hypothesis that both groups show similar changes in rate and pattern of neuronal activity in the GPi (Fig. 13.4, first and second column, respectively). The mean discharge rates in the PGD patient (43 spikes/s) and the CD patient (39 spikes/s) were similar to one another. Furthermore, burstiness (measured as the percentage of spikes located within bursts) in the PGD patient was 20 % compared to 33 % in the CD patient, while the burst duration (0.06 burst duration in

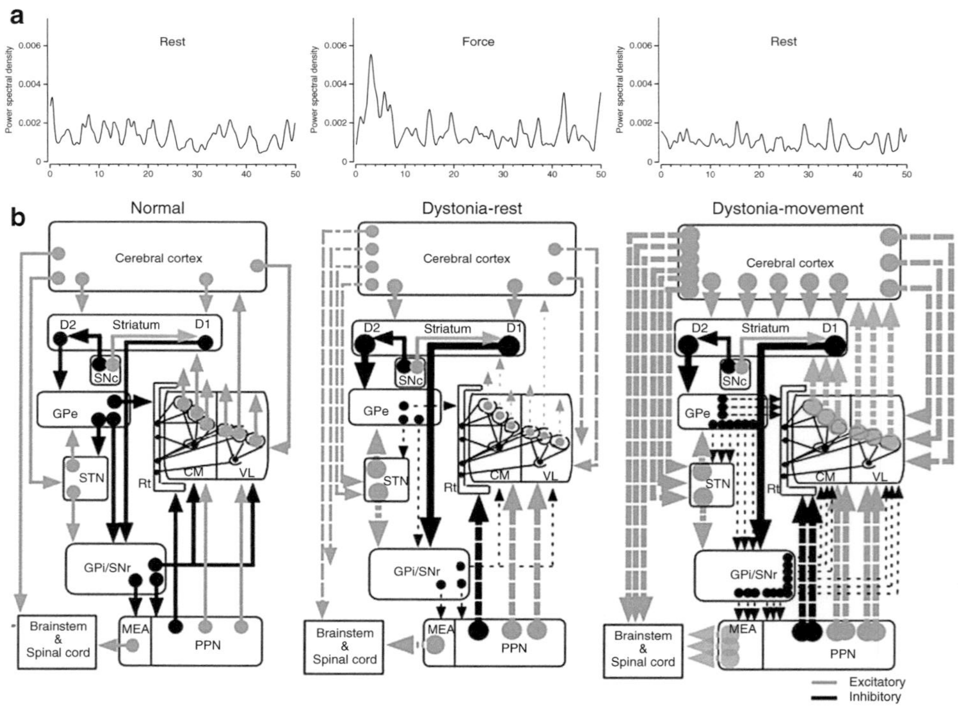

Fig. 13.3 Mechanism underlying dystonic activity during rest and during movement. (**a**) Change in low-frequency spectral power during subsequent periods of rest (*left*), force control (*middle*), and rest (*right*) indicates increased oscillatory activity in GP activity in patient with PGD during movement. (**b**) Model for primary dystonia. Consistent with previous reports, thalamic and pallidal activity in dystonia is reduced at rest (*middle panel*) compared to normal activity (*left panel*). During movement, pallidal activity is further reduced (*right panel*), leading to an increase in thalamic activity and the development of uncontrolled synchrony throughout the subcortical-cortical network. *CM* centromedian, *VL* motor thalamus, *Rt* reticular nucleus of the thalamus, *PPN* pedunculopontine nucleus, *MEA* midbrain extrapyramidal area, *GPe* globus pallidus pars externa, *GPi* globus pallidus pars interna, *STN* subthalamic nucleus, *SNr* substantia nigra pars reticulata, *SNc* substantia nigra pars compacta. Multiple lines of *different lengths* exiting a nucleus represent *asynchronous* neuronal activity, whereas *multiple broken lines of the same length* illustrate *altered patterns of synchronous* activity. The width of the lines depicts the amount of neuronal activity (Figure adapted from previous publication by Vitek [4])

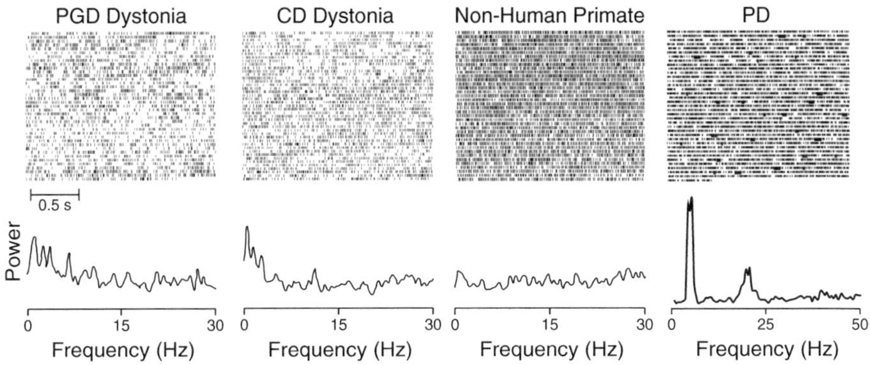

Fig. 13.4 Pallidal GPi activity in PGD (*left*) and CD (*middle*) patients and normal monkey (*right*). The raster display (*top row*) and power spectrum (*bottom row*) are shown for each cell. Also shown, activity in GPi in PD patient with manifestations of tremor, bradykinesia, and rigidity

seconds) and mean number of spikes (8 spikes/s) within a burst was the same. In contrast to dystonia patients, normal NHP rates were higher (84 Hz), burstiness was less (7 %), and burst characteristics were different with shorter burst durations (0.03 s) and the number of spikes in a burst was smaller (6 spikes/burst) (Fig. 13.4, third column). Finally, the power spectra (bottom row) show large peaks at low frequency (1–5 Hz) in the PGD and CD patients that were not present in the normal NHP. GPi neurons from both dystonia groups demonstrated elevated power throughout much of the frequency range below 6 Hz. Although these data are taken from a limited sample, they are consistent with previous observations that low-frequency synchronized oscillations may be more closely associated with dystonia and causal to its development than previously thought. Similar findings have been noted in PD (Fig. 13.4, last column). Persistent bursting or oscillations reported within the BGTC [37–41, 43, 48, 49, 96] may result in maladaptive learning leading to system-wide changes over time. These findings are in support of the hypothesis that different phenotypes of primary dystonia arise from the same pathophysiological changes occurring in somatotopically distinct regions of the pallidum.

13.4 Network Properties

Although widely considered purely a disorder of the BG dystonia is now being viewed as a more system-wide dysfunction of multiple brain circuits [70, 99]. CB and BG motor circuits are often considered in isolation of each other in both normal and dysfunctional motor control. In the network model of dystonia, CB and BG dysfunction is relayed via separate subcortical pathways to the cerebral cortex, resulting in a final expression of dystonia [100]. However, recognition of the interaction within and across these pallidothalamocortical and cerebellothalamocortical circuits and their respective roles in the pathophysiology of dystonia is increasing. The functional segregation between BGTC and CBTC circuits has come into question and possible

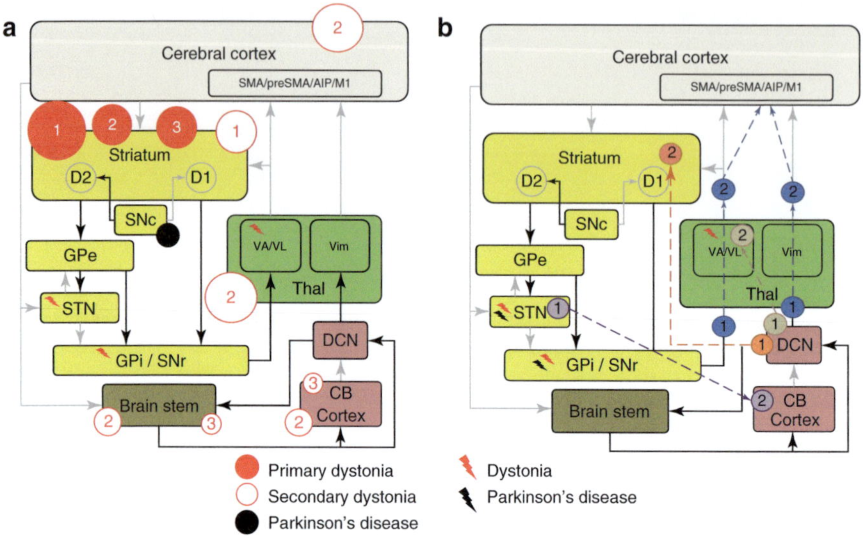

Fig. 13.5 Sources and pathways related to the pathophysiology of PD and dystonia. (**a**) PD is associated with a depletion of dopaminergic neurons in the SNc (*black circle*). Most forms of dystonia (*red filled and open circles*) are largely associated with BGTC dysfunction but have also been linked to the CBTC motor circuit. Primary dystonias, including PGD (*1*), focal dystonias (*2*), and genetic forms (*3*), have no known cause but are generally associated with striatal dysfunction. Secondary dystonias (*open circles*) have a known cause or recognized pathological condition, including drug-related tardive dystonias (*1*), structural lesions (*2*), and other rare familial forms (*3*). (**b**) BGTC and CBTC interactions and pathways. Lines represent inhibitory (*black*) or excitatory (*gray*) projections. The basal ganglia (BG, *yellow*) include the striatum (dopamine receptor subtypes represented as D1 and D2), *GPe* and *GPi* (globus pallidus pars externa and interna, respectively), *SNc* and *SNr* (substantia nigra pars compacta and reticulata, respectively), and *STN* (subthalamic nucleus). The thalamus, *VA/VL* ventroanterior/ventrolateral, *CB* cerebellum, *Vim* ventral intermediate nucleus, *DCN* deep cerebellar nuclei, *SMA* supplementary motor cortex, *AIP* anterior intraparietal area, and *M1* primary motor cortex. Disynaptic targets from both the BG and CB converging within cortical areas (*blue*). Disynaptic pathways linking CB output from the dentate nucleus to the BG via striatal inputs (*orange*), linking BG output from the STN to the CB (*purple*), and projections from the lateral CB nucleus to the thalamus (*brown*)

anatomical substrates have linked BG and CB involvement in dystonia (Fig. 13.5). Specifically, disynaptic targets from both the BG and CB converging within cortical areas, including the supplementary motor area (SMA) and pre-SMA [100], the anterior intraparietal area (AIP) [101], and primary motor cortex (M1) [102], support the view that the CB and BG communicate via separate thalamic relays to the cortex [103]. In the network model of dystonia, CB or BG dysfunction is relayed via separate subcortical pathways to the cerebral cortex, resulting in a final expression of dystonia. However, direct cerebral-cerebellar involvement may not be necessary and other subcortical connections may link CB and BG circuits. Disynaptic pathways linking CB output from the dentate nucleus to the BG via striatal inputs [104] and linking BG output from the STN to the CB [103] have been shown in the NHP and projections from the lateral CB nucleus to the thalamus in rat [105]. There is also considerable evidence for brainstem involvement in the pathophysiology of dystonia. In humans, secondary forms of CD are largely linked to lesions in the CB or within brainstem afferents to the CB rather than the BG [106]. Dystonia in a rare familial form of ataxia is also limited to pathology of the CB and brainstem [107]. While DYT1 dystonia does not explicitly show signs of neurodegeneration or cell loss within the CB or BG, brainstem inclusions have been noted [108]. Furthermore, improvements in CD were noted after the treatment of causative tumors of the posterior fossa [109]. The extent to which pallidothalamocortical and cerebellothalamocortical substrates are involved, or whether it is a deficiency in the integration between overlapping BG and CB circuits that leads to some forms of dystonia, remains unclear.

Consistent with the anatomical substrates, there is increasing evidence that altered CB output can drive changes in BGTC circuits [99, 110–113]. While these studies show the importance of BG dysfunction in the development of dystonia, they also provide evidence that abnormal activity in structures outside the BG cortical-subcortical circuits can disrupt normal BG function. Circuit abnormalities underlying the development of dystonia may be preferentially related to deficiencies in cerebellothalamocortical fiber tracks rather than neurodegeneration or cell loss [70, 114–117]. Importantly, phenotypic variability in affected DYT1/DYT6 gene carriers has been linked to thalamocortical projection abnormalities subjacent to somatotopic regions of the Rolandic cortex related to the affected limb [70]. Furthermore, non-manifesting carriers were distinguished by additional, tandem micro-lesions in the more distal thalamocortical tract segments of the CBTC motor pathway. These findings are consistent with the hypothesis that, in non-manifesting carriers, distal lesions in the CBTC motor pathway effectively disrupt or override the pathological activity from the more proximal segment, thereby reducing disease penetrance [70, 117].

The vast majority of PGD patients show improvement with GPi lesions or DBS in the STN or GPi, suggesting a major role of the BGTC circuit. While many patients respond very well to DBS, we do not know why some patients respond better than others. Mechanism by which DBS improves dystonia is also unclear. Ineffective DBS may occur as the result of more system-wide alterations occurring in these patients that are not affected by DBS at the targeted site. As we learn more about the circuitry and pathways involved in patients with dystonia, it is becoming clear that future models of dystonia must take into account network-wide alterations. Network alterations in dystonia may help address the delayed therapeutic response of patients with dystonia to pallidotomy and DBS as well as address changes in synaptic plasticity and altered receptive fields in both pallidothalamocortical and cerebellothalamocortical circuits. While the relative contribution of BG and CB pathways in the development of dystonia remains unclear, it is increasingly apparent that we can no longer ignore the CB when developing pathophysiological models of dystonia.

13.5 Discussion

Divergent hypotheses exist concerning the pathophysiological basis for the development of primary dystonia. One perspective is that different phenotypes of primary dystonia arise from the same pathophysiological changes occurring in somatotopically distinct regions of the pallidum [66] while others argue that these different phenotypic forms are due to distinct pathophysiological changes in the pallidum. We suggest that, at least in PGD, dysfunction begins in the striatum and preferentially involves the "direct" pathway in the development of uncontrolled increases in synchronization in GP. The "indirect" pathway might account for the changes in firing rates/patterns and in receptive fields. In dystonia, GP activation is decreased via inhibition from the "direct" pathway (GPi) and "indirect" pathway (GPe), allowing widened receptive fields that might originate in the striatum. We view the underlying pathophysiology in dystonia to be similar in PD. However, in PD we believe that there is a greater disruption in the "indirect" pathway. Our model postulates that striatal overactivity in both the "direct" and "indirect" pathways accounts for the observed changes in mean discharge rate and altered patterns of neuronal activity as well as the increased incidence of somatosensory responses of GPi. The degree to which changes in firing rates/patterns and broadened receptive fields contribute to dystonia is unclear and may be secondary to the development of uncontrolled increases in synchronization. We propose a "population" theory, suggesting that a progressive increase in the amount and degree of uncontrolled synchrony of spontaneous neuronal activity in the pallidum could eventually involve a critical population of neurons within GPi, triggering similar synchronous activity throughout the BGTC. As the pathology incorporates more cells and wider regions of GPi, dystonia may become more persistent, occurring with less stimulation, under stress, or at rest. As such, patients may not manifest dystonia until a certain critical number of neurons are "affected." In PGD, the pathology may involve many somatotopic regions and thus affects many body parts, while in focal dystonia, the dystonic body part is likely determined by pathology in the somatotopic region of the pallidum associated with the body part. Differences in phenotypic forms of (primary) dystonia would therefore arise from the same, rather than distinct, pathophysiological profile in which the differential distribution of neurons with the same profile within the sensorimotor GPi determines the phenotypic distribution of dystonic symptoms.

References

1. Fahn S, Elton RL. The UPDRS Development Committee, United Parkinson disease rating scale. In: Fahn S, Marsden CD, Clne D, Goldstein M, editors. Recent developments in Parkinson's disease, vol. 2. Floral Park: Macmillan; 1987. p. 293–304.
2. DeLong MR. Primate models of movement disorders of basal ganglia origin. Trends Neurosci. 1990;13:281–5.
3. Albin RL, Young AB, Penney JB. The functional anatomy of basal ganglia disorders. Trends Neurosci. 1989;12:366–75.
4. Vitek JL. Pathophysiology of dystonia: a neuronal model. Mov Disord. 2002;17:S49–62.
5. Hendrix CM, Vitek JL. Toward a network model of dystonia. Ann N Y Acad Sci. 2012; 1265:46–55.
6. Vitek JL, Giroux M. Physiology of hypokinetic and hyperkinetic movement disorders: model for dyskinesia. Ann Neurol. 2000;47:S131–40.
7. Kultas-ilinsky K, Ilinsky IA. Fine structure of the magnocellular subdivision of the ventral anterior thalamic nucleus (V Amc) of Macaca mulatta: II. Organization of nigrothalamic afferents as revealed with EM autoradiography. J Comp Neurol. 1990;294:479–89.

8. Alexander GE, Crutcher MD, DeLong MR. Basal ganglia-thalamocortical circuits: parallel substrates for motor, oculomotor, "prefrontal" and "limbic" functions. Prog Brain Res. 1990;85:119–46.
9. Mink JW, Thach WT. Basal ganglia intrinsic circuits and their role in behavior. Curr Opin Neurobiol. 1993;3:950–7.
10. Mink JW. The basal ganglia: focused selection and inhibition of competing motor programs. Prog Neurobiol. 1996;50:381–425.
11. Uno M, Yoshida M. Monosynaptic inhibition of thalamic neurons produced by stimulation of the pallidal nucleus in cats. Brain Res. 1975;99:377–80.
12. Alexander GE, DeLong MR, Strick PL. Parallel organization of functionally segregated circuits linking basal ganglia and cortex. Annu Rev Neurosci. 1986;9:357–81.
13. Smith Y, Wichmann T, DeLong MR. The external pallidum and the subthalamic nucleus send convergent inputs onto single neurons in the internal pallidal segment in the monkey: anatomical organization and functional significance. In: Percheron G, McKenzie FS, Feger J, editors. The basal ganglia IV – new ideas and data on structure and function. New York: Plenum Press; 1994. p. 51–61.
14. Nambu A, Tokuno H, Takada M. Functional significance of the cortico–subthalamo–pallidal 'hyperdirect' pathway. Neurosci Res. 2002;43:111–7.
15. Nambu A. Seven problems on the basal ganglia. Curr Opin Neurobiol. 2008;18:595–604.
16. Tepper JM, Abercrombie ED, Bolam JP. Basal ganglia macrocircuits. In: James EDA, Tepper M, Bolam JP, editors. Progress in brain research, vol. 160. Amsterdam: Elsevier; 2007. p. 3–7.
17. Nambu A, Tokuno H, Hamada I, Kita H, Imanishi M, Akazawa T, et al. Excitatory cortical inputs to pallidal neurons via the subthalamic nucleus in the monkey. J Neurophysiol. 2000;84:289–300.
18. Goto S, Lee LV, Munoz EL, Tooyama I, Tamiya G, Makino S, et al. Functional anatomy of the basal ganglia in X-linked recessive dystonia-parkinsonism. Ann Neurol. 2005;58:7–17.
19. Crittenden JR, Graybiel AM. Basal Ganglia disorders associated with imbalances in the Striatal striosome and matrix compartments. Front Neuroanat. 2011;5:59.
20. Tokuno H, Chiken S, Kametani K, Moriizumi T. Efferent projections from the striatal patch compartment: anterograde degeneration after selective ablation of neurons expressing μ-opioid receptor in rats. Neurosci Lett. 2002;332:5–8.
21. Gerfen CR. The neostriatal mosaic: compartmentalization of corticostriatal input and striatonigral output systems. Nature. 1984;311:461–4.
22. Bonsi P, Cuomo D, Martella G, Madeo G, Schirinzi T, Puglisi F, et al. Centrality of striatal cholinergic transmission in basal ganglia function. Front Neuroanat. 2011;5:6.
23. Koós T, Tepper JM. Dual cholinergic control of fast-spiking interneurons in the neostriatum. J Neurosci. 2002;22:529–35.
24. Pisani A, Bernardi G, Ding J, Surmeier DJ. Re-emergence of striatal cholinergic interneurons in movement disorders. Trends Neurosci. 2007;30:545–53.
25. Calabresi P, Centonze D, Gubellini P, Pisani A, Bernardi G. Acetylcholine-mediated modulation of striatal function. Trends Neurosci. 2000;23:120–6.
26. Duvoisin RC. Cholinergic-anticholinergic antagonism in parkinsonism. Arch Neurol. 1967;17:124–36.
27. Barbeau A. The pathogenesis of Parkinson's disease: a new hypothesis. Can Med Assoc J. 1962;87:802–7.
28. Fahn S. High dosage anticholinergic therapy in dystonia. Neurology. 1983;33:1255–61.
29. Graybiel AM. Neurotransmitters and neuromodulators in the basal ganglia. Trends Neurosci. 1990;13:244–54.
30. Contant C, Umbriaco D, Garcia S, Watkins KC, Descarries L. Ultrastructural characterization of the acetylcholine innervation in adult rat neostriatum. Neuroscience. 1996;71:937–47.
31. Smith Y, Villalba R. Striatal and extrastriatal dopamine in the basal ganglia: An overview of its anatomical organization in normal and Parkinsonian brains. Mov Disord. 2008;23: S534–47.

32. Sidibé M, Smith Y. Thalamic inputs to striatal interneurons in monkeys: synaptic organization and co-localization of calcium binding proteins. Neuroscience. 1999;89:1189–208.
33. Lapper SR, Bolam JP. Input from the frontal cortex and the parafascicular nucleus to cholinergic interneurons in the dorsal striatum of the rat. Neuroscience. 1992;51:533–45.
34. DeBoer P, Abercrombie ED. Physiological release of striatal acetylcholine in vivo: modulation by D1 and D2 dopamine receptor subtypes. J Pharmacol Exp Ther. 1996;277:775–83.
35. Pakhotin P, Bracci E. Cholinergic interneurons control the excitatory input to the striatum. J Neurosci. 2007;27:391–400.
36. Zhou FM, Wilson CJ, Dani JA. Cholinergic interneuron characteristics and nicotinic properties in the striatum. J Neurobiol. 2002;53:590–605.
37. Zhuang P, Li Y, Hallett M. Neuronal activity in the basal ganglia and thalamus in patients with dystonia. Clin Neurophysiol. 2004;115:2542–57.
38. Starr PA, Rau GM, Davis V, Marks Jr WJ, Ostrem JL, Simmons D, et al. Spontaneous pallidal neuronal activity in human dystonia: comparison with Parkinson's disease and normal macaque. J Neurophysiol. 2005;93:3165–76.
39. Hutchison WD, Lang AE, Dostrovsky JO, Lozano AM. Pallidal neuronal activity: implications for models of dystonia. Ann Neurol. 2003;53:480–8.
40. Vitek JL, Chockkan V, Zhang JY, Kaneoke Y, Evatt M, DeLong MR, et al. Neuronal activity in the basal ganglia in patients with generalized dystonia and hemiballismus. Ann Neurol. 1999;46:22–35.
41. Chang EF, Turner RS, Ostrem JL, Davis VR, Starr PA. Neuronal responses to passive movement in the globus pallidus internus in primary dystonia. J Neurophysiol. 2007;98: 3696–707.
42. Lenz F, Byl N, Garonzik I, Lee J, Hua S. Microelectrode studies of basal ganglia and VA, VL and VP thalamus in patients with dystonia: dystonia-related activity and sensory reorganization. In: Kultas-Ilinsky K, Ilinsky I, editors. Basal ganglia and thalamus in health and movement disorders. New York: Kluwer Academic/Plenum Publishers; 2001. p. 225–37.
43. Tang JK, Moro E, Mahant N, Hutchison WD, Lang AE, Lozano AM, et al. Neuronal firing rates and patterns in the globus pallidus internus of patients with cervical dystonia differ from those with Parkinson's disease. J Neurophysiol. 2007;98:720–9.
44. Merello M, Cerquetti D, Cammarota A, Tenca E, Artes C, Antico J, et al. Neuronal globus pallidus activity in patients with generalised dystonia. Mov Disord. 2004;19:548–54.
45. Baron MS, Vitek JL, Bakay RA, Green J, Kaneoke Y, Hashimoto T, et al. Treatment of advanced Parkinson's disease by posterior GPi pallidotomy: 1-year results of a pilot study. Ann Neurol. 1996;40:355–66.
46. Lozano AM, Kumar R, Gross RE, Giladi N, Hutchison WD, Dostrovsky JO, et al. Globus pallidus internus pallidotomy for generalized dystonia. Mov Disord. 1997;12:865–70.
47. Lozano AM, Lang AE, Galvez-Jimenez N, Miyasaki J, Duff J, Hutchinson WD, et al. Effect of GPi pallidotomy on motor function in Parkinson's disease. Lancet. 1995;346:1383–7.
48. Sanghera MK, Grossman RG, Kalhorn CG, Hamilton WJ, Ondo WG, Jankovic J. Basal ganglia neuronal discharge in primary and secondary dystonia in patients undergoing pallidotomy. Neurosurgery. 2003;52:1358–70; discussion 1370–3.
49. Magarinos-Ascone CM, Regidor I, Gomez-Galan M, Cabanes-Martinez L, Figueiras-Mendez R. Deep brain stimulation in the globus pallidus to treat dystonia: electrophysiological characteristics and 2 years' follow-up in 10 patients. Neuroscience. 2008;152:558–71.
50. Starr PA, Turner RS, Rau G, Lindsey N, Heath S, Volz M, et al. Microelectrode-guided implantation of deep brain stimulators into the globus pallidus internus for dystonia: techniques, electrode locations, and outcomes. Neurosurg Focus. 2004;17:E4.
51. Suarez J, Verhagen-Metman L, Reich S, Dougherty P, Hallett M, Lenz F. Pallidotomy for hemiballismus: efficacy and characteristics of neuronal activity. Ann Neurol. 1997;42: 807–11.
52. Kumar R. Methods for programming and patient management with deep brain stimulation of the globus pallidus for the treatment of advanced Parkinson's disease and dystonia. Mov Disord. 2002;17 Suppl 3:S198–207.

53. Tisch S, Zrinzo L, Limousin P, Bhatia KP, Quinn N, Ashkan K, et al. Effect of electrode contact location on clinical efficacy of pallidal deep brain stimulation in primary generalised dystonia. J Neurol Neurosurg Psychiatry. 2007;78:1314–9.
54. Alterman RL, Miravite J, Weisz D, Shils JL, Bressman SB, Tagliati M. Sixty hertz pallidal deep brain stimulation for primary torsion dystonia. Neurology. 2007;69:681–8.
55. Kupsch A, Benecke R, Muller J, Trottenberg T, Schneider GH, Poewe W, et al. Pallidal deep-brain stimulation in primary generalized or segmental dystonia. N Engl J Med. 2006;355:1978–90.
56. Vidailhet M, Vercueil L, Houeto JL, Krystkowiak P, Benabid AL, Cornu P, et al. Bilateral deep-brain stimulation of the globus pallidus in primary generalized dystonia. N Engl J Med. 2005;352:459–67.
57. Aagaard P, Simonsen EB, Andersen JL, Magnusson P, Dyhre-Poulsen P. Increased rate of force development and neural drive of human skeletal muscle following resistance training. J Appl Physiol. 2002;93:1318–26.
58. Sanger TD, Pascual-Leone A, Tarsy D, Schlaug G. Nonlinear sensory cortex response to simultaneous tactile stimuli in writer's cramp. Mov Disord. 2002;17:105–11.
59. Sanger T, Tarsy D, Pascual-Leone A. Abnormalities of spatial and temporal sensory discrimination in writer's cramp. Mov Disord. 2001;16:94–9.
60. Byl NN. Focal hand dystonia may result from aberrant neuroplasticity. Adv Neurol. 2004; 94:19–28.
61. Byl N. Sensory dysfunction associated with repetitive strain injuries of tendinitis and focal hand dystonia: a comparative study. J Orthop Sports Phys Ther. 1996;23:234–44.
62. Byl N, Merzenich M, Cheung S, Bedenbaugh P, Nagarajan S, Jenkins W. A primate model for studying focal dystonia and repetitive strain injury: effects on the primary somatosensory cortex. Phys Ther. 1997;77:269–84.
63. Tinazzi M, Priori A, Bertolasi L, Frasson E, Mauguiere F, Fiaschi A. Abnormal central integration of a dual somatosensory input in dystonia. Evidence for sensory overflow. Brain. 2000;123(Pt 1):42–50.
64. Tempel LW, Perlmutter JS. Abnormal vibration-induced cerebral blood flow responses in idiopathic dystonia. Brain. 1990;113(Pt 3):691–707.
65. Lenz FA, Jaeger CJ, Seike MS, Lin YC, Reich SG, DeLong MR, et al. Thalamic single neuron activity in patients with dystonia: dystonia-related activity and somatic sensory reorganization. J Neurophysiol. 1999;82:2372–92.
66. Vitek JL, Bakay RA, Hashimoto T, Kaneoke Y, Mewes K, Zhang JY, et al. Microelectrode-guided pallidotomy: technical approach and its application in medically intractable Parkinson's disease. J Neurosurg. 1998;88:1027–43.
67. Taha JM, Favre J, Baumann TK, Burchiel KJ. Characteristics and somatotopic organization of kinesthetic cells in the globus pallidus of patients with Parkinson's disease. J Neurosurg. 1996;85:1005–12.
68. Topp KS, Byl NN. Movement dysfunction following repetitive hand opening and closing: anatomical analysis in owl monkeys. Mov Disord. 1999;14:295–306.
69. Garraux G, Bauer A, Hanakawa T, Wu T, Kansaku K, Hallett M. Changes in brain anatomy in focal hand dystonia. Ann Neurol. 2004;55:736–9.
70. Argyelan M, Carbon M, Niethammer M, Uluğ AM, Voss HU, Bressman SB, et al. Cerebellothalamocortical connectivity regulates penetrance in dystonia. J Neurosci. 2009;29:9740–7.
71. Asanuma K, Ma Y, Okulski J, Dhawan V, Chaly T, Carbon M, et al. Decreased striatal D2 receptor binding in non-manifesting carriers of the DYT1 dystonia mutation. Neurology. 2005;64:347–9.
72. Carbon M, Su S, Dhawan V, Raymond D, Bressman S, Eidelberg D. Regional metabolism in primary torsion dystonia: effects of penetrance and genotype. Neurology. 2004;62:1384–90.
73. Ghilardi MF, Carbon M, Silvestri G, Dhawan V, Tagliati M, Bressman S, et al. Impaired sequence learning in carriers of the DYT1 dystonia mutation. Ann Neurol. 2003;54:102–9.
74. Agostino R, Berardelli A, Curra A, Manfredi M. The performance of rapid arm movements in Parkinson's disease – a review. In: Battistin L, Scarlato G, Caraceni T, Ruggieri S, editors. Parkinson's disease. Philadelphia: Lippincott-Raven Pub; 1996. p. 135–46.

75. Cohen LG, Hallett M. Hand cramps: clinical features and electromyographic patterns in a focal dystonia. Neurology. 1988;38:1005–12.
76. Berardelli A, Hallett M, Rothwell JC, Agostino R, Manfredi M, Thompson PD, et al. Single-joint rapid arm movements in normal subjects and in patients with motor disorders. Brain. 1996;119(Pt 2):661–74.
77. Lestienne F. Effects of inertial load and velocity on the braking process of voluntary limb movements in man. Exp Brain Res. 1979;35:407–18.
78. Prodoehl J, Corcos DM, Vaillancourt DE. Effects of focal hand dystonia on visually guided and internally guided force control. J Neurol Neurosurg Psychiatry. 2006;77:909–14.
79. Prodoehl J, MacKinnon CD, Comella CL, Corcos DM. Rate of force production and relaxation is impaired in patients with focal hand dystonia. Parkinsonism Relat Disord. 2006;12:363–71.
80. Nowak DA, Rosenkranz K, Topka H, Rothwell J. Disturbances of grip force behaviour in focal hand dystonia: evidence for a generalised impairment of sensory-motor integration? J Neurol Neurosurg Psychiatry. 2005;76:953–9.
81. Odergren T, Iwasaki N, Borg J, Forssberg H. Impaired sensory-motor integration during grasping in writer's cramp. Brain. 1996;119:569–83.
82. Graybiel AM, Aosaki T, Flaherty AW, Kimura M. The basal ganglia and adaptive motor control. Science. 1994;265:1826–31.
83. Albin RL, Cross D, Cornblath WT, Wald JA, Wernette K, Frey KA, et al. Diminished striatal [123I] iodobenzovesamicol binding in idiopathic cervical dystonia. Ann Neurol. 2003;53:528–32.
84. Saka E, Iadarola M, Fitzgerald DJ, Graybiel AM. Local circuit neurons in the striatum regulate neural and behavioral responses to dopaminergic stimulation. Proc Natl Acad Sci U S A. 2002;99:9004–9.
85. Pisani A, Martella G, Tscherter A, Bonsi P, Sharma N, Bernardi G, et al. Altered responses to dopaminergic D2 receptor activation and N-type calcium currents in striatal cholinergic interneurons in a mouse model of DYT1 dystonia. Neurobiol Dis. 2006;24:318–25.
86. Quartarone A, Morgante F, Sant'Angelo A, Rizzo V, Bagnato S, Terranova C, et al. Abnormal plasticity of sensorimotor circuits extends beyond the affected body part in focal dystonia. J Neurol Neurosurg Psychiatry. 2008;79:985–90.
87. Ruge D, Cif L, Limousin P, Gonzalez V, Vasques X, Hariz MI, et al. Shaping reversibility? Long-term deep brain stimulation in dystonia: the relationship between effects on electrophysiology and clinical symptoms. Brain. 2011;134:2106–15.
88. Peterson DA, Sejnowski TJ, Poizner H. Convergent evidence for abnormal striatal synaptic plasticity in dystonia. Neurobiol Dis. 2010;37:558–73.
89. Lovinger DM, Partridge JG, Tang KC. Plastic control of striatal glutamatergic transmission by ensemble actions of several neurotransmitters and targets for drugs of abuse. Ann N Y Acad Sci. 2003;1003:226–40.
90. Bonsi P, Martella G, Cuomo D, Platania P, Sciamanna G, Bernardi G, et al. Loss of muscarinic autoreceptor function impairs long-term depression but not long-term potentiation in the striatum. J Neurosci. 2008;28:6258–63.
91. Jun D, Guzman JN, Tkatch T, Songhai C, Goldberg JA, Ebert PJ, et al. RGS4-dependent attenuation of M4 autoreceptor function in striatal cholinergic interneurons following dopamine depletion. Nat Neurosci. 2006;9:832–42.
92. Ruge D, Tisch S, Hariz MI, Zrinzo L, Bhatia KP, Quinn NP, et al. Deep brain stimulation effects in dystonia: time course of electrophysiological changes in early treatment. Mov Disord. 2011;26:1913–21.
93. Hebb MO, Chiasson P, Lang AE, Brownstone RM, Mendez I. Sustained relief of dystonia following cessation of deep brain stimulation. Mov Disord. 2007;22:1958–62.
94. Coubes P, Cif L, El Fertit H, Hemm S, Vayssiere N, Serrat S, et al. Electrical stimulation of the globus pallidus internus in patients with primary generalized dystonia: long-term results. J Neurosurg. 2004;101:189–94.
95. Gernert M, Bennay M, Fedrowitz M, Rehders JH, Richter A. Altered discharge pattern of basal ganglia output neurons in an animal model of idiopathic dystonia. J Neurosci. 2002; 22:7244–53.

96. Liu X, Yianni J, Wang S, Bain PG, Stein JF, Aziz TZ. Different mechanisms may generate sustained hypertonic and rhythmic bursting muscle activity in idiopathic dystonia. Exp Neurol. 2006;198:204–13.
97. Silberstein P, Kuhn AA, Kupsch A, Trottenberg T, Krauss JK, Wohrle JC, et al. Patterning of globus pallidus local field potentials differs between Parkinson's disease and dystonia. Brain. 2003;126:2597–608.
98. Bittar RG, Yianni J, Wang S, Liu X, Nandi D, Joint C, et al. Deep brain stimulation for generalised dystonia and spasmodic torticollis. J Clin Neurosci. 2005;12:12–6.
99. Neychev VK, Fan X, Mitev VI, Hess EJ, Jinnah HA. The basal ganglia and cerebellum interact in the expression of dystonic movement. Brain. 2008;131:2499–509.
100. Akkal D, Dum RP, Strick PL. Supplementary motor area and presupplementary motor area: targets of basal ganglia and cerebellar output. J Neurosci. 2007;27:10659–73.
101. Clower DM, Dum RP, Strick PL. Basal ganglia and cerebellar inputs to 'AIP'. Cereb Cortex. 2005;15:913–20.
102. Hoover JE, Strick PL. The organization of cerebellar and basal ganglia outputs to primary motor cortex as revealed by retrograde transneuronal transport of herpes simplex virus type 1. J Neurosci. 1999;19:1446–63.
103. Bostan A, Strick P. The cerebellum and basal ganglia are interconnected. Neuropsychol Rev. 2010;20:261–70.
104. Hoshi E, Tremblay L, Feger J, Carras PL, Strick PL. The cerebellum communicates with the basal ganglia. Nat Neurosci. 2005;8:1491–3.
105. Ichinohe N, Mori F, Shoumura K. A di-synaptic projection from the lateral cerebellar nucleus to the laterodorsal part of the striatum via the central lateral nucleus of the thalamus in the rat. Brain Res. 2000;880:191–7.
106. LeDoux MS, Brady KA. Secondary cervical dystonia associated with structural lesions of the central nervous system. Mov Disord. 2003;18:60–9.
107. Le Ber I, Clot F, Vercueil L, Camuzat A, Viémont M, Benamar N, De Liège P, Ouvrard-Hernandez AM, Pollak P, Stevanin G, Brice A, Dürr A, Camuzat A, Viémont M, Benamar N, De Liège P, Ouvrard-Hernandez AM, Pollak P, Stevanin G, Brice A, Dürr A. Predominant dystonia with marked cerebellar atrophy: a rare phenotype in familial dystonia. Neurology. 2006;67:1769–73.
108. McNaught KSP, Kapustin A, Jackson T, Jengelley TA, JnoBaptiste R, Shashidharan P, et al. Brainstem pathology in DYT1 primary torsion dystonia. Ann Neurol. 2004;56:540–7.
109. Krauss JK, Seeger W, Jankovic J. Cervical dystonia associated with tumors of the posterior fossa. Mov Disord. 1997;12:443–7.
110. Ratcheson RA, Li CL. Effect of dentate stimulation on neuronal activity in the caudate nucleus. Exp Neurol. 1969;25:268–81.
111. Li CI, Parker LO. Effect of dentate stimulation on neuronal activity in the globus pallidus. Exp Neurol. 1969;24:298–309.
112. Nieoullon A, Cheramy A, Glowinski J. Release of dopamine in both caudate nuclei and both substantia nigrae in response to unilateral stimulation of cerebellar nuclei in the cat. Brain Res. 1978;148:143–52.
113. Calderon DP, Fremont R, Kraenzlin F, Khodakhah K. The neural substrates of rapid-onset Dystonia-Parkinsonism. Nat Neurosci. 2011;14:357–65.
114. Carbon M, Kingsley PB, Su S, Smith GS, Spetsieris P, Bressman S, et al. Microstructural white matter changes in carriers of the DYT1 gene mutation. Ann Neurol. 2004;56:283–6.
115. Carbon M, Kingsley PB, Tang C, Bressman S, Eidelberg D. Microstructural white matter changes in primary torsion dystonia. Mov Disord. 2008;23:234–9.
116. Delmaire C, Vidailhet M, Wassermann D, Descoteaux M, Valabregue R, Bourdain F, et al. Diffusion abnormalities in the primary sensorimotor pathways in writer's cramp. Arch Neurol. 2009;66:502–8.
117. Uluğ AM, Vo A, Argyelan M, Tanabe L, Schiffer WK, Dewey S, et al. Cerebellothalamocortical pathway abnormalities in torsinA DYT1 knock-in mice. Proc Natl Acad Sci. 2011;108:6638–43.

14 Neurorehabilitation in Dystonia

Jorge Hernandez Franco and Raymond L. Rosales

14.1 Introduction

Current pharmacologic treatment of dystonia is based mainly on a combination of anticholinergic drugs, benzodiazepines, muscle relaxants, anticonvulsants, and local chemodenervation injections.

Although different surgical interventions for the treatment of dystonia have been studied, the most commonly used approaches are selective peripheral denervation of muscles responsible for abnormal movements and, in severe cases, bilateral pallidal deep-brain stimulation of the globus pallidus internus or the subthalamic nucleus [1–3].

The use of intrathecal baclofen for the management of generalized dystonia [4] and cervical dystonia [5] has also been described.

While only few investigators may have evaluated the effects of physical therapy, in a cohort of 220 patients with isolated idiopathic spasmodic torticollis, followed over 14 years in a retrospective study, 98 % of the patients received medications, and 85 % received simultaneous physiotherapy consisting of active cervical exercises, motor-activating electrical current, and muscle feedback [1].

The need of a complementary treatment approach such as physiotherapy, biofeedback, and the use of sensory tricks is probably due to the modest symptomatic benefit provided by oral medications [6] and the attendant adverse event profile of these drugs at higher doses in later stages of the dystonic problem (e.g., dry mouth, cognitive disturbance, drowsiness, diplopia, glaucoma, and urinary retention) [7]. In a case report, Ramdharry [8] outlines the rehabilitation of a patient who was able to reduce

J.H. Franco (✉)
Department of Neurological Rehabilitation, National Institute of Neurology and Neurosurgery, Mexico City, Mexico
e-mail: jhfranco@medicapolanco.com

R.L. Rosales, MD, PhD
Department of Neurology and Psychiatry,
The Royal and Pontifical University of Santo Tomas, Manila, Philippines
e-mail: rlrosales@mnl.ust.edu.ph

P. Kanovsky et al. (eds.), *Dystonia and Dystonic Syndromes*,
DOI 10.1007/978-3-7091-1516-9_14

the effective dose of BoNT injections when physiotherapy management was added to a long-term pharmacological regimen. The exercise program consisted mainly of strengthening exercises of the muscles opposing the direction that the dystonic muscles pulled in, postural head and trunk exercises while moving the arms in a task, and soft tissue mobilization of painful muscular areas of the neck and shoulders.

The aim of this chapter is to highlight the non-pharmacological strategies to neurorehabilitation for cervical dystonia and task-specific dystonia including musician's dystonia. These strategies will be categorized and detailed as follows:

(a) Physical therapy exercises
(b) Sensory stimulation
(c) Biofeedback
(d) Physical agents
(e) Orthotic devices

14.2 Physical Therapy Intervention

14.2.1 Cervical Dystonia

Many approaches have been proposed for the treatment of cervical dystonia. However, a few investigators have examined the effects of physical therapy interventions with randomized controlled trials, and the studies that have been done were poorly controlled and had small samples sizes.

A personalized intensive physiotherapy program has been used to treat many people with cervical dystonia in Paris. This program developed by Jean Pierre Bleton [9], based on his expertise and not on specific research [8], involves realignment of the muscles using mirrors, reducing tone and stretching dystonic muscles, and reinforcing corrective muscles [10] promoting the activation of the muscles groups opposing the dystonic posture.

The muscles recognized as responsible for a pathological posture are stretched and relaxed. These antagonist muscles are the focus of the rehabilitation plan.

Classically, the steps to follow [11] are stated below:

- To stop abnormal movements by specific relaxation techniques
- To reinforce the corrective muscles by bringing about the head to the opposite side to the torticollis attitude
- To avoid muscular overflow and co-contractions by precise and analytic activation of the corrective muscles
- Then, to replace spasms by voluntary and adapted movements of the head

Bleton described two forms of cervical dystonia: a myoclonic form and a tonic form. In the myoclonic form, emphasis is placed on immobilizing the head and in the tonic form on rehabilitating corrective muscles [11].

Reducing pathological head movement is a principal aim for the myoclonic form. The rehabilitation program attempts to achieve muscle relaxation and near-total head immobility. The position of the eyes plays an important role. Directing the line of sight to the side opposite to the torticollis reduces the spasm and promotes correction of the cervical dystonia. Controlled immobility is repeated several times a day in front of a mirror to allow self-monitoring. The stabilizing role of the deep cervical muscles, particularly the semispinalis muscles, must be developed. The need to reinforce the activity of the muscles correcting the cervical dystonia is more appropriate for tonic variety.

The physiotherapy program tries to focus selectively on the inactive muscles and not to overflow onto other muscles. The supine position, with the torso slightly raised, permits muscle relaxation, allowing for easier voluntary contraction of corrective muscles. At a more advanced stage, the rehabilitation of deficient muscles and the search for postural adjustments are carried out while sitting and in front of a mirror for dual monitoring by the therapist and the dystonic patient. Progression continues to monitoring while standing and finally when walking.

For the pain in some patients, Bleton [9] suggests that muscle relaxation is obtained by local stretching and adapted head placement and eventually with ultrasound, heat packs, or hydrotherapy. Patients also are encouraged to correct the pathological posture at all times by organizing their activities of daily living in such a way as to hold their heads on the opposite side to the deformation. They are encouraged to be active and practice sports such as walking, swimming, tai chi, or yoga. Motor education, along with the specific rehabilitation of corrector muscles, allows a functional balance to be struck between dystonic muscles that have been weakened by injections and corrective muscles strengthened by the exercises.

The exercises proposed for spasmodic torticollis (ST) are [12] as follows:

14.2.1.1 Rotational Torticollis

Rotation with Hands Behind Your Head

This exercise consists of turning the head completely in the direction opposite the ST, fingers crossed behind the head at the occipital bone. Throughout the duration of this exercise, the chest and arms remain immobile. Only the head turns. Rotation is to be made to the right or left (the execution of a rotation of the head in the opposite direction of where it turns spontaneously).

Cervical Rotation with Extension

This exercise consists of raising the arms above the head, the hands holding a light weight of about 2 lb (1 kg) while at the same time turning the head in the correcting direction. The goal is the extension of the whole spinal column, accompanied by a correction of ST.

Cervical Rotation with the Shoulders in a Fixed Position

This exercise consists of moving the head into the correcting position while at the same time keeping the chest straight and the shoulders lowered by holding a stick in the hands.

Cervical Rotation and Correction of the Asymmetry of the Shoulders

This exercise consists of correcting the discrepancy in the height of the shoulders, while at the same time working against the ST. To do this, one raises the arm on the side of the lower shoulder and lowers the other arm. During this movement of the arms, the head turns to the side that corrects the ST. This exercise is more effective if the arms are rotated inward.

Exercise for Maintaining a Static, Stable Head Position

If the ST is accompanied by a head tremor, one is trained to keep it immobile even if it is only for a few seconds. In order to do this, one can use the sensory trick that stabilizes the ST. When the head stops moving, one removes the hand and then maintains this immobility for as long as possible. In order to achieve better control of the spasm, one can perform this exercise in front of a mirror.

It is possible to use a biofeedback machine in order to learn how to conquer the spasm. As improvement is achieved, the help of the hand or the biofeedback machine can be eliminated trying to maintain the stability of the head at all times.

14.2.1.2 Laterocollis

Relaxing a Lateral Tilt

The main goal is stretching the cervical muscles that determine the pathological tilted posture. It consists of bringing the head into a correcting position with the help of one's hand. In order to avoid compensating by raising the shoulder, keep the opposite hand on the crosspiece of the chair. Important: cervical traction must be done slowly without jerks or sudden movements.

Active Correction of a Lateral Tilt

The goal of this exercise is to strengthen the muscles that correct the pathological cervical tilt. It consists of tilting the head toward the correct position while at the same time reversing the asymmetry in the height of the shoulders. To do this, one raises the arm on the side of the lower shoulder and lowers the other arm. The exercise is more effective if the arms are rotated inward.

Horizontal Displacement of the Head

This disorder frequently causes a displacement of the axis of symmetry of the head compared to the axis of symmetry of the body.

The displacement of the head can be measured with a ruler and two strings weighted with lead. One string is placed along the axis of symmetry of the face, the other along the axis of symmetry of the body. Then it is only a question of reading on the ruler the number of inches or centimeters of displacement.

The purpose of this exercise is to move the head into the position opposite to that of the displacement. To achieve this, the head and body are moved in opposite directions. This movement is easier to do when the arms are stretched out horizontally. When you have improved, you can do this exercise with your arms at your sides in order to achieve better control of the correction. This exercise can be done in front of a mirror.

14.2.1.3 Coordination Exercises

Coordination of the Eyes and Neck

The goal of this exercise is to keep the head stable and in a neutral position while at the same time moving the eyes. It consists of having the eyes follow a swinging object while avoiding any sign of ST.

Coordination of the Neck and Shoulders

The goal of this exercise is to keep the head stable and in a neutral position, while at the same time moving the arms.

Exercise with clubs: keeping the head stable and in a neutral position, one alternately raises and lowers the left and right hands holding gymnastic clubs.

Exercise with a jump rope: while the head is kept in a stable position, one swings a jump rope to the right and left of the body.

14.2.1.4 Coordination of the Neck and the Rest of the Body

The goal of these exercises is to keep the head stable and in a neutral position while doing activities with the whole body.

Exercise marching in place: while keeping the head stable and in a neutral position, one marches in place exaggerating the height of the knees and the arm swing.

Exercise walking: one keeps the head stable while walking. This exercise is easier to do while controlling the position of the head, while one looks at the mirror placed in front.

Caution: if the spasm reappears while walking, one avoids holding the head with the hand; rather, one strives to maintain the correction using only the cervical muscles.

Given the lack of randomized evidence to support this approach and the considerable time and cost for such intensive physiotherapy, a pilot trial was done to establish the effectiveness of the Bleton technique. Specifically, the pilot trial aimed to establish whether this specific physiotherapy program for primary cervical dystonia improved the patient's outcome in terms of neck position, pain, disability, and impact on life compared to a standard physiotherapy approach. The randomized trial was done and published by the Dystonia Society, United Kingdom [10]. The two randomized trials were given by trained physiotherapists up to once a week for 24 weeks, with a control group receiving a standard physiotherapy advice program for neck problems.

Patients were randomized by telephoning a central randomization service. The primary outcome was the change in total TWSTR score between baseline and 6 months (the end of the intervention) and was evaluated by a clinician who was blinded to the treatment group. Secondary outcomes included the Cervical Dystonia Impact Profile-58 subscore (CDIP-58). There were 110 patients randomized (55 in each group), but 6-month outcomes were only available for 84 patients.

In the 2 months prior to the trial, the physiotherapist went to Paris for 6 weeks to be trained and accredited by Monsieur Bleton in his technique. After initial assessment, the physiotherapist developed a personalized regimen according to the Bleton protocol. Exercises also focus on strengthening the underactive muscles in the neck that are being under-recruited due to the overactive pull of the dystonic muscles in order to regain muscle balance. Conventional physiotherapy starts with the patient lying on his/her back, followed by a sitting position, then standing and walking and finally with their main functional activities. Patients were encouraged to adapt their environmental surroundings at work and home in order to work their head and neck in the opposite direction to their natural pull from dystonic muscles, for example, positing themselves with their TV or computer on the side opposite to their cervical dystonia. Patients received their treatment in an outpatient physiotherapy department up to once a week for 45 min per session.

The control group was offered standard physiotherapy advice for neck and back problems. The advice was standardized between the two centers and included advice about posture, sleeping position, relaxation, standard neck exercises to increase the range of neck movement, as well as shoulder, thoracic spine, lumbar spine, and core stability exercises. The emphasis was on self-management, and again patients were encouraged to perform their exercises at home and record this with a diary. Patients were offered follow-up physiotherapy sessions every 2–4 weeks for 24 weeks. The patients and their general practitioners were informed that no other neck physiotherapy was permitted for their dystonia during the 1-year trial period.

Adherence to both physiotherapy regimens was good. There was no difference between the groups in the total TWSTR score at 6 months. Both groups showed large improvements compared to baseline in the TWSTR and CDIP-58 scores which were sustained at 12 months.

The authors concluded that the specialized physiotherapy program developed by Bleton was no more beneficial than a standard program of the neck physiotherapy advice and home exercises [10]. Therefore, there is no justification for widespread use of the technique in the UK, and a larger trial does not seem merited. It is unclear whether the improvement from baseline seen with both forms of physiotherapy was a direct treatment effect, or a placebo effect. Both physiotherapy interventions were safe.

The following rehabilitation programs were also developed for research studies:

A physiotherapy program included two initial sessions of 60 min of passive myofascial elongation maneuvers and deep massage of the cervical muscles; the target muscles were the sternocleidomastoid, scalene complex, levator scapulae, and upper trapezius [13, 14].

The subsequent sessions comprised 20 min of passive myofascial elongation maneuvers; 20 min of active stretching to increase the range of motion of muscles, tendons, and ligaments; and 20–40 min of exercises to improve postural control while maintaining correct head-trunk alignment, balance, and strengthening of the axial musculature.

Sessions were done on a daily schedule, lasting 60–90 min each for 2 weeks.

In another study, Ramdharry [8] designed his exercise approach starting with a strengthening program of the deep neck flexor muscles that oppose the posture resulting from the dystonic muscle activity. Pain in the surrounding muscles was managed with soft tissue mobilization of the upper fibers of the trapezius muscle.

Strategies were taught to further reduce effort such as light touch stimulation of the side of the face, similar to a sensory trick. The physiotherapy exercises progressed to a standing position, maintaining head and neck in the midline while moving the arms during a task, for example, putting on a shirt or reaching to a shelf.

Other exercises included an indirect rotation, for example, looking at a fixed point keeping the head still, then stepping to the left on the spot, and a combined movement, for example, looking at points in the room that required a mixture of rotation flexion or side flexion.

A transversus abdominis muscle strengthening program was done in order to address core stability problems that were probably secondary to maintaining a long-term abnormal posture of the upper spine. Static hold exercises were added to his physiotherapy regimen. In addition, visual tracking tasks were given which required rapid movement in different directions to encourage more automatic movement during visual pursuit.

14.2.1.5 Combined Neck Exercises and Botulinum Toxin Injections

Bleton [9] states also that the muscles able to correct misalignment are more effective when the dystonic muscles are weakened by botulinum toxin (BoNT) injections to target areas. Thus, initiating a physiotherapy program as soon after the injection is provided would reinforce the effects of the strengthening exercises.

In summary, rehabilitation with BoNT injections can include various active and passive exercises aimed at reducing the dystonic posturing of the neck, returning the head to a normal position, and preventing long-term muscle shortening and contracture. Possible techniques include neck range of motion exercises, deep neck flexor strengthening exercises, soft tissue mobilization, strengthening exercises to oppose the direction of the neck pull, deep massage of the cervical structures, passive myofascial elongation maneuvers, cervical and axial strengthening and stretching exercises, and balance and posture retraining. Experts recommend that exercises are performed in various positions (e.g., in supine, sitting, and standing positions) and during tasks specific to activities of daily life (e.g., while watching the television, using a computer, or driving).

As idiopathic cervical dystonia is a long-term condition, experts predict that exercises would have to be performed regularly over an extended period of time and may be most effective when incorporated into the usual daily activities of the patient. Again, these theories have not been tested scientifically [15]. Meanwhile, a tailored BoNT program will be combined, as potentially new dystonic muscles become active, while previous ones may have been "toxin silenced."

14.2.2 Focal Hand Dystonia and Musician's Dystonia

The task-specific focal hand dystonia arising from occupations includes variants like writer's cramp, musician's cramp, and sports' cramp. Primary sensory modalities are intact, although impaired spatial or temporal discrimination may be identified.

Hand movement requires a degree of fine motor control which entails the precise activation of the hand area in the sensorimotor homunculus and inhibition of other uninvolved muscles. In healthy subjects, input from one muscle facilitates motor output to that muscle ("homotopic" effect), whereas it reduces motor output to the other muscles ("heterotopic" effect). In professional musicians, proprioceptive input from muscles that are functionally related such as the abductor pollicis brevis and the first dorsal interosseous facilitates motor output to these muscles, while it suppresses the output to functionally unrelated ones, such as the abductor digiti minimi [16].

In focal hand dystonia there is evidence of lack of inhibition at multiple levels in the central nervous system. This lack of inhibition with simultaneous contraction of both agonist and antagonist muscles is integral in the development of writer's cramp.

The major contributing factor in the development of focal hand dystonia appears to be the prolonged, repetitive use of the hand [17, 18]. The hands are represented in the primary somatosensory cortex in high resolution, and receptive fields are small and sharply differentiated, not including more than one finger [19, 20]. It is known that through repeated use, this representation in the somatosensory cortex is malleable through the process of sensory learning called neuroplasticity. Among trained musicians, there is enlarged cortical representation of the hands in the somatosensory cortex and auditory domains which demonstrates this normal plasticity [21]. In focal hand dystonia, repetitive sensory stimulation during the execution of the skilled manual tasks might lead to maladaptive sensorimotor plasticity in susceptible individuals. This maladaptive sensorimotor plasticity leads to changes in the representation of the digits within the somatosensory and motor cortices of the brain. As a result, the brain is unable to distinguish between near simultaneous sensory inputs to the cortex, disrupting sensory feedback to the motor system and consequently fine motor movements. Magnetic resonance imaging showed that representational areas in the brain seem to fuse among these patients with this dystonia [22]. Restrictions in musculoskeletal structure and performance have been reported in patients with focal hand dystonia, producing limitations in flexibility and strength.

In idiopathic focal dystonia, the muscle spindles become stiff, and their elastic properties vary as they are stretched or what is termed as "spindle thixotropy" [23, 24]. Compared to normal individuals, the muscle spindles become stiffer in dystonia but then become more elastic after they are overstretched. Thus, it is the inconstancy of elastic properties of the dystonic muscle spindle that leads to the motor subroutine corruption. These peripherally induced mechanisms may elucidate two phenomena: (a) why idiopathic focal dystonia symptoms tend to affect skilled and heavily practiced movements and (b) why sometimes dystonic symptoms evolve with time [25, 26].

The pattern of focal hand dystonia varies and may involve various combinations of distal and proximal muscles, flexors and extensors, and supinators or pronators and is dependent on which groups of muscles are more often used by the sufferer. For instance, in writer's cramp, the dominant writing hand is more commonly affected. Finger flexors and wrist extension and flexion are commonly affected [27]. Writer's cramp may be classified as simple or complex. In the complex type, other acts of dexterity, such as buttoning clothes or handling cutlery, are impaired. Writer's cramp is often associated with other dystonias, most frequently cervical dystonia.

Several sensory and motor interventions have been used therapeutically in focal hand dystonia to relieve the presence of involuntary contractions [28–30]. However, the mechanism responsible for any benefit remains unknown since no study has yet established a correlation between intervention-induced neurobiological effects and behavioral improvement. Conservative therapies are primarily empiric, and no single treatment modality had yet proven to be effective with predictability.

Although BoNT injections or baclofen can decrease dystonic cramping, the medications do not improve somatosensory differentiation. Conservative intervention strategies based on the principles of neuroplasticity such as constraint-induced movement therapy can contribute to the patient's movement learning process [30].

Researchers assume that the basic mechanisms of motor learning function can be achieved through intensive practice.

In one pilot study five professional musicians were studied with "sensorimotor retuning" therapy, which consisted of immobilization by splint(s) of one or more of the digits other than the focal dystonic finger. The focal dystonic finger was required to carry out repetitive exercises in coordination with one or more of the other digits for 1.5–2.5 h daily (depending on patient fatigue) over a period of eight consecutive days under therapist supervision.

The patients subsequently continued practicing the exercises with the splint for 1 h a day at home in combination with progressively longer periods of repertoire practice without the splint.

All patients showed improvement without the splint at the end of treatment, and two of the patients even resumed concert performances.

Nancy N. Byl et al. [31] has published several studies suggesting guidelines for the non-pharmacological management of focal hand dystonia, including a therapy home program.

In a general overview, the patient is asked to think positively about recovery and to try to stop all activities that caused abnormal finger movements and to practice stress-free hand movements at target and nontarget tasks. The patient is also educated about the potential for recovery based on the capacity for adaptation in the nervous system. He is asked to try and manage personal and professional stress and to participate in a fitness program (walking, running, riding a bicycle, swimming, or doing a workout in the gym). The program includes a learning-based memory program designed to improve memory and a so-called learning-based sensorimotor training (LBSMT).

The LBSMT program consists of imagining normal movement, trying to stop the abnormal movement, learning to interface the hand with the target instrument, and then creating attended, progressive, graded, repetitive sensory, sensorimotor, and fine motor movements on nontarget and then small parts of the target task, to finally practicing the complete performance on the target task.

14.3 Sensory Stimulation

14.3.1 Cervical Dystonia

A special characteristic of the dystonias is the response to certain "sensory tricks," whereby doing something else apart from the task may alleviate the dystonia. Many patients with cervical dystonia are able to temporarily decrease or abolish their spasms following a specific maneuver called "geste antagonistique." Any clinical maneuver that modulates afferent sensory and efferent motor pathways could be used as a sensory trick in patients with cervical dystonia. Although the central mechanism responsible for this response is unknown, the modification of abnormal motor activity by enhanced proprioceptive or other sensory input may play a role [27]. In fact, Rosales and Dressler [24] included BoNT as a form of a proprioceptive "sensory trick."

Vibratory and electrical stimulation to muscles or mixed and cutaneous nerves was applied to patients with cervical dystonia in order to modify the movement patterns [32]. Vibration was delivered by an air-driven vibrator oscillating at a frequency of 150 Hz with a 2–3 mm displacement. The vibrator was applied for up to 10 s to the forehead, occiput, cervical spine, paraspinal and shoulder girdle muscles, masseteric muscles, suprahyoid muscles, and remaining neck flexors and extensors. Then, electrical stimulation was delivered sequentially to mixed and cutaneous nerves, including the supraorbital nerves, mental nerves of the chin, greater occipital nerves, facial nerves, spinal accessory nerves (branch to upper trapezius muscles in myotomy cases), and branches of the cervical plexus. Stimulus parameters included square waves of 0–30 mA, with an adjustable pulse width of 40–200 ~ts. Pulse rates of 1, 2, 4, 10, 20, and 30 pulse/s (pps) were tested at each nerve site. Stimulation was maintained for up to 10 s, at sensory and motor threshold stimulus intensity.

Cutaneous electrical stimulation applied to digits of the hand has also been reported to reduce spasmodic activity in torticollis patients [33].

Karnatah [34] also studied the effect of neck muscle vibration on cervical dystonia but with a longer time interval of 15 min with good results.

A visual-sensory trick was studied in a case report with good results, even though the effects were transient [35]. The patient was instructed to look at a point on a blackboard to exclude the possibility of voluntary correction of posture when she saw herself reflected on a mirror, which also resulted in a transient relief of symptom to the same degree as looking at a mirror.

When the patient closed her eyes or terminated visual fixation to a specific target, her dystonic neck movements recurred within 10 s. When she was instructed to move her eyes while staring at a target to evaluate the role of visual fixation, dystonic contraction soon developed. The authors considered the possibility that direct visual sensory input (looking at images reflected in mirrors, fixating the gaze) influenced connections between the occipital and parietal cortexes, that is, the visual and somatosensory cortexes, and subsequently affected neck muscle contraction by higher order sensorimotor integration through compensation for impaired proprioceptive information from the head. These findings imply the role of the visual system as a compensating mechanism for impaired vestibular and proprioceptive systems.

14.3.2 Focal Hand Dystonia

Byl [36] developed a sensory discrimination rehabilitation program to restore the subject's sensory processing accuracy. Supervised physical therapy sessions and home 1- to 1.5-h sessions were held one or two times a week. Repetitive non-simultaneous sensory stimulation retraining tasks involved stimulation of the mechanoreceptors, muscle afferents, and Golgi tendon organs. The sensory tasks were designed to restore the sensitivity of the individual receptive fields of the affected hand and fingers, with the objective of redifferentiating the somatosensory map of the hand. Graphesthesia, localization, and stereognosis tasks were performed by the patient in the supine, sitting, and prone positions with the arms in positions that were different from those used during the target tasks.

The sensory discrimination activities included (1) identifying sensory stimulation of the skin with varying textures and temperatures; (2) identifying various objects and designs drawn on the hand and fingers; (3) reading Braille with one finger and playing games with Braille cards (with the eyes closed); (4) identifying shapes, letters, and numbers pressed or drawn on the skin; (5) playing dominoes and determining the match with the eyes closed; (6) identifying matched pairs of objects either in a game format or by matching objects drawn from a bag or a box full of rice, beans, or noodles; (7) identifying raised letters and numbers; (8) searching for shapes to place in a matched opening; (9) discriminating and matching coins, beads, buttons, shapes, and small animal figures; and (10) asking the patient to locate the point on the skin where they felt they were touched.

Sensorimotor retraining activities were integrated with the sensory retraining. For example, biofeedback was used to help patients eliminate the co-contraction of the flexors and extensors and to quiet unnecessary motor contraction. Patients were asked to hold and palpate the target instruments but not to play the instrument until the involuntary movements were controlled.

As part of the program, patients were asked to do sensory and motor exercises by placing a mirror between the affected and unaffected sides, making the affected hand look like the mirror image. The patient was asked to place the hand in a functional position without hyperextension of the interphalangeal joints. Finally, he was asked to perform fine motor activities concentrating on making the affected hand perform just like the mirror image of the unaffected side.

In addition, the patient was asked to spend at least 30 min a day imagining normal sensory processing, normal motor control, and effective, efficient execution of the target task.

14.4 Biofeedback

14.4.1 Cervical Dystonia

Biofeedback may be defined as the technique of using equipment to reveal to human beings some of their internal physiological events, normal and abnormal, in the form of visual and auditory signals in order to teach them to manipulate these otherwise involuntary events by manipulating the displayed signals [37].

The clinical observation of tactile self-induced feedback used by patients with torticollis to control their head movements supports the idea that external feedback may be used as a therapeutic mode. Patients often use maneuvers touching with their hands the chin, neck, or nose, developing a variety of gestures called "geste antagonistique" or counter-pressure phenomenon [38].

External audiovisual feedback of integrated electromyographic (EMG) muscle activity is used as a therapeutic tool in the treatment of cervical dystonia and other focal dystonias.

Prior to treatment abnormal motor activity is analyzed and primary and secondary goals are established: control of involuntary spasmodic activity or increased range of motion of head and neck and strengthening of atrophied muscles.

The patient receives the treatment initially in a sitting position with three to five 45-min sessions a week. Surface electrodes are placed in the targeted muscles. The patient is instructed to initially be aware of the EMG of the target hypertrophied spasmodic muscle. The next goal is trying to decrease the amplitude of the EMG activity below a target line. When these goals are achieved, the patient is asked to move the head in the opposite direction of the spasmodic activity, trying to maintain spasmodic activity at a low level. Then simultaneous feedbacks of both agonist and antagonist (spasmodic) muscles are presented to the patient in order to maintain a balanced activity.

Visual feedback is withdrawn and the patient practices using auditory feedback and concomitant mirror viewing of his head position. If primary goals are achieved in the sitting position, the next step is trying to control head position with ongoing feedback while walking in a treadmill. During the treatment, the patient is asked to do home exercises only with the use of a mirror.

Martin [39] recommends a complementary homework consisting of four tasks:

The first task was to practice exercises consisting of moving the neck in three basic directions (rotation, lateral flexion, flexion/extension), initially as single movements and then in combinations (so that they were carrying out diagonal movements, spiral movements, etc.).

It was hoped that using the neck muscles to perform fairly complex movements might help patients reestablish control of neck muscles and that the exercises would strengthen neck muscles which had atrophied.

The second task included relaxing in front of a mirror. The aim of this task was for patients to practice the skills they were learning during the EMG feedback training. The mirror gave some feedback as to their performance both in terms of how straight their heads were and how tense their sternomastoid muscles were.

A third task consisted of instructions to patients to relax whenever they felt tense, doing relaxation exercises to counteract feelings of tension and anxiety.

The fourth task was aimed at reducing asymmetrical muscular habits.

14.4.2 Focal Hand Dystonia

A brief comment of biofeedback for the management of focal hand dystonia has already been mentioned in the sensory stimulation part.

14.5 Physical Agents

14.5.1 Focal Hand Dystonia

Single-case studies have reported that high-frequency transcutaneous electrical nerve stimulation (TENS) on forearm flexor muscles decreased the excitability over stimulated muscles and increased the excitability of the antagonist muscles lasting approximately 40 min, improving cervical dystonia and dystonic tremor [40]. It has been hypothesized that TENS may remodulate excitatory and inhibitory relationships in the central nervous system (CNS) between agonist and antagonist muscles, mainly through the activation of muscular large-diameter afferent fibers, promoting a reshaping of reciprocal excitatory and inhibitory functions between agonist and antagonist muscles in the CNS that are severely impaired in dystonia.

In his study, Tinazzi [40] stimulated the forearm flexor muscles, placing the negative electrode over the flexor carpi ulnaris and radialis muscles and the positive electrode 4 cm distally over the same muscles. The current was an asymmetric rectangular biphasic waveform and had a frequency of 50 Hz with a pulse width of 250 μs. The stimulation intensity was below the pain threshold and muscular contraction. Electrical stimuli were administered throughout each 20-min session in 2-s trains (100 stimuli/train) separated by 2-s pauses.

Lowering the temperature of the human skeletal muscle increases instantaneous muscle stiffness and prolongs relaxation time, blocking muscle spindle activity and providing temporary relief of dystonic symptoms [41]. An altered sensory feedback from superficial skin receptors might also help as a "sensory trick," improving writing in cases of writer's cramp, providing a temporary relief.

Pohl [41] immersed the hand and forearm muscles in 15 °C cold water for 5 min, decreasing writing time and improving writing ability in patients with writer's cramp.

14.6 Orthotic Devices

14.6.1 Focal Hand Dystonia

Zeuner [29] designed a motor training model for patients with writer's cramp. The patient's fingers and wrist were splinted using thermoplastic material. The finger that trained was not splinted, while the remaining four fingers were splinted to reduce co-contraction. A plastic splint for the tip of the training finger with a pen attached to it was used for the training. Patients were instructed to train each finger individually by making clockwise and counterclockwise circles on a piece of paper for 4–8 weeks daily.

Priori et al. [42] suggested an alternative approach for the management of focal hand dystonia called "inactivity-dependent neuroplasticity." He evaluated the effects of 4 weeks' immobilization by a plastic splint in eight patients with idiopathic occupational focal dystonia of the upper limb with good results in a follow-up at 6 months.

The rationale for this approach is that fatigue transiently reduces and reshapes motor cortical areas through motor inactivity, inducing a reduction in afferent input and causing a restriction in motor neuron firing activity.

Nihal Tas [43] demonstrated an improvement in writing ability of dystonic patients while using a static hand orthosis.

The writing device supported and immobilized the segments of the hand (thumb, fingers, or wrist) that interfere with writing and allow the proximal large muscles to control the coordination of movement for writing, substituting the action of the distal muscles that show co-contraction. A place for grasping the pen was made in the device.

We present a case (figure) that serves as an example of a combined management of writer's cramp dystonia using BoNT and an orthotic device.

We evaluated a 30-year-old housewife with writer's cramp diagnosed 1 year ago affecting only the right hand. Her thumb as well as her second and third digits would hyperflex at the proximal and distal interphalangeal joints, while her wrist hyperflex. She has no family history of any movement disorder and was diagnosed with idiopathic vasculitis 1 year before, presenting moderate dysarthria and writer's cramp dystonia as sequelae. The right hand digits' activity during writing was videotaped and analyzed. AboBTXA (abobotulinum A) was injected as follows:

Flexor digitorum sublimis fingers 2 and 3, 75 units each; flexor digitorum profundus, fingers 2 and 3, 75 units each. Flexor pollicis brevis 50 units; adductor pollicis 50 units; flexor carpi radialis 100 units.

After the injection, the patient was instructed to use an orthotic device on the third digit 12 h during the day for 3 weeks, as this finger presented a higher hyperflexion activity in comparison to the thumb and second finger. Writing improved and she did not present wrist hyperflexion anymore. The Writer's Cramp Rating Scale [44] grade showed a favorable change from 11 to 6 points.

The patient reported a moderate decrease in grip strength during her daily activities.

Use of orthotic device immediately, after BoNT injection

Writing ability three weeks after the BoNT injection

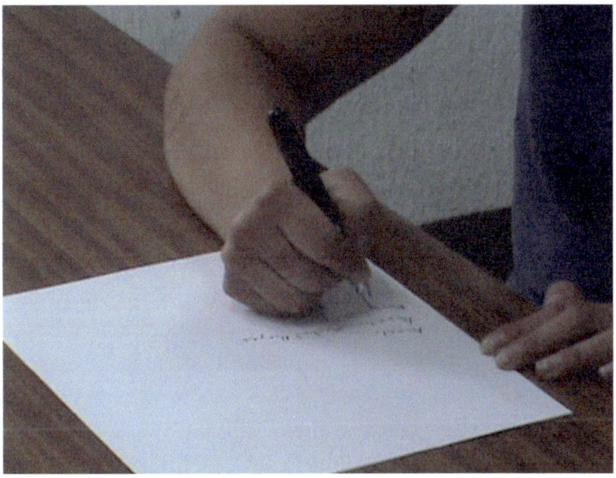

14.7 Vestibular Rehabilitation

Some studies have described changes in vestibular function and perception of body orientation in patients with cervical dystonia [45].

Müller studied static posturography and dynamic equilibrium in ten patients with tonic cervical dystonia with a fixed postural deviation and 20 patients with cervical dystonia with phasic head movements. Seventeen age-matched volunteers formed the control group. The results showed no abnormalities in static posturography in the study group, but dynamic balance was impaired in the phasic cervical dystonia group.

Maintenance of balance involves multiple peripheral sensory inputs, central integrating pathways, and efferent outputs, which could affect the vestibular input due to repetitive involuntary and mainly horizontal head oscillations in phasic cervical dystonia.

Vestibular rehabilitation (VR) can be defined [46] as the recovery of the balance deficits by creating new patterns of reaction to different daily balance situations.

VR exercises can be divided in three main strategies: habituation, which tries to reduce avoidance of certain positions; adaptation, where the therapist teaches the unaffected balance receptors to undertake the function of the damaged ones; and substitution, by teaching other sensory systems to compensate for the vestibular impairment.

Describing all VR exercises is beyond the scope of this chapter.

Han and colleagues [47] designed an exercise-based vestibular rehabilitation therapy program. The goals include the following:

1. To enhance gaze stability
2. To enhance eye movements
3. To enhance postural stability
4. To improve vertigo
5. To improve activities of daily living

14.7.1 Cervical Dystonia and Dysphagia

Besides focusing in movement therapy as a non-pharmacological approach in the treatment of cervical and focal dystonia, several studies have shown dysphagia in cervical dystonias such as spasmodic dysphonia and spasmodic torticollis before focal BoNT injections. In his review of 42 patients with cervical dystonia, Marsden observed dysphagia in six patients [48].

Ertekin [49] studied oropharyngeal swallowing with the aid of submental muscle EMG. His major finding was that the neurogenic nature of dysphagia was more prominent and longer lasting than the mechanical dysphagia. The same author [49] thought of the neurogenic cause of dysphagia in cervical dystonia patients, based on the following conclusions:

- Head or neck movement is not essential for dysphagia in dystonic patients.
- In clinically affected patients, stabilizing the head and neck was not effective in changing the patient's dysphagia clinically.
- No EMG relation was observed between swallowing and EMG activity of the sternocleidomastoid muscle.
- Videofluoroscopic evaluations of patients with cervical dystonia generally show a delayed initiation of swallowing and pharyngeal residue, which are regarded as neurogenic dysfunction abnormalities in oropharyngeal assessments.
- Cricopharyngeal sphincter muscle EMG during swallowing was not normal in patients with cervical dystonia. EMG findings were compatible with the hyperreflexive/hypertonic nature of the upper esophageal sphincter observed in suprabulbar involvement of the bulbar swallowing center in the central nervous system.
- The cause of neurogenic dysphagia in cervical dystonia is unknown.

Conclusion

Neurorehabilitation for focal dystonias, specifically the cervical and the task-specific limb dystonias, includes a regimen of physical therapy exercises, sensorimotor stimulation, biofeedback, and physical and orthotic devices. The effectiveness of these approaches can potentially be enhanced with BoNT injections to targeted dystonic muscles that need to be abolished to balance motor control and achieve dexterity. Vestibular neurorehabilitation in relation to perception of body orientation may be an added novel approach that may facilitate motor control in dystonia. All these approaches are largely geared toward achieving a normal head and limb position, preventing pain, posturing, long-term muscle shortening, and contracture.

References

1. Rondot P, Marchand MP, Dellatolas G. Spasmodic torticollis-review of 220 patients. Can J Neurol Sci. 1991;18:143–51.
2. Kiss ZH, Doig K, Eliasziw M, et al. The Canadian multicenter trial of pallidal deep brain stimulation for cervical dystonia: preliminary results in three patients. Neurosurg Focus. 2004;17(1):E5.

3. Cohen-Gadol AA, Ahlskog J, Matsumoto JY, et al. Selective peripheral denervation for the treatment of intractable spasmodic torticollis: experience with 168 patients at the Mayo clinic. J Neurosurg. 2003;98:1247–54.
4. Albright AL, Barry MJ, Shafton DH, Ferson SS. Intrathecal baclofen for generalized dystonia. Dev Med Child Neurol. 2001;43:652–7.
5. Dykstra DD, Mendez A, Chappuis D, et al. Treatment of cervical dystonia and focal hand dystonia by high cervical continuously infused intrathecal baclofen: a report of 2 cases. Arch Phys Med Rehabil. 2005;86:830–3.
6. Adler CH, Kumar R. Pharmacological and surgical options for the treatment of cervical dystonia. Neurology. 2000;55(12 Suppl 5):S9–14.
7. Jankovic J, Beach J. Long – term effects of tetrabenazine in hyperkinetic movement disorders. Neurology. 1997;48:358–62.
8. Ramdharry G. Case report: physiotherapy cuts the dose of botulinum toxin. Physiother Res Int. 2006;11:117–22.
9. Bleton JP. Spasmodic torticollis. Handbook of rehabilitative physiotherapy. Paris: Frison-Roche; 1994.
10. Counsel C, Sinclair H, Fowlie J, Tyrrell E, Derry N, Meager P, et al. A randomised trial comparing a specialized physiotherapy programme versus standard physiotherapy advice in patients with idiopathic cervical dystonia. UK, The Dystonia Society: 2007–2010. http://www.dystonia.org.uk/index.php/professional-research/61-physiotherapy-research-study.
11. Bleton JP. Physiotherapy of focal dystonia: a physiotherapist's personal experience. E J Neurol. 2010;17 Suppl 1:107–12.
12. Bleton JP. Spasmodic torticollis of cervical dystonia. Guide for functional retraining produced by L'Hospital Sainte-Anne centre Raymond Garcin. http://perso.orange.fr/jeanpierre.gadbois/fasciculekinebleton.pdf.
13. Smania N, Corato E, Tinazzi M, et al. The effect of two different rehabilitation treatments in cervical dystonia: preliminary results in four patients. Funct Neurol. 2003;18:219–25.
14. Tassorelli C, et al. Botulinum toxin and neuromotor rehabilitation: an integrated approach to idiopathic cervical dystonia. Mov Disord. 2006;21(12):2240–3.
15. Novak I, Campbell L, Boyce M. Botulinum toxin assessment, intervention and aftercare for cervical dystonia and other causes of hypertonia of the neck: international consensus statement. EFNS Eur J Neurol. 2010;17 Suppl 2:94–108.
16. Rosenkranz K, Butler K, Williamon A, Rothwell JC. Regaining motor control in musician's dystonia by restoring sensorimotor organization. J Neurosci. 2009;29(46):14627–36.
17. Soland VL, Bhata KP, Marsden CD. Sex prevalence of focal dystonias. J Neurol Neurosurg Psychiatry. 1996;60(2):204–5.
18. Bessman SB. Dystonia update. Clin Neuropharmacol. 2000;23(5):239–51.
19. Jankovic J. Parkinson's disease and movement disorders. Philadelphia: Lippincott and Williams; 2007. p. 322–4.
20. Bara-Jimenez W, Catalan MJ, Hallett M, Gerloff C. Abnormal somatosensory homunculus in dystonia of the hand. Ann Neurol. 1998;44(5):828–31.
21. Pantev C, Engelien A, Candia V, Elbert T. Representational cortex in musicians. Plastic alterations in response to musical practice. Ann N Y Acad Sci. 2001;930:300–14.
22. Elbert T, Candia V, Altenmuller E, Rau H, Sterr A, Rockstroh B, et al. Alteration of digital representations in somatosensory cortex in focal hand dystonia. Neuroreport. 1998;9(16):3571–5.
23. Proske U, Morgan DL, Gregory JE. Thixotropy in skeletal muscle and in muscle spindles: a review. Prog Neurobiol. 1993;41(6):705–21.
24. Rosales RL, Dressler D. On muscle spindles, dystonia and botulinum toxin. Eur J Neurol. 2010;17 Suppl 1:71–80.
25. Grunewald RA. Progression of dystonia: learning from distorted feedback? J Neurol Neurosurg Psychiatry. 2007;78(9):914.
26. Rosset-Llobet J, Candia V, Fabregas S, Ray W, Pascual-Leone A. Secondary motor disturbances in 101 patients with musician's dystonia. J Neurol Neurosurg Psychiatry. 2007;78(9):949–53.

27. Raymond L. Rosales (2012) Dystonia, spasticity and botulinum toxin therapy: rationale, evidences and clinical context, dystonia - The many facets, Prof. Raymond Rosales (Ed.), ISBN: 978-953-51-0329-5. Available from: http://www.intechopen.com/books/dystonia-the-many-facets/dystonia-spasticity-and-botulinum-toxin-therapy-rationale-evidences-and-clinical-context
28. Zeuner KE, Bara-jimenez W, Noguchi PS, Goldstein SR, Dambrosia JM, Hallett M. Sensory training for patients with focal hand dystonia. Ann Neurol. 2002;51:593–8.
29. Zeuner KE, Shill HA, Sohn YM, Molloy FM, Thornton BC, Dambrosia JM, Hallett M. Motor training as treatment in focal hand dystonia. Mov Disord. 2005;20:335–41.
30. Candia V, Elbert T, Altenmüller E, Rau H, Schafer T, Taub E. Constraint-induced movement therapy for focal hand dystonia in musicians. Lancet. 1999;353(2):42.
31. Byl NN, Nagajaran S, McKenzie AL. Effect of sensory discrimination training on structure and function in patients with focal hand dystonia: a case series. Arch Phys Med Rehabil. 2003;84:1505–14.
32. Leis AA, Dimitrijevic MR, Delapasse JS, Sharkey PC. Modification of cervical dystonia by selective sensory stimulation. J Neurol Sci. 1992;110:79–89.
33. Cleeland CS. Behavioral techniques in the modification of spasmodic torticollis. Neurology. 1973;23:1241–7.
34. Karnath HO, Konczak J, Dichgans J. Effect of prolonged neck muscle vibration on lateral head tilt in severe spasmodic torticollis. J Neurol Neurosurg Psychiatry. 2000;69:658–60.
35. Lee CN, Eun MY, Kwon DY, Park MH, Park KW. Visual sensory trick in patient with cervical dystonia, Case report. Neurol Sci. 2012;33:665–7.
36. ByI NN, McKenzie A. Treatment effectiveness for patients with a history of repetitive hand use and focal hand dystonia: a planned, prospective follow-up study. J Hand Ther. 2000;13(4): 289–301.
37. Basmajian JV. Biofeedback, principles and practice for clinicians. 3rd ed. Baltimore: Williams & Wilkins; 1989. p. 1–4.
38. Korein J, Brudny J. Integrated EMG feedback in the management of spasmodic torticollis and focal dystonia: a prospective study of 80 patients. Res Publ Assoc Res Nerv Ment Dis. 1976;55:385–426.
39. Martin PR. Spasmodic torticollis: investigation and treatment using EMG feedback training. Behav Ther. 1981;12(2):247–62.
40. Tinazzi M, Farina S, Bhatia K, Fiaschi A, Moretto G, Bertolasi L, Zarattini S, Smania N. TENS for the treatment of writer's cramp dystonia: a randomized, placebo controlled study. Neurology. 2005;64:1946–8.
41. Pohl C, Happe J, Klockgether T. Cooling improves the writing performance of patients with writer's cramp. Mov Disord. 2002;17(6):1341–4.
42. Priori A, Pesenti A, Capellari A, Scarlato G, Barbieri S. Limb immobilization for the treatment of focal occupational dystonia. Neurology. 2001;57:405–9.
43. Taş N, Karataş GK, Sepici V. Hand orthosis as a writing aid in writer's cramp. Mov Disord. 2001;16(6):1185–9.
44. Wissel J, Kabus C, Wenzel R, et al. Botulinum toxin in writer's cramp: objective response evaluation in 31 patients. J Neurol Neurosurg Psychiatry. 1996;61:172–5.
45. Müller J, Ebersbach G, Brenneis C, Brady L, Poewe W. Disturbances of dynamic balance in phasic cervical dystonia. J Neurol Neurosurg Psychiatry. 1999;67:807–10.
46. Georgescu M, Stoian S, Mogoantă CA, Ciubotaru GV. Vestibulary rehabilitation – election treatment method for compensating vestibular impairment. Rom J Morphol Embryol. 2012;53(3):651–6.
47. Han BI, Song HS, Kim JS. Vestibular rehabilitation therapy: review of indications, mechanisms, and key exercises. J Clin Neurol. 2011;7(4):184–96.
48. Marsden CD, Harrison MJ. Idiopathic torsion dystonia (dystonia musculorum deformans). A review of forty-two patients. Brain. 1974;97(4):793–810.
49. Ertekin C, Aydogdu I, Seçil Y, Kiylioglu N, Tarlaci S, Ozdemirkiran T. Oropharyngeal swallowing in craniocervical dystonia. J Neurol Neurosurg Psychiatry. 2002;73(4):406–11.

Dystonia in Art: The Impact of Psychiatric and Neurological Disease on the Work of the Sculptor F. X. Messerschmidt

15

Michal Maršálek

15.1 Introduction

Not many other pieces of art are accompanied by such a lack of clarity and so many speculations as the set of "character heads" of the Austrian sculptor Franz Xaver Messerschmidt. Even 230 years after his death, there is still no consensus as to what the expressions of these sculptures really mean. The reason why we cannot define a correct answer may be the fact that we fail to pose the correct question. Ernst Kris [1], who provided the most elaborated interpretation of the meaning of the character heads we have seen thus far, is undoubtedly right when he, as an art historian and psychoanalyst, insists that "the question, why many heads comprise only unrelated elements of expression instead of meaningful expression, should be asked again". The answer to his question extends beyond the borders of art history. The task of this chapter is to add a contemporary medical view to the existing opinions and to offer a new interpretation of the sense of the character heads on that basis.

Following an overview of the basic data about the artist's life and work, I will attempt, in my own analysis, to answer three questions:

1. Did F. X. Messerschmidt suffer from a mental illness?
2. What do the character heads depict?
3. Why did he create the character heads?

M. Maršálek
National Psychiatric Hospital Bohnice, Prague, Czech Republic
e-mail: marsalek@bohnice.cz

P. Kanovsky et al. (eds.), *Dystonia and Dystonic Syndromes*,
DOI 10.1007/978-3-7091-1516-9_15

15.2 Curriculum Vitae of F. X. Messerschmidt

Franz Xaver Messerschmidt was born on February 6, 1736, in the town of Wiesensteig near Stuttgart as the fourth of seven children sired by the local tanner Johann Georg Messerschmidt. Both parents were exceptionally vital. Messerschmidt's father was 66 at the time of the artist's conception. His mother, a descendent of the well-known sculptor family Straub, was 30 years younger. Johann died at 77, while his wife lived to the age of 97. Franz Xaver inherited both good physical health and a talent for arts. He completed his sculptor's apprenticeship with his uncle in Munich and in Graz and successfully graduated from the prestigious Vienna Academy of Fine Arts. He became a protégé of the court painter Martin van Meytens, the director of the Academy. Even during his studies, he excelled in depicting his models quickly and precisely. By the 1760s, he was becoming a sought-after court portraitist, and his sculpture of Maria Theresa as a Hungarian queen is considered one of the best art pieces of Vienna rococo to this day. In this period, he inclined more and more towards the newly arising artistic concept of classicism which he had the chance to became familiar with during his several months' stay in Rome in 1765 [1–4]. In 1769, he bought a house with a homestead, garden and a summer residence in Vienna. He became an assistant of Professor Schletterer and was assured of inheriting the post after the professor's retirement. At 33 years old, he was respected and financially secure, and the quality of his early work guaranteed that he would be taken for an important artist even in the future. But there was a sudden breakdown in his promising career. In 1774, he was not appointed professor, apparently due to certain "symptoms of insanity", and had to retire prematurely. In 1771, he began to lose all his commission work, and finally he had to sell his residence, and in May 1775, he left Vienna in financial crisis [1–3, 5]. After 2 years of existence in his native Wiesensteig and in Munich, he moved to Pressburg (today's Bratislava) to the house of his younger brother, the sculptor Johann Adam Messerschmidt, in 1777. In 1780, he bought his own house in Zuckermandel, a district below the castle of Pressburg. There he lived and created in solitude, enjoying the reputation of an eminent artist but also of a capricious eccentric. He was able to earn his living on occasional commission work until his sudden death (probably of pneumonia) in August 1783 [1, 3, 6]. After his death, a set of strange busts was found in his house. The busts were classicist in style and depicted exclusively men's heads, grimacing in bizarre fashions. According to a contemporary report, he started to work on those "character heads", as they were later improperly labelled, in 1771 [7]. He never sold those statues. He dedicated most of his creative activity in his last 13 years of life to their creation from metal or alabaster. At present, we are familiar with the appearance of 56 such heads [1, 3].

15.3 Interpretation of Character Heads

The approaches of art historians often interpret Messerschmidt's character heads as a reaction to the social situation of the era of the late Enlightenment, as a reaction to his personal situation or as an attempt to deal with different art styles, past or contemporary. The character heads were taken for a satire aimed at his contemporaries. This is where the label "Hogarth of sculpture" comes from. It never became clear, however, what the subject of this putative satire actually was. What's more, it is difficult to conceive of an artist who dedicates 13 years of life to the creation of caricatures which he does not show to anybody. Others conclude that the expressions of the heads represent venting of aggression against circumstances [8]. Oberhaidacher [9] sees the heads in context of art trends of his era, as an extension of the naturalistic tendencies of German gothic or as an anticlassicism, which, in the era of the late Enlightenment, emphasised a subjective approach, disharmony and irrational elements of the human psyche [5, 6, 10]. Some of the above-mentioned influences certainly could have been elemental in the form and content of the work, but their influences are not sufficient for a full understanding of Messerschmidt's late work. The formal aspect of the work was influenced by Roman and Egyptian art, with which Messerschmidt came into contact in 1765 [2–4]. The character heads were first considered to be systematic depictions of human characters in the physiognomic frame, but this opinion was later questioned [1–3, 6, 11, 12]. Physiognomy was a fashionable discipline of the end of the eighteenth century dealing with the relationship between the physical appearance and the character of a personality. The basic work of its founder Johann Kaspar Lavater, *Physiognomische Fragmente zur Befőrderung der Menschenkenntnis und Menschenliebe*, was published in four books in the years 1775–1778, at least 4 or 5 years after the first character heads were created by Messerschmidt. Besides, Lavater's doctrine dealt with the firmer features of the face, while Messerschmidt worked exclusively with the variable, muscular component of expression. What's more, the set of character heads does not represent a systematic typology; many sculptures repeat an expression almost mechanically [1, 3]. Albert Ilg [13] in his first monograph on Messerschmidt pronounced a hypothesis that in some statues, the artist depicted persons under hypnosis, whom he had seen during sessions with a renowned Viennese physician and magnetiser Franz Anton Mesmer. This view has recently been further elaborated upon by Krapf [12, 14], according to whom the two men had known each other many years and were friends. Even Messerschmidt lived in Mesmer's house after his return from Rome (1765–1770). This assertion, however, does not correspond to reality [15]. Messerschmidt certainly portrayed Mesmer and made a fountain for his garden in 1770. Nothing more is known, though, about their relationship. The first hypnotic experiment was performed by Mesmer in August 1774, and the whole method was fully developed by him in Paris in 1778, when Messerschmidt had already been living in Pressburg and working on his character heads for at least 6 years. Krapf also surmised that some of the statues had magnets (as used in Mesmer's séances) in their mouths but was unequivocally debunked by several authors [15–17]. Another presumption [12, 14] that the string tied around the head and neck of two statues was a reference to the so-called baquet, a string by which the hypnotised were connected in group séances, is clearly untrue based simply on timing; the method was first used by Mesmer in Paris in 1778 [15, 17]. Pfarr surmised in 2001 that Mesmer, as a member of several secret societies, could have influenced Messerschmidt through his learning of animal gravity.

As far as medical interpretation of Messerschmidt's late work is concerned, the key position is held by the analyses of Ernst Kris who first suggested in 1932 that the sculptor suffered from schizophrenia and the creation of his character heads was the result of this disease [1]. Some other psychiatrists [18] and psychologists [19] were of the same opinion, while other psychiatrists considered some other types of mental disease less damaging to the personality, either a paraphrenia [20] or a paranoia [21]. These authors based their opinions mainly on the detailed report of a Berlin representative of the Enlightenment, Friedrich Nicolai, who visited Messerschmidt in Pressburg in 1781 and described the visit 4 years later in the book *Beschreibung einer Reise durch Deutschland und die Schweiz im Jahre 1781* [7]. The opinion that Messerschmidt suffered from lead poisoning [22] can also be dismissed as groundless, as no typical symptoms of lead intoxication were ever described [6, 15]. The psychiatrist Peters [23] does not accept that Messerschmidt suffered from any mental disease, and some authors argue against the objectivity of Nicolai's report and thus its reliability as evidence of delusions and hallucinations [10, 23, 24]. Some of them especially reject Kris's psychoanalytical interpretation of Messerschmidt's work [24, 23].

15.4 Author's Analysis

15.4.1 Did F. X. Messerschmidt Suffer from Mental Illness?

15.4.1.1 The Symptoms of the Mental Illness

Kris [1] described the signs of psychotic disease mostly according to a report by Nicolai [7]. Messerschmidt told Nicolai that ghosts visited him in the night and caused pain in his right lower abdomen and thigh. His conviction that he really saw the ghosts was irrefutable. What's more, anybody who doubted their existence became his enemy according to Nicolai. His persecution by the ghosts was a long-term affair. The delusions of this persecution crystallised. First, the sculptor could not understand why he was being persecuted despite his flawless life. Later he began to understand that the ghost envied his deep knowledge of body proportions (he showed Nicolai an old Italian book about proportions. Besides this, a paper torso of an Egyptian figure was hanging on the wall which apparently served as a standard of correct proportions). At another point in time, he confided that the worst of all was the Spirit of Proportions, who wanted to kill him because of his deep knowledge of the secret of proportions. They kept pinching each other and Messerschmidt almost died of fear. Finally he won, when he depicted the ghost in a statue. The Spirit of Proportions later disappeared leaving only an odorous cloud behind him. Further statements of Messerschmidt will be mentioned later.

The convictions and experiences so far cited were assessed as hallucinations and persecution delusions not only by Kris [1] but also by other authors [18, 19, 21] who concluded that Messerschmidt suffered from a mental illness. Others do not agree with him. Rudolf and Margot Wittkower [24] consider the possibility of erroneous interpretation of Nicolai's description, which in itself, according to them, is already an interpretation of Messerschmidt's views. (Nicolai's vague handling of reality – within the scope of artistic licence – was proven anew by Pötzl-Malikova [25]. She compared Nicolai's description of Messerschmidt's studio – which responded to the notion of Messerschmidt as an unappreciated artist – with the relatively rich furnishings of the house as described in the actual list of inventory after the artist's death). The Rudolf and Margot Wittkower (1963) also point out the context of the period, in which belief in supernatural forces was common for many people. The bizarre visions of F. X. Messerschmidt could, in the eighteenth century, fit into the context of contemporary conventions, while in the twentieth century, it was no longer widely acceptable. The degree to which it is bizarre in itself, though, is not sufficient to distinguish delusion from mere superstition. Only the irrefutability of the vision and its influence on the behaviour of the subject are specific for delusion.

Further descriptions fit the picture of psychosis well too. The report of State Chancellor Wenzel Prince Kaunitz for the Kaiserin dated December 1774 includes a statement that in the last 3 years Messerschmidt had demonstrated "some mental confusion" [3]. According to the same report, the condition improved, and Messerschmidt could start working again, but "he still suffered occasionally from an unhealthy imagination". Messerschmidt felt persecuted by the professors' board of the Vienna Academy, which could have some real basis. At the beginning of the 1770s, he was, however, very paranoid even towards his brother Johann Adam, whom he accused in a letter from June 14, 1770, of coming to see him only to get hold of his property and of threatening him with a sword when FXM was ill [3]. It is not clear whether the above-mentioned situation really happened or whether it was a product of the artist's imagination. A quote from a letter to his brother from January 1777 asserting that all Germans were against him can be judged as an exaggeration of a certain truth, albeit a strange one. The idea that the artist was labelled mad thanks to the unconventional concepts of his art is blatantly false as it was the behaviour of the artist which was questioned, not his work [26]. Kris's interpretation of Messerschmidt's physical pains as hallucinations was questioned as well. He could have suffered from bodily disease which could have caused him pains, which he could then have interpreted through the lens of contemporary superstition. Messerschmidt was always described as a man bouncing with health. The fight with the Spirit of Proportions with the mutual pinching does not fit with this assumption at all. What's more, he saw the ghost, which can hardly be explained in any way other than a hallucinatory experience. The concept that Messerschmidt suffered from psychosis was from a psychiatric point of view rejected mainly by Peters [23], who in a thorough analysis questioned Kris's psychoanalytical explanation of the artist's motivation. Even if we reject the psychoanalytical explanation completely, it changes nothing in the concluded diagnosis.

If we summarise all the data gathered, the first references to a changed behaviour in the artist can be traced at least to 1771. Paranoia related to many spheres is particularly evident. The condition developed (the worst of it evidently being at the beginning); the sense of persecution crystallised and became irrefutable; the changes continued over many years; they waxed and waned in intensity and were influential on behaviour. This is all very characteristic of delusions, in this case of a paranoid-persecutory delusion. Visual, bodily and olfactory hallucinations were identified with high rates of probability.

15.4.1.2 The Course of the Mental Illness

The assessment of a mental illness in different times and different milieu can be affected by many misunderstandings concerning the symptoms of the disease. On the other hand, the most important characteristic, i.e. the impact of the disease on the quality of life, can easily be assessed even after 200 years. According to Heinrichs [19], Messerschmidt's problems, whatever they may have been, had caused a sudden social decline towards his middle age (over the years 1771–1777, he suffered the loss of nearly all commission work, and he retired, sold up his property and left the capital). A rupture and a decline in life process such as this can only be caused by psychosis, dementia or a severe drug addiction. There is no evidence in Messerschmidt's life of the development of dementia or drug addiction. The ascertained hallucinations and delusions, the dynamics of the disease and its social impact bear witness to the fact that Messerschmidt in his last 13 years of life suffered from a psychosis, very probably of the schizophrenic spectrum. It was not, however, a typically developing case, which is often characterised by a progression of cognitive changes and personality disintegration. Despite the fact that the consequences of the disease were milder, the decline of life quality was evident and irreversible. A certain portion of patients nevertheless suffer a milder course of the disease, and undoubtedly they achieve sporadic outstanding successes even while the disease is in progress.

15.4.1.3 Formally Recognisable Signs of Psychotic Illness in the Creation of the Character Heads

Reitmann [27] observes that the characteristic feature of the creative activity of people with schizophrenia is a conspicuous lack of expression, because those with the illness put the face together as a mosaic. They are able to recognise and depict details, but they do not reach a higher synthesis – they do not capture the expression. They can see the face as a cluster of lines and forms, which does not evoke any corresponding emotion in them. It is not a question of impaired perception but of an impaired final review. Glandien [21] observes that Messerschmidt's busts bear incomprehensible expressions and are composed of parts, but he refuses the idea that they do not constitute a final unit. At the same time, he refutes the presence of stereotypy and ornamental symbolism on the character heads, features which were understood as typical in schizophrenic creativity by Meurer-Keldenich (1979) (quoted according to [21]. Kris's analysis [1], however, finds some features of exaggerated ornamentation in the accenting of a single element out of context (multiplication of wrinkles or exceedingly decorative hair styling) (Fig. 15.1). These are the signs of stereotypy in the creative expression, which correspond to the features of creation of patients with schizophrenia according to Rennert [28] (ornamental stereotypy, perseverance). Kris [1] further mentions the stereotypic use of several expressions in a series. It is of course impossible to assume the mental illness of an artist or to make a diagnosis only according to the features of his work. Such signs

Fig. 15.1 Stereotypy in the excessive multiplication of the wrinkles (F. X. Messerschmidt: Childish Weeping. After 1770. Bratislava City Museum. Reproduced with permission)

can only be important in cases where mental illness has previously been proven. In this case the signs do not bear witness to any deep deterioration of the creative forces, but certain visual signifiers that are typical for schizophrenia can be detected in Messerschmidt's work.

15.4.1.4 The Probability of Genesis of a Mental Disease

Two important risk factors for the development of schizophrenia can be proven for F. X. Messerschmidt. He was born on February 6, 1736. The fact that children born in winter months have an 8 % higher probability for later development of schizophrenia was ascertained in more than 250 studies (review Torrey et al. [29]). The probability of malnutrition or infection of the foetus in the critical period is higher in winter. Those schizophrenias are not genetically loaded and their course is often milder. The father of F. X. Messerschmidt was 67 when his son was born. The probability of the development of schizophrenia increases sharply with paternal age at the time of conception and does not depend on the maternal age. The descendants of fathers older than 50 years have a threefold higher probability of developing schizophrenia compared to those born to fathers younger than 25 years [30]. Children born to fathers another 10 years older have not been explored so far, but a further steep increase of the risk can be presumed. Today these findings are confirmed by three meta-analyses of controlled studies (e.g. Miller et al. [31]). The same results were gained in a study performed in Iran, where children born to fathers in their younger and older age in large Iranian families were studied [32]. Older fathers are actually more inclined to have epigenetic changes of the genetic information in the development of sperm even without any previous genetic disposition to schizophrenia. According to the current knowledge, the daughters of older fathers are more affected, but both genders have a higher severity of both positive symptoms and autism compared with other patients suffering from schizophrenia. On the other hand, the symptoms often remit when treated and the family is not genetically affected [33]; thus, the prognosis seems better. Coming back to

Messerschmidt, the epigenetic risk of schizophrenia is clear in his case and we should not be too surprised by its development.

15.4.1.5 Conclusion

F. X. Messerschmidt had a higher risk of developing schizophrenia due to two unfavourable epigenetic factors related to his birth, and he did indeed develop a psychosis, at the age of 33, probably on the schizophrenic spectrum with little personality impairment. Within the framework of the disease, he suffered from partially systematised paranoid-persecutory and grandiose delusion and from hallucinations affecting several senses till the end of his life. There was no clear cause for the start of the problems and the disease caused a substantial decline of his quality of life even if the problems did not worsen over time. More important than the ambition to categorise the diagnosis of the disease according to the newest criteria is the fact that Messerschmidt suffered from psychosis, but the disease did not substantially affect his creative abilities.

15.4.2 What Do the Character Heads Depict?

15.4.2.1 Survey of Opinions

All the heads (aside from two exceptions) are life-size, frontally oriented, and strictly symmetrical. They represent men of different ages and of several different types including heads with distinct features of the artist. The expressions of most of the statues are undecipherable; they make a hostile impression and do not express any particular emotion. They are notably strained, with a forcefully open or clamped mouth or eyes often found. Some of the figures have heads in forced anteflexion; others have the neck craned forward. The elements of the expression repeat stereotypically and non-systematically. Some heads are decorated by bizarre embellishments. The ornamentation (mannerisms, stereotypy) on certain heads is outstanding, changing the face into a grimace. Kris [1] presumes that the expression of many of the heads reflects a psychotic experience and that the stylisation into a mask simultaneously has a magical-defensive function (intimidation of the ghosts). Grimacing with a magical function is also pointed out by Glandien [21] and Krauss [18]. Glandien adds that the sculptor systematically created a series of his heads to retain his identity. The expressions of the heads were interpreted psychoanalytically by Kris. Kris [1] in general sees a pent-up aggression behind all the expressions of the character heads and points out the predominantly negative and aggressive tendencies in the expressions – as does Pötzl-Malikova [3]. Some of the motives are repeated mechanically and in a stereotyped manner and they certainly do not constitute a compact unity [1, 3, 18]. Kris [1] wrote about the repetition of the same motif: "We get the impression that Messerschmidt was affixed to some expressive constellations. All the variations just confirm that insurmountable limits existed for him… The mouth plays a definite role…." Messerschmidt himself told Nicolai that 64 principal proportions existed and they contained the fundamental secrets known by the old Egyptians. He creates the sculptures to give the secret of proportions over to mankind. He himself did not stick to the rule very much. This statement fits well within the frame of schizophrenic idle philosophising without a clear concept. Krapf [2, 14] thought that some of the heads described real situations from Mesmer's hypnosis. The Spirit of Proportions, according to him, could be a stylisation of the old Egyptian god Thoth, transformed into a Greek Hermes Trismegistus, a god with a bird's head, the forefather of all esoteric doctrines avowed by secret societies. Messerschmidt's connection with these societies is very probable [10, 12, 14, 18, 34], and already according to Nicolai [7], the sculptor was under their influence, this including his interest in old Egyptian doctrines. Egyptomania of the end of the eighteenth century effected Messerschmidt strongly both theoretically and practically (the bizarre headdress of the statue "A Grievously Wounded Man", resembling the pharaoh's headdress). The hairless figures resemble the old Egyptian statues very much, too [1, 3]. Even Messerschmidt's teaching of proportions relates to it. All the character heads created in classicist style have retained high artistic qualities. All have a striking lack of expression. It is as if the single expressive elements were randomly composed. A combination of expressions as seen in a mirror was presumed by Oberhaidacher [9] and also by Kirchner [35]. That is how Messerschmidt really worked according to the only available report [7]. To analyse the expressions of the character heads, the FACS (Facial Action Coding System) [36] was used by Pfarr [6]. The method categorises facial movements according to muscular positioning and intensity and is used to assess the character and force of emotion expressed by the face. Pfarr used this method to verify that the particular expressions of the heads are emotionally inconsistent. Contrary to Kris, however, he does not accept it as a consequence of a mental disease but presumes that Messerschmidt intentionally changed the muscular configurations of the character heads in much the same way as in counterpoint baroque music and that those constructions were influenced by the esoteric theories of the day. According to Krauss [18], the artist depicted schizophrenic stereotypies (grimacing). These psychomotor manifestations are rare nowadays, evidently thanks to modern pharmacotherapy, but up to the second half of the twentieth century, they were relatively often observed in schizophrenic patients. They were regarded as a conscious reaction to hallucinatory experiences by Krauss [18]. Rendered as statues, they were supposed to intimidate the ghosts, according to him. The same opinion was held by Pötzl-Malikova [3] (Messerschmidt thought that by creating the statues, he gave form to proportions, which intimidated the ghosts) and by Glandien [21], who used ethological terminology (Messerschmidt created the grimaces on the statues to increase his volume and frighten the ghosts that way).

15.4.2.2 Author's Interpretation

Leaving aside stereotypy and mannerisms, the expressions of the heads resemble one of the motor disorders of the extrapyramidal system – dystonia. The dystonias are neurological affections of the extrapyramidal system of the brain, which can differ in origin but which present with similar symptoms. Typical for dystonias are extended tonic contractions of different muscle groups, which, in psychiatric patients, are mainly expressed in the face (predominantly in the region of the mouth)

Fig. 15.2 "Closed position" of the expression on one face (F. X. Messerschmidt: An Arch-Rascal. After 1770. Bratislava City Museum. Reproduced with permission)

Fig. 15.3 "Open position" of the expression on one face. The same character as in Fig. 15.2 is depicted (F. X. Messerschmidt: The Yawner. After 1770. Museum of Fine Arts, Budapest. Reproduced with permission)

and neck. They present as forceful opening and closing of the mouth and eyes, grimacing, poking the tongue out and turning the head and neck to either side, forward or back, in different constellations, which then repeat stereotypely. Either a myoclonic component with short dystonic motions or the myostatic form with a stiffly retained posture prevails. A selection of Messerschmidt's character heads plus corresponding images of patients suffering from dystonia (of the same muscle groups as the statues) illustrates the link. The selected pairs very clearly display similarities in facial expression between the sculptures and the patients (compare Fig. 15.2 with Fig. 15.4, Fig. 15.3 with Fig. 15.5, Fig. 15.6 with Fig. 15.7 and Fig. 15.8 with Fig. 15.9). Dystonias exactly correspond with Kris's notion [1] that the muscle groups of the character heads are contracted in a random constellation in the face. That is why they do not formulate any actual expression but merely resemble them. Dystonic motions are involuntary and often painful, which is why the patient is strongly aware of them. They react sensitively to any noxious stimulus by instantly worsening, but they cannot be directly controlled at will. Psychic strain and different activating manoeuvres (counting by memory, voluntary movement or painful stimuli) can provoke the movements or enhance them. They grow weaker once the patient calms down. It can be sometimes difficult to distinguish dystonia from

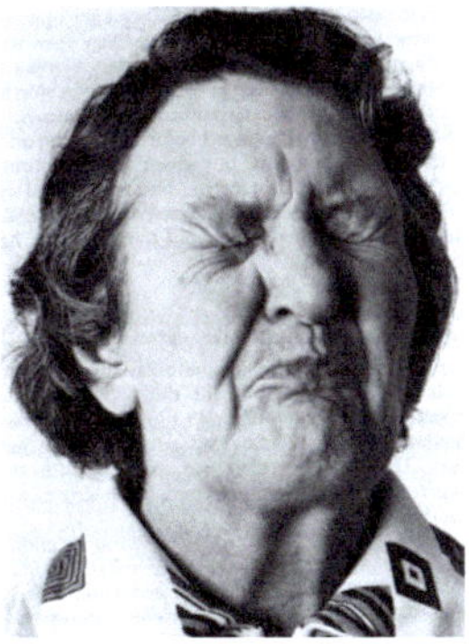

Fig. 15.4 The expression of a patient suffering from idiopathic dystonia (Meige's syndrome) bears great resemblance to the expression of the statue in Fig. 15.2 (Reproduced from Marsden [37], with permission from BMJ Publishing Group Ltd)

Fig. 15.5 The expression of a patient suffering from idiopathic dystonia (Meige's syndrome) bears great resemblance to the expression of the statue in Fig. 15.3 (Reproduced from Marsden [37], with permission from BMJ Publishing Group Ltd)

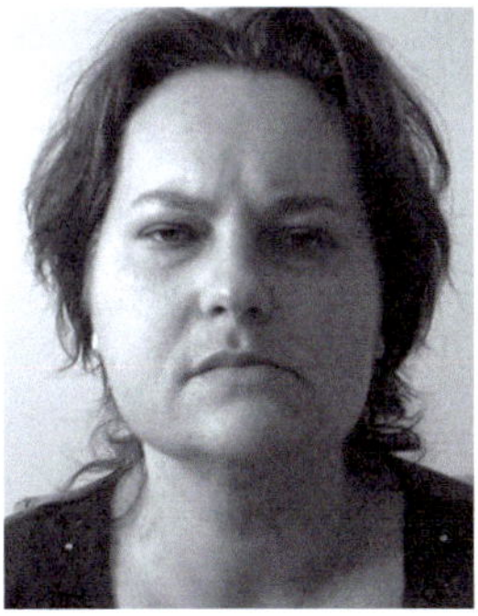

Fig. 15.6 A mild form of antipsychotic-induced tardive dystonia resembles the expression of the statue in Fig. 15.7 (Photo by the author. Reproduced with permission of a patient)

catatonic stereotypy and mannerisms. The movement disorder is simpler, though, and the patient can easily distance themselves from it and take the observer's standpoint. From the neurological point of view, dystonias are understood as a variant of dyskinesias, and they are commonly explored in patterns which are identical for both dystonias and dyskinesias.

Fig. 15.7 F. X. Messerschmidt: The Melancholic. After 1770 (Bratislava City Museum. Reproduced with permission)

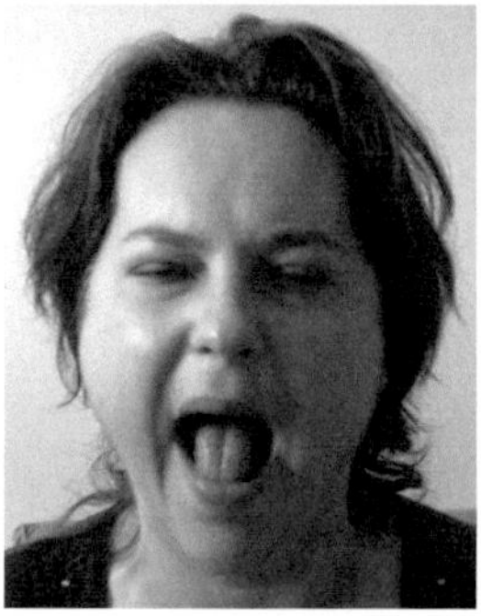

Fig. 15.8 In a severe form, antipsychotic-induced tardive dystonia resembles the expression of the statue in Fig. 15.9 (Photo by the author. Reproduced with permission of a patient)

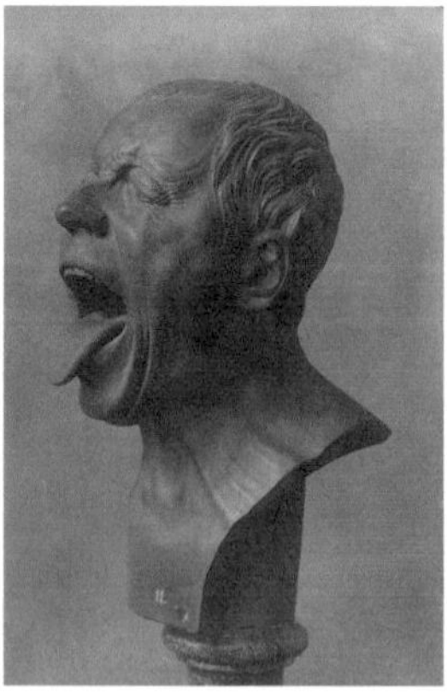

Fig. 15.9 F. X. Messerschmidt: A Man Vomiting. After 1770 (Bratislava City Museum. Reproduced with permission)

In the set of character heads, not only can exact depictions of particular dystonias be identified but also several pairs depicting the same person, expressing either an open or a closed expressive constellation, two marginal positions of a single dystonic movement (specifically, vigorously open or closed eyes or mouth) (Figs. 15.2 and 15.3).

These pairs make an impression of one motion pattern, as if the two most characteristic frames of a film take had been chosen (see also [38]). Two extreme variants of one expressive constellation can be seen even on the pair of portraits of the Spirit of Proportions.

It would be pertinent to ask where Messerschmidt could have witnessed dystonias. He depicted them in different spells and different places. Therefore the idea that he could have portrayed a specific acquaintance suffering from dystonia can be excluded. Similarly he could not have visited a place where more people with dystonia would have been concentrated. The only option for him to record dystonias repeatedly would be the case if he himself had suffered from them. Therefore I agree with most of the authors that Messerschmidt's heads were created as auto-portraits. This means that he would have had to have been suffering from a neurological disease besides psychosis. Would that have been in any way possible? I will try to give evidence that it was indeed possible, even very probable. Despite the fact that dystonias of different origins are relatively rare, they appear comparatively often in patients with schizophrenia (in about 2–94 % of acute [39] and 1–4 % of chronic [tardive] forms [40]. These dystonias arise, however, as side effects of treatment with antipsychotics – and antipsychotics were discovered in the 1950s. Why should Messerschmidt have suffered from dystonias 180 years before those side-effective treatments were discovered? The opinion that antipsychotics induce extrapyramidal symptoms was recently revised, influenced by studies performed in countries like Morocco or India, where patients such as this were found who had never been treated by psychopharmacological agents. Meta-analysis of six controlled studies of antipsychotic-naïve patients with schizophrenia and of six studies of their non-ill relatives of the first degree proved a significantly higher prevalence of extrapyramidal disorders in them compared to within the healthy population. Tardive dyskinesias (among which tardive dystonias are categorised) and parkinsonism were tracked [41]. Similar results were obtained in studies of healthy twin siblings of patients with schizophrenia [42]. The same disorders in the frame of schizophrenia were described in the literature of the first half of the twentieth century, when no antipsychotics yet existed [43–45]. All those findings support a hypothesis that extrapyramidal disorders including dystonias are integral components of the clinical picture of schizophrenia and that treatment with antipsychotics only accelerates their manifestation [46, 47]. If Messerschmidt suffered from schizophrenia, the probability of a concurrence of some extrapyramidal disorder including dystonias was relatively high. The sculptor then depicted them on the character heads.

The number of motions described by Messerschmidt is strictly limited and many theoretically possible combinations do not occur within the set. This corresponds to Kris's observation [1] who explained it as the artist's concentration on a limited number of sexual themes. But it also fits well into the concept that Messerschmidt depicted only those dystonic movements (or eventually their combinations) from which he himself suffered.

A forcefully closed mouth is very common; however; it was found in 36 character heads [18]. Its comparative frequency could be explained by another of Messerschmidt's delusions, i.e. that the animals understand ghosts better than we and therefore know well why the lips should be tightly closed [1, 18].

15.4.3 Why Did F. X. Messerschmidt Create the Character Heads?

The fact that Messerschmidt suffered both from schizophrenia and dystonia is not enough to explain why he depicted his dystonias so persistently. It is appropriate to come back to Nicolai's report [7] with awareness that some of the Messerschmidt views are described fragmentarily and the connection between them is only loose. They should be therefore considered in the context of his actions. Messerschmidt had mentioned to Nicolai, among other things, the mysterious relationship of bodily proportions. He said that "all things in the World have their just proportions and all effects correspond to their causes. Everything in the World is governed by proportions and whoever invokes in himself the proportions which correspond to or are superior to those of another, will of necessity bring forth effects which will correspond to the effect of the other, or will of necessity be superior to it". The proportions represent the relationship between the dimensions of an object expressed by their ratio. If proportion is constant, we can change the size of an object without changing its shape. We can tell that two objects have the same proportions according to their identical shape or in other words, if they look the same. In essence, Messerschmidt says: if somebody manages to look the same as someone else, he will exercise power over him. One way to do it is to change one's face and make it resemble the other's. The other way to do it is to change the face of the other so as he looks the same as you. The result would be the same: they would both have the same appearance, that is, proportion. Messerschmidt said too that the specific proportions that occur in the head of a man also repeat themselves throughout the entire body [7]. The proportions of the face correspond to certain points on the body. It means that any change in the proportions of the face can subsequently affect the corresponding parts of the body of the same person. In practice, however, Messerschmidt demonstrated to Nicolai that the system could work the other way around – he pinched himself and then modelled the grimaces elicited in front of a mirror. The assertion that he had solved the secret of proportions and mastered the ghosts that way appears as another of Messerschmidt's delusions, this time an inventive one; it could have been derived from Mesmer's views as Pfarr [10] claims, or it could reflect the teachings of Vitruvius which had been popular for several centuries [1]. What is important is that Messerschmidt believed in the system and stuck to it.

As dystonias arise spontaneously, they are commonly interpreted by patients with schizophrenia as being induced by some outside hostile energy. The possibility that Messerschmidt perceived the spontaneous dystonias as being caused by the ghosts is quite probable. His concept even gives a clear motive for their actions. Inducing their facial proportions upon Messerschmidt's face, they achieve the result of Messerschmidt's face having the same proportions as theirs; thus, the ghosts could exercise power over him. On the other hand, if Messerschmidt did the same, if he went on to depict the dystonic grimaces induced by the ghost, he would be able to finally master it. This is why, in my opinion, he created sculptures with the grimaces observed on his own face. The depictions would have had to be as exact as possible, as the rules of magic say that the more precisely we name or depict a ghost, the more we know about its secrets and the more easily we can master it. The character heads would then be the most precise pictures of the sculptor's own dystonia and the motivation for their creation would have been the struggle with imaginary foes. Creation of the heads was a long and laborious process lasting far longer than one dystonic seizure. To be able to capture them properly, the sculptor could have elicited them repeatedly by pinching different parts of the body. The painful stimulus acted as an activating manoeuvre which brought on or enhanced the dystonias. The character heads could not only halt the ghost's attack at the moment when the ghost's appearance was captured but could also intimidate the ghosts preventively, if the sculptor created all the 64 mysterious proportions. It could be objected that it was not dystonia but simple painful grimaces that Messerschmidt elicited. These would all look much alike though, and a large variety of the sculptures' expressions (most of them not resembling painful grimace at all) would never have come into existence. Many authors [3, 18, 21] are of the same opinion as Kris and consider the creation of most of the character heads to be a magic weapon in the struggle with the ghosts. I agree with them in general but I believe that the war with the ghosts was mostly waged in accordance with my description. The struggle, aided by his depiction of the enemy, was described by Messerschmidt himself when he spoke about his victory over the Spirit of Proportions [7]. Some other heads could have evolved even on the basis of schizophrenic grimacing as supposed by Krauss [18]. There could have been more battle strategies against the ghosts, of course.

The hypothesis that Messerschmidt depicted dystonias could explain the sense of most of the heads, including those which suggest physiological functions like yawning or retching (both those heads perfectly fit a picture of severe facial dystonia). It would even explain the changes in neck position, which were not taken so much into account in most of the analyses. It would give a reason as to why the depicted muscle constellations did not correspond to any particular emotion. Further explanations follow from the fact that it was a magical struggle situation, which is why he worked so vigorously, why the statues never left the studio. Even the strictly frontal position of the heads is in accordance with the combative hypothesis. If the artist were to beat these ghosts, he had to face them, as in classical African art, which is magical too, not aesthetic.

Let me give a final summary of my thesis: in the last 13 years of his life, Messerschmidt suffered from psychosis on the schizophrenic spectrum with a milder course. In the scope of this illness, he for one thing felt endangered by the ghosts and for another suffered from dystonia of the head and neck. Thanks to his simultaneous belief that objects of the same proportions can mutually influence each other, he perceived his spontaneous dystonias as an attempt on the part of the ghost to replicate their features on his face and to exercise power over him in that way. He chose the same magical defence against it, and by depicting his real dystonias as proportions of the ghosts' faces, he defeated them. Or at least he thought so. Moreover, he could have elicited dystonias by an activating manoeuvre (pinching) and then using their depiction in his struggle with the ghosts. The answer to the old question as to which emotions are expressed by the character heads is simple, from that point of view. The character heads do not express any emotion. They represent a constellation of the muscles in the head and neck, which can sometimes be reminiscent of emotions.

After all the analyses of the artist's activities, one final question remains unspoken so far: could the magical creative activity have been of any real help to Messerschmidt? We do not know the answer. But we can presume that it could have helped in two ways. The first relief might have been in the fact that he had found his own, albeit erratic system in the psychotic chaos. He identified the enemy and its weak points and could use against it what he had mastered best – his art. And he knew that he had mastered that weapon excellently. He could have told himself: there is no end to the labyrinth but I know that my way is the right one. Besides giving him the ability to live with psychosis, real improvement of the dystonias could have arisen as another effect of the work, as these afflictions often sensitively react to the calm brought by amelioration. Short-term improvement of dystonias is commonly described even with treatment by placebo. Radical temporary improvement of the extrapyramidal disorders is exceptional but possible, too. One of my female patients with schizophrenia and severe antipsychotic-induced dystonia of the mouth and neck lost this cosmetically hampering disorder for several weeks after she had happily fallen in love. We do not know if any of those mechanisms was of any help to Messerschmidt, but we know for sure that he never gave up his struggle. If his dystonias abated from time to time, it could have been a confirmation for him that the steps he had taken had been correct.

Conclusion

The result of Messerschmidt's endeavour is a unique creative work of high artistic quality unparalleled in art history. As a magically conceived creative activity, it stands beyond the narrative concept of all European art till the beginning of the twentieth century and goes back somewhere very far, to the roots of the human psyche, where thought was still identical to action. The psychiatric diagnosis is only a means of understanding some of the links. From the human point of view, we can observe the very unique attempt of an extraordinarily strong personality to struggle with his own insanity. Messerschmidt keeps struggling and his art takes its own path.

Acknowledgements I would like to thank Prof. Maria Pötzl-Malikova for the valuable comments from an art historian's point of view and for the help with procuring otherwise inaccessible documents. I would further like to thank Dr. Evzen Nespor for preparing the English version of the text and Melanie Steynberg BA. B.Ed for revising the final version. The author also gratefully acknowledges the helpful comments and suggestions of the reviewers, which have significantly improved the presentation.

References

1. Kris E. Die "Charakterköpfe"des Franz Xaver Messerschmidt: Versuch einer historischen und psychologischen Deutung. Jahrbuch der Kunsthistorischen Sammlungen in Vienna (new ser), 4; 1932. p. 169–228.
2. Krapf M. Franz Xaver Messerschmidt's life and work. The career of a difficult one. In: Krapf M, editor. Franz Xaver Messerschmidt 1736–1783 (exh. cat.). Vienna: Hatje Cantz Publishers; 2003. p. 15–32.
3. Pötzl-Malikova M. Franz Xaver Messerschmidt. Wien-München: Jugend und Volk; 1982.

4. Scherf G. Messerschmidt's portraits and heads: a singular art. In: Pötzl-Malikova M, Scherf G editors. Franz Xaver Messerschmidt 1736–1783. From neoclassicism to expressionism. (exh. cat.), Louvre éditions. Paris: Officina libraria; 2010. p. 31–40.
5. Pötzl-Malikova M. The life and work of Franz Xaver Messerschmidt. In: Pötzl-Malikova M, Scherf G, editors. Franz Xaver Messerschmidt 1736–1783. From neoclassicism to expressionism. (exh. cat.), Louvre éditions. Paris: Officina libraria; 2010. p. 17–29.
6. Pfarr U. Franz Xaver Messerschmidt Menschenbild und Selbstwahrnehmung. Neue Frankfurter Forschungen zur Kunst 2. Berlin: Gebr. Mann; 2006.
7. Nicolai F. Beschreibung einer Reise durch Deutschland und die Schweiz im Jahre 1781, Berlin und Stettin 1785. In: Pötzl-Malikova M, editor. Franz Xaver Messerschmidt, Wien-München: Jugend und Volk,.Dok. XLIV, 1982. p. 145–148. In: Pötzl-Malikova M, Scherf G, editors. Franz Xaver Messerschmidt 1736–1783. From neoclassicism to expressionism. (exh. cat.), Louvre éditions, Paris, Neue Galerie New York, Officina libraria; 1785. p. 207–11.
8. Pakosta F. Mein Bekenntnis zu Franz Xaver Messerschmidt. Tendenzen. 1976;17(106):41–4.
9. Oberhaidacher J. Kunstgeschichte oder Psychologie? Zu dem Charakterköpfen des Franz Xaver Messerschmidt. Osterreichische Zeitschrift für Kunst und Denkmalpflege 38, [1–2], Vienna; 1984. p. 25–42.
10. Pfarr U. Ernst Kris on F. X. Messerschmidt A valuable stimulus for new research? American Imago, studies in psychoanalysis and culture. 2001 58, [1], p. 445–61.
11. Bückling M. The truth about man. In: Bückling M, editor. The fantastic heads of Franz Xaver Messerschmidt, (exh. cat.). München: Liebieghaus Frankfurt/Main; 2006. p. 213–40.
12. Krapf M. The "character heads": from the simple heads conformable with nature to the convulsive works. In: Krapf M, editor. Franz Xaver Messerschmidt 1736–1783 (exh. cat.). Vienna: Hatje Cantz Publishers; 2003. p. 33–48.
13. Ilg A. Franz Xaver Messerschmidt's Leben und Werke. Leipzig: Prag; 1885.
14. Krapf M. Patrons and friends, with special regard to the role of the "magnetizer" F. A. Mesmer. In: Krapf M, editor. Franz Xaver Messerschmidt 1736–1783 (exh. cat.). Vienna: Hatje Cantz Publishers; 2003. p. 49–60.
15. Pötzl-Malikova M. Zur aktuellen Situation in der Messerschmidt – Forschung. Osterreichische Z Kunsthist Denkmalpflege. 2003;57(2):259–64.
16. Maué C. Franz Xaver Messerschmidt, 1736–1783. Frühneuzeit Info. 2003;14(1):169–78 (exhibition reviews).
17. Pfarr U. Franz Xaver Messerschmidt (1736–1783). Kunstchronik. 2003;56(1):9–14.
18. Krauss P. Die "Charakterköpfe"des Franz Xaver Messerschmidt aus psychiatricher Sicht. In: Konrad AH, editor. Hohenstaufen, Veröfentlichungen des Geschichts und Altertumsvereins Göppingen 13, Weissenhorn; 1986. p. 114–134.
19. Heinrichs RW. Historical origin of schizophrenia. Two early madmen and their illness. J Hist Behav Sci. 2003;39(4):349–63. Fall.
20. Turcek M, Malikova M. Franz Xaver Messerschmidt – a contribution to evaluation of the pathological elements of his creation (in Slovak). Bratisl Lek Listy. 1962;42:289–98. 1, [5].
21. Glandien O. Franz Xaver Messerschmidt (1736–1783) – Ausdrucksstudien und Charakterköpfe. Arbeiten der Forschungstelle des Instituts für Geschichte und Medizin der Universität zu Köln, Band 20, Köln; 1981. p. 78–105.
22. Behr H-G, Grohmann H, Hagedorn B-O. Charakter-Köpfe: Der Fall Messerschmidt. Wieverrückt darf Kunst sein? Weinheim: Beltz; 1983.
23. Peters UH. Physiognomic als Mittel des Psychoanalyse: Ernst Kris deutet Franz Xaver Messerschmidt. In: Schmölders C, Gilman SL, editors. Gesichter der Weimarer Republik: Eine Physiognomische Kulturgeschichte. DuMont: Cologne; 2000. p. 262–79.
24. Wittkower R, Wittkower M. Genius, madness, and melancholy. In: Wittkower R, Wittkower M, editors. Born under Saturn. London; 1963. p. 124–32.
25. Pötzl-Malikova M. Zwei unbekannte Dokumente zum Leben und Werk Franz Xaver Messerschmidts, Ars. 1996 1–3, p. 215–24.
26. Pötzl-Malikova M. Neuerscheinungen über Franz Xaver Messerschmidt. Kunstchronik. 2007;60(9–10):429–39 (reviews).

27. Reitmann F. Facial expression in schizophrenic drawings. J Ment Sci. 1939;85:264–72.
28. Rennert H. Die Merkmale schizophrener Bildnerei. Jena: Fischer; 1962.
29. Torrey EF, Miller J, Rawkungs R, Yolken RH. Seasonality of births in schizophrenia and bipolar disorder: a review of the literature. Schizophr Res. 1997;28(1):1–38.
30. Malaspina D, Harlap S, Fennig S, Heiman D, Nahon D, Feidman D, Susser ES. Advancing paternal age and the risk of schizophrenia. Arch Gen Psychiatry. 2001;58(4):361–7.
31. Miller B, Messias E, Miettunen J, Alaräisänen A, Marjo-Riita J, Koponen H, Räsänen P, Isohanni M, Kirkpatrick B. Meta-analysis of paternal age and schizophrenia risk in male versus female offspring. Schizophr Bull. 2011;37(5):1039–47.
32. Naserbakht M, Ahmadkhaniha H-R, Mokri B, Smith CL. Advanced paternal age is a risk factor for schizophrenia in Iranians. Ann Gen Psychiatry. 2011;10:15.
33. Rosenfield PJ, Kleinhaus K, Opler M, Perrin M, Learned N, Gotz R, Stanford A, Messinger J, Harkavy-Friedman J, Malaspina D. Later paternal age and sex differences in schizophrenia symptoms. Schizophr Res. 2010;116(2–3):191–5.
34. Yonan M. The man behind the mask? Looking at Franz Xaver Messerschmidt. Eighteenth-Century Studies. 2009 42, [3] p. 31–451.
35. Kirchner T. Franz Xaver Messerschmidt and the construction of expressions. In: Bückling M, editor. The fantastic heads of Franz Xaver Messerschmidt, (exh. cat.). München: Liebieghaus Frankfurt/Main; 2006. p. 266–81.
36. Ekman P, Friesen W. Facial action coding system: a technique for the measurement of facial movements. Palo Alto: Consulting Psychologists Press; 1978.
37. Marsden CD. Blepharospasm-oromandibular dystonia syndrome (Brueghel's syndrome). A variant of adult-onset torsion dystonia? J Neurol Neurosurg Psychiatry. 1976;38:1204–9.
38. Schmid T. 49 Köpfe. Die Grimassen-Serie des Franz Xaver Messerschmidt. Zurich: T. Schmid; 2004.
39. Chen JJ, Swope DM. Movement disorders. In: Tisdale JE, Miller DA, editors. Drug-induced diseases. Prevention, detection, and management. 2nd ed. Bethesda: American Society of Health-System Pharmacists, Special Publishing; 2010. p. 211–35.
40. Marsalek M. Tardive drug-induced extrapyramidal syndromes. Pharmacopsychiatry. 2000;33 Suppl 1:14–33.
41. Koning JPF, Tenback DE, van Os J, Aleman A, Kahn RS, van Harten PN. Dyskinesia and parkinsonism in antipsychotic-naive patients with schizophrenia, first-degree relatives and healthy controls: a meta-analysis. Schizophr Bull. 2010;36(4):723–31.
42. Koning JP, Kahn RS, Tenback DE, van Schelven LJ, van Harten PN. Movement disorders in nonpsychotic siblings of patients with nonaffective psychosis. Psychiatry Res. 2011;188(1):133–7.
43. Bleuler E. Lehrbuch der Psychiatrie. Berlin: Julius Springer Verl; 1923.
44. Kraepelin E. Dementia praecox and paraphrenia. Chicago: Chicago Medical Books Co; 1919.
45. Reiter P. Extrapyramidal motor-disturbances in dementia praecox. Acta Psychiatr Neurol Scand. 1926;1:287–309.
46. Tenback DE, van Harten PN. Epidemiology and risk factors for (tardive) dyskinesia. Int Rev Neurobiol. 2011;98:211–30.
47. Whitty PF, Owoeye O, Waddington JL. Neurological signs and involuntary movements in schizophrenia: intrinsic to and informative on systems pathobiology. Schizophr Bull. 2009;35(2):15–424.

Index

P. Kanovsky et al. (eds.), *Dystonia and Dystonic Syndromes*,
DOI 10.1007/978-3-7091-1516-9

MIX
Papier aus verantwortungsvollen Quellen
Paper from responsible sources
FSC® C105338

If you have any concerns about our products,
you can contact us on
ProductSafety@springernature.com

In case Publisher is established outside the EU,
the EU authorized representative is:
Springer Nature Customer Service Center GmbH
Europaplatz 3, 69115 Heidelberg, Germany

Printed by Libri Plureos GmbH
in Hamburg, Germany